D. M. SHAPIRO

8 February 2012

CORPORATE SOCIAL RESPONSIBILITY IN THE MINING INDUSTRIES

Corporate Social Responsibility Series

Series Editor:
Professor David Crowther, London Metropolitan University, UK

This series aims to provide high quality research books on all aspects of corporate social responsibility including: business ethics, corporate governance and accountability, globalization, civil protests, regulation, responsible marketing and social reporting.

The series is interdisciplinary in scope and global in application and is an essential forum for everyone with an interest in this area.

Also in the series

Perspectives on Corporate Social Responsibility
Edited by David Crowther and Lez Rayman-Bacchus
ISBN 0 7546 3886 3

Corporate Social Performance: A Stakeholder Approach
Stuart Cooper
ISBN 0 7546 4174 0

Nonprofit Trusteeship in Different Contexts
Rikki Abzug and Jeffrey S. Simonoff
ISBN 0 7546 3016 1

Human Values in Management
Edited by Ananda Das Gupta
ISBN 0 7546 4275 5

Ethical Boundaries of Capitalism
Edited by Daniel Daianu and Radu Vranceanu
ISBN 0 7546 4395 6

Corporate Social Responsibility in the Mining Industries

NATALIA YAKOVLEVA
ESRC Centre for Business Relationships, Accountability, Sustainability and Society (BRASS), Cardiff University, UK

ASHGATE

Published by
Ashgate Publishing Limited
Gower House
Croft Road
Aldershot
Hampshire GU11 3HR
England

Ashgate Publishing Company
Suite 420
101 Cherry Street
Burlington, VT 05401-4405
USA

Ashgate website: http://www.ashgate.com

British Library Cataloguing in Publication Data
Yakovleva, Natalia
Corporate social responsibility in the mining industries. - (Corporate social responsibility series)
1. Social responsibility of business 2. Mineral industries - Social aspects 3. Mineral industries - Environmental aspects
4. Social responsibility of business - Russia (Federation)
5. Mineral industries - Social aspects - Russia (Federation)
I.Title
658.4'08

Library of Congress Cataloging-in-Publication Data
Yakovleva, Natalia.
Corporate social responsibility in the mining industries / by Natalia Yakovleva.
p. cm. -- (Corporate social responsibility series)
Includes bibliographical references and index.
ISBN 0-7546-4268-2
1. Social responsibility of business. 2. Mineral industries--Social aspects. 3. Mineral industries--Environmental aspects 4. Social responsibility of business--Russia (Federation) 5. Mineral industries--Social aspects--Russia (Federation) I. Title. II. Series.

HD60.Y34 2005
622'.068'4--dc22

2005009755

ISBN 0 7546 4268 2

Printed and bound in Great Britain by MPG Books Ltd. Bodmin, Cornwall

Contents

List of Figures

List of Figures

List of Tables

List of Abbreviations

AIDS	Acquired Immune Deficiency Syndrome
ALROSA	Almazy Rossii-Sakha
BHP	Broken Hill Proprietary Limited
CSO	Central Selling Organisation
CSR	Corporate Social Responsibility
EIA	Environmental Impact Assessment
EMS	Environmental Management Systems
GDP	Gross Domestic Product
GRI	Global Reporting Initiative
HIV	Human Immunodeficiency Virus
HTML	Hyper Text Markup Language
ICME	International Council on Metals and Minerals
IIED	International Institute for Environment & Development
IMF	International Monetary Fund
ISO	International Standard Organization
NGO	Nongovernmental Organization
OECD	Organisation for Economic Cooperation and Development
OHSAS	Occupational Health and Safety Assessment Series
oz	troy ounce
PDF	Portable Document Format
RF	Russian Federation
RS(Y)	Republic of Sakha (Yakutia)
RSFSR	Russian Soviet Federated Socialist Republic
SAPI	Sakhainvestprovalmaz
UK	United Kingdom
UN	United Nations
UNCED	United Nations Conference on Environment and Development
UNCTAD	United Nations Conference on Trade and Development
UNEP	United Nations Environmental Programmes
US	United States
USA	United States of America
USSR	Union of Soviet Socialist Republics
WBCSD	World Business Council for Sustainable Development
WCED	World Commission on Sustainable Development
WWF	World Wide Fund for Nature

Introduction

Business is a critical factor in the world's economic and social development. Business interactions with society go beyond economic impacts on nations that include supply of goods and services, generation of profits, provision of jobs, tax payments and trade development. Consequences of business operations affect the interests of society not only through products and services supplied, but also through impacts on the natural environment, impacts on the livelihood of people and impacts on traditions, culture and social constructions. As business grows in size and expands geographically, more groups of society are affected by business operations. These affected groups feel that business has an obligation to share its profits and benefits with the society as the society helps to sustain its existence and operations. Today, society expects business to act in the way that would satisfy society's interests and share benefits with those members of the society who are not business shareholders.

Business is working towards increasing its positive inputs into society through making voluntary contributions to education, health, social services, arts and culture; and aligning its practices in accordance with societal norms and expectations in the areas of employee relations, impact on the environment, supply chain relations, etc. that were previously considered to be spheres of economic and legal consideration. A concept of corporate social responsibility (CSR) suggests that business has an obligation to contribute to social progress beyond economic transactions of a firm. Corporate social responsibility implies that business should consider interests of those affected by business operations and respond to their desires and concerns. Corporate social responsibility explores the relationship between business and society in the context of satisfying society expectations, achieving society goals and bringing benefits to society that in many cases implies voluntary actions, which are prescribed by moral and philanthropic values rather than by legal requirements.

This book critically assesses the nature of corporate social responsibility in the mining industries. Mining is one of the world's major industrial sectors, which in some cases holds a dominant position in the socio-economic development of many nations and is rapidly expanding into remote areas of the world due to the liberalization of investment regimes in developing countries and economies in transition. Mining industries significantly influence societies in which they operate. Metals and minerals mining industries contribute to regional and national economies in terms of providing budget and export revenues, employment and infrastructure development. It is demonstrably responsible for causing a wide range of negative environmental and social impacts at local, regional and global levels such as disruption of river flows, degradation of land and forest resources, impacts on livelihood of local communities near mines and disturbance of traditional lifestyles of indigenous people. Due to the severity and diversity of impacts caused by mining industries and a wide range of stakeholder concerns and expectations related to mineral extraction, the mining industries represent an interesting example for the study of business and society relationships.

Mining companies are now faced with challenges to improve their environmental and social performances, to ensure integration of the concept of sustainable development and to deliver on their corporate social and environmental responsibilities in reaction to increasing pressure from international organizations, national governments, consumers, employees, local communities, NGOs and the wider public. However, improvement of environmental and social performances in the mining industries can be hampered by ineffective government systems, the absence of appropriate regulations, and economic, political and social problems, which are particularly present in developing countries and countries with transition economies.

Previous studies in the area of environmental and social responsibilities in the mining industries explored the extent of environmental and social impacts caused by mining operations in developed and developing countries (Sengupta, 1993; Warhurst, 1999; McMahon and Remy, 2001); resolution of conflicts between mining companies and stakeholders after industrial incidents and violation of human rights or rights of indigenous people (Shrivastava, 1995; Dixon and Dillon, 1990; Hilson, 2002); problems associated with and provision for mine closure (Warhurst and Mitchell, 2000); and the nature of corporate social disclosure of mining companies that originate in developed countries (Tilt and Symes, 1999; Warhurst, 1998).

This book assesses the role of the mining industries in progress towards sustainable development, analyses modern approaches to corporate social responsibility adopted by mining companies, and evaluates policies and practices of mining companies within four key areas of corporate social responsibility: environmental protection, health and safety, employee relations and community development. The book addresses the current developments in corporate conduct of global mining companies, especially within the global gold and diamond mining industries and with specific reference to mining companies in a transition economy, the Russian Federation.

In the Russian Federation, which is the world's largest transition economy, there has been a significant emphasis on the extractive industry in its economic development (Schroeder, 1998), especially in the regions of East Siberia and the Russian Far East where the extractive industry is particularly strongly developed. The extractive industry in the Russian Federation has undoubtedly inherited a legacy of environmental misuse related to mineral resources exploitation in the former Soviet Union (Lluardo, 1995), but is undergoing a transformation in the context of transition from planned to market economy, privatization, liberalization of trade, democratization of society and integration into the world economy.

The Republic of Sakha, which is the largest territorial unit of the Russian Federation and located in north-east Siberia, has an economy which is highly dependent on mining, particularly on precious metals and minerals mining such as gold and diamonds (Yakovleva et al, 2000) and presents an interesting case for the investigation of interactions between the mining industry and communities in which it operates due to several key concerns. Firstly, the mining industry in the Republic of Sakha is a dominant sector of the regional economy, which substantially contributes to budgets and export revenues, employment, provision of technical infrastructure and social welfare. Secondly, the mining industry and accompanying infrastructure development have caused a variety of negative environmental impacts including the disruption and pollution of river flows, disturbance of land and forest resources, harmful use of 'peaceful' nuclear explosions, negative effects of traditional lifestyles of indigenous peoples and the health of the population in the mining area. Thirdly, privatization and economic changes in the Russian Federation have led to the transformation of the minerals industry involving structural, economic and financial implications. Fourthly, the rise of sovereignty has led to center-periphery disputes relating to distribution of benefits from mineral resource exploitation. Fifthly, innovative approaches have helped towards solving environmental, social and economic problems concerning the mineral extraction.

The book offers a comprehensive review of existing literature on the character and range of environmental and social concerns related to the mining industries and influences on the mining industries for the improvement of their social and environmental performances. Using qualitative analysis, the book presents a comprehensive account of corporate strategies and tools adopted by the mining companies in response to pressures from stakeholders. The book draws from the analysis of corporate social disclosure of global gold and diamond mining companies in Annual Reports, Environmental Reports and corporate websites for 1998-2000. It also incorporates an analysis of three case studies from the Republic of Sakha, including: gold mining company Aldanzoloto; diamond mining company ALROSA; and foundations for delivery of socio-economic and environmental initiatives in the diamond province, SAPI Foundation and Target Fund. Primary data for the case studies involved questionnaires, in-depth and semi-structured interviews with representatives from the mining companies, governmental agencies and local authorities, and the analysis of internal corporate and government documents.

The book is presented in 6 chapters, which are organized in two parts. Part one is dedicated to the review of major societal concerns related to economic, social and environmental impacts of mining industries and their role in contributing towards sustainable development. Part one investigates strategies and tools available for mining companies to mitigate adverse environmental and social impacts caused by their operations. This first part also explores sectoral differences and provides a detailed insight into specific stakeholder concerns and industry response in a case of global diamond and gold mining industries. Part one unites Chapters 1, 2, 3 and 4.

Chapter 1 presents a brief review of the origin and evolution of the concept of corporate social responsibility and its relevance to the mining industries; this chapter also outlines methods for assessment of corporate social responsibility. Chapter 2 explores the role of the mining industries in economic development and provides a review of impacts caused by the mining industries on communities and the environment. This chapter reveals the specific nature of corporate social responsibility in the mining industries and addresses a range of issues concerned. Chapter 2 also provides an analysis of tools used by national and international communities to influence the operations of the mining corporations towards the improvement of their environmental and social performance. Further, this chapter highlights key approaches available for mining companies in order to respond to stakeholder concerns. Chapter 3 addresses the importance of sectoral differences and specifics that affect the level of corporate social responsibility in various mining industries such as competition, sector size, ownership and market structure. This chapter presents an in-depth analysis and comparison of corporate environmental and social strategies adopted in two mining industries, gold and diamond mining, and is based on the analysis of environmental and social disclosure of the top global gold and diamond mining companies in their Annual Reports, Environmental Reports and corporate websites.

Chapter 4 concentrates on local geographical specifics affecting the level of corporate social responsibility within the mining industries that include political, economic, legal and cultural factors. The chapter outlines the contextual framework for case study mining companies in the Republic of Sakha and presents three levels within which corporate social responsibility of the case study mining companies is formed: national level (the Russian Federation); regional level (the Republic of Sakha); and local level (mining areas in the Republic of Sakha). Chapter 4 focuses on the significance of local concerns in the formation of the CSR strategies within the mining industries and emphasizes the importance of developing corporate strategies that tackle specific problems or issues created in the mining areas.

The second part of the book provides an investigation and discussion of four key areas of corporate social responsibility, which are considered to be crucial in order for mining industries to demonstrate their responsibilities towards their operations, namely environmental protection, health and safety, employee relations and community development. The nature of corporate strategies in these key areas of corporate social responsibility is analyzed using the findings from the case study of gold and diamond mining companies in the Republic of Sakha. Chapter 5 illustrates specific issues which mining companies face while implementing CSR strategies in the areas of environmental protection, health and safety and employee relations. Chapter 6 presents a detailed analysis of three models for community development in the mining industries. The first model is a development of company-led community programs, which are explored using the evidence from case study companies in the Republic of Sakha. The second model is the establishment and delivery of community initiatives via a corporate philanthropic foundation, the analysis of which is based on the evidence from corporate foundations established by various international mining corporations. The third model for community development is a partnership between government, mining industry and civil society organizations formed to address community problems in mining areas; this model is analyzed using the case study of the SAPI Foundation and Target Fund.

Conclusions summarize the findings from each chapter, the case study companies and foundations. They discuss the ways in which mining companies incorporate corporate social concerns in their management structures and the variety of approaches for integration of CSR issues in business strategies. Here the discussion on practices and different degrees of considering stakeholder concerns in business operations of mining companies is presented. Finally, the book discusses the environmental and social performance of case study companies from the Republic of Sakha in comparison with international practices and analyses the companies in the context of corporate social responsibility and stakeholder approach.

This book demonstrates that mining companies clearly recognize the importance of environment, health and safety, employee and community issues and are starting to make systematic actions to mitigate negative impacts and address stakeholder concerns by adopting proactive strategies, which include the establishment of environmental management systems, regular reporting on environmental and social performance, delivery of community initiatives, engagement in charity giving and improvements in the treatment of employees. The book shows that a variety of tools are available for mining companies to integrate the notion of corporate social responsibility into their business operations. However, mining companies have different levels of integration of corporate social responsibility strategies and it is evident that mining companies in the Russian Federation in many cases lag behind their Western counterparts. Although Russian companies show a distinct willingness to improve their social and environmental practices and have already adopted some successful initiatives, economic difficulties, lack of general management excellence, remains of old Soviet systems thinking, lack of state support for sustainability in the industry and insufficient legislation prevent fuller development of CSR in the Russian mining sectors. Nevertheless, the Republic of Sakha demonstrates a remarkable example of how revenues from mining industry could be invested and utilized for socio-economic and environmental benefit in the mining area.

The case of the SAPI Foundation and Target Fund represents a rare example of an evolving partnership between mining industry, government and nonprofit organization, created to help solve multifaceted problems associated with mining operations in the region. The analysis of these foundations shows that it is difficult to sustain a true tri-sector partnership due to failures in involving stakeholders and demonstrating accountability of the foundation's operations.

PART I

MINING AND THE ENVIRONMENT

Chapter 1

CSR in Mining: Origin and Concept

Origin and Concept of Corporate Social Responsibility

Companies have to meet a variety of responsibilities assigned to them by law, shareholders other stakeholders and the society at large in order to undertake their duties legitimately. Such responsibilities are categorized into four broad types (Carroll, 1979; Brummer, 1991; Peattie, 1992):

- economic responsibility requires a company to be productive and produce goods and services that are desirable by society;
- legal responsibility requires a company to follow set legal requirements to pursue its business;
- moral and ethical responsibility requires a company to follow the acknowledged ethical norms and values;
- social (sometimes referred to as philanthropic) responsibility requires a company to be proactively involved in practices that benefit the society beyond its economic, legal and ethical responsibilities.

According to the view that a firm is a purely economic institution (Friedman, 1970), the overriding economic goal of a firm is to maximize its return to investors via enhancing the value of the firm's stock and increasing the dividends paid to each stockholder. However, corporate social responsibility (CSR) stems from the notion that companies relate to society and cause impacts that go beyond simple marketplace transactions and that companies serve a wider range of human values that cannot be captured by solely focusing on economic values (Walden and Schwartz, 1997; Buchholz, 1998; McWilliams and Siegel, 2001).

The concept of corporate social responsibility goes beyond the narrow economic, technical and legal requirements of the firm, beyond profitable production of goods and services and is increasingly being seen as helping to solve important social and environmental problems, especially those that the companies have helped to create. One important thesis that contributed towards the development of CSR concept is a view that firms can no longer be seen as purely as private institutions but as social institutions. Firms are expected to operate in full understanding of the general welfare of society and to share benefits from their economic activities with society. Therefore, firms retain their social role within the society by responding to society's needs and giving society what it wants (Wood, 1991; Frederick et al, 1992; Walden and Schwartz, 1997; Balabanis et al, 1998; Buchholz, 1998).

The roots of CSR may be found as early as the beginning of 20th century (see Table 1.1). Carroll (1999) calls Howard Bowen, who published a book *Social Responsibilities of the Businessman* in 1953, a father of corporate social responsibility. However, it was not until the 1960s that corporate social responsibility became a much more fully developed concept (Halal, 2001). The concept of corporate social responsibility is derived from several concepts and theories such as legitimacy theory, concepts of social contract and public responsibility, stakeholder theory, business ethics and corporate citizenship (Walden and Schwartz, 1997).

Table 1.1 Development of organization's corporate social responsibility

Year	Type	Characteristics
1859-1910	Corporation is responsible to owners and managers only	A firm has primarily an economic responsibility to contribute to the society by making profit and produce goods and services, while labor is a commodity to be bought and sold.
1900-1950	Corporation is responsible to owners, managers and employees	A firm's consideration of employees grows more than a mere factor of production as the firm has an obligation to provide a stimulating work environment and recognize employee rights as promulgated by employee unions.
1945-1965	Corporation is responsible to owner, manager, employee and other target environment	A firm must supply complete product information to consumers and not engage in deceptive practices. Firms must not be abusive in their pricing policies, and market unsafe products.
Since 1960	Corporation is responsible to owner, manager, employee and other target environment and public at large	A firm must not degrade the environment, must provide opportunities to minority groups, and actively work to promote social justice. A firm is a social institution as much as a legal and economic institution.

Source: Adapted from Kolk, 'Economics of Environmental Management' (2000).

Legitimacy theory According to theory of organizational legitimacy, a business is operating under a mandate from the society, which could be withdrawn if a business is seen not to be doing the thing the society expects of it (Woodward et al, 2001). Corporate social responsibility is seen as a contractual obligation a company has towards society. The notion of the social contract implies that a company operates in society via an implied social contract. It is society that has permitted a company to use natural and human resources and has given it the right to perform its productive functions to attain its power status (Donaldson, 1983; Balabanis et al 1998). However, society's license for a company to operate is not permanent and organizational survival and company's growth are based upon this type of social contract. Therefore, a company must constantly evolve and adapt to the changing needs and expectations of society (Walden and Schwartz, 1997) and must meet the test of legitimacy and seek society's approval (Patten, 1991).

Legitimacy theory stresses the way company management reacts to community expectations and may include pollution prevention and remediation of the physical environment, assurance of health and safety of employees and consumers and those who reside in the communities where products are manufactured and wastes are dumped, as well as responsibility in relation to consequences of unemployment through technological innovation or plant closure (Patten, 1991). Legitimacy theory implies that given the growth in community awareness and concern, that companies will take measures to ensure their activities and performances are acceptable to society and meet society's wider and growing expectations (Walden and Schwartz, 1997; Wilmshurst and Frost, 2000).

Public responsibility One of the driving principles of corporate social responsibility is a public responsibility, by which businesses are responsible for the outcomes relating to their primary and secondary areas of involvement with the society (Wood, 1991; Kolk, 2000). Businesses have often been forced to respond to various social issues that have been considered to be consequences of their activities (Buchholz, 1998). Bowen (1953) emphasizes that business behavior and methods of business operation must fall within the guidelines set by society. From this perspective, CSR is viewed as an obligation for a firm to follow policies and make decisions that would be desirable to objectives and values of the society at large (Bowen, 1953; Carroll, 1979).

Stakeholder theory Stakeholder theory is tightly linked with legitimacy theory. A company through its policies and operations can impact upon various stakeholder groups, including: consumers, suppliers, government, competitors, communities, employees and shareholders. The company may encounter demands from these stakeholders to devote resources to meet its responsibilities (Buchholz, 1998; McWilliams and Siegel, 2001). Henriques and Sadorsky (1999) classify corporate stakeholders into four groups: (a) organizational (including employees, customers, shareholders, suppliers); (b) community (local residents and special interest groups); (c) regulatory (municipalities, regional and central governments, regulatory system) and (d) media stakeholders. Stakeholder theory emphasizes the necessity for companies to consider the needs, interests and influence of those affected by their policies and operations. Stakeholder management involves taking the concerns of these various individuals and groups into account in a process of decision-making, so they can be satisfied to some extent, at least the most of them (Buchholz, 1998).

The central point of CSR is stakeholder management. As shown in Table 1.1 the range of stakeholders that companies target increased over the years. The interests of all corporate stakeholders need to be given consideration by the company, and if their concerns are disregarded they may damage or halt the company's operations. Justification for the stakeholders' involvement in corporate decision-making is that they are affected by the company's operations. Stakeholders' capabilities and power could be turned for or against the company. Therefore the company needs to respond to stakeholder demands (Frederick et al, 1992).

Ethics Business ethics theory, which is often referred to as 'normative theory' concerns holding managers and their companies responsible for implementing ethical principles in their organizations and using moral reasoning in their decisions, in the formulation of policies and strategies, and the general direction of their enterprises (Buchholz, 1998). In the CSR context, managers perform as moral actors and are obliged to exercise managerial discretion, which is available to them, in all domains of social responsibility towards socially responsible outcomes (Wood, 1991; Kolk, 2000). Ethics-driven view of CSR implies that corporate actions are not only made in the name of self-interest that is proposed by social and stakeholder obligations, but in favor of ethical values such as equality, liberty and fairness (Maignan and Ferrell, 2004).

Corporate citizenship The concept of CSR is closely linked to the wider concept of corporate citizenship. According to Carroll and Buchholtz (1999), a company becomes a good corporate citizen when it meets its fourth type of responsibility, a philanthropic responsibility, contributes resources to the community and improves quality of life. Corporate citizenship suggests a parallel between companies and individual citizens, and that companies are endowed with both rights and responsibility in the conducting of their business. Like individual citizens companies are expected to make voluntary contributions to help maintain the overall well being of the society that sustains them (Logan, 1998).

There is no universally accepted definition of CSR; various definitions are based on different business and social theories:

> It refers to the obligations of businessmen to pursue those policies, to make those decisions, or to follow those lines of actions which are desirable in terms of the objectives and values of our society (Bowen, 1953, p. 6).

> The social responsibility of business encompasses the economic, legal, ethical, and philanthropic expectations placed on organizations by society at a given point in time (Carroll and Buchholtz, 1999, p. 35).

> Corporate social responsibility is the continuing commitment by business to behave ethically and contribute to economic development while improving the quality of life of the workforce and their families as well of the local community and society at large (Watts et al, 1998, p. 3).

CSR generally refers to business decision-making linked to ethical values, compliance with legal requirements, and respect for people, communities and the environment. CSR is defined as operating a business in a manner that meets or exceeds the ethical, legal, commercial and public expectations that society has of business. CSR is viewed as a comprehensive set of policies, practices and programmes that are reviewed by corporate top managers (Business for Social Responsibility, online).

Corporate social responsibility implies compliance plus the active development and implementation of mainstream business strategy, supported by technological and organisational innovation, to prevent negative social impacts and optimise social benefits from the outset. It also involves, through responsible management, the mitigation on an ongoing basis, of negative effects, if and when they occur (Warhurst, 1998, p. 82).

Corporate social responsibility means that a corporation should be held accountable for any of its actions that affect people, their communities, and the environment (Frederick et al, 1992, p. 30).

Different definitions and concepts underlying CSR reveal several essential principles upon which CSR is based, they include:

- *Accountability* According to the CSR concept, businesses are held responsible for actions that have an effect on their stakeholders. Accountability implies that companies are obliged to demonstrate the reasonableness of their actions to a community of others through the activity of giving accounts (Shearer, 2002).
- *Responsiveness* In some sense, legitimacy theory and stakeholder theory suggests a degree of reaction from businesses to social concerns and demands. Responsiveness implies a capacity of business to acknowledge social pressures and make actions to respond. Responsiveness also suggests a two-way relationship between business and society (Frederick et al, 1992).
- *Proactive corporate behavior* Being proactive reflects the degree to which corporate behavior is planned in anticipation of emerging economic, political, social, cultural and environmental trends and in the absence of crisis conditions. Therefore, businesses have a capacity to anticipate and prevent negative consequences using proactive measures (Brummer, 1991; Burke and Logsdon, 1996).
- *Voluntarism* Voluntarism of corporate actions indicates the scope of discretionary decision-making by a company in the absence of regulatory or economic pressure. CSR often implies voluntary actions which aim to bring social good that not only go beyond economic interests of a company but which are predominantly not required by law (Walden and Schwartz, 1997; McWilliams and Siegel, 2001). In the context of CSR, those companies that exceed minimum standards set by legislation exhibit voluntarism (Brummer, 1991; Burke and Logsdon, 1996).

There are two contradicting views on the role of voluntarism in CSR. The first view supported by Carroll and Buchholtz (1999) suggests that CSR consists of all four types of corporate responsibilities: economic, legal, ethical and social, which suggests that CSR refers to both types of corporate operations: operations towards compliance with legislation requirements and voluntary operations towards social benefit not stipulated by law or economic requirements. The second view considers that CSR lies outside of legal and economic responsibilities of the firm, because CSR starts when the law ends and it calls on the firm to consider interests of stakeholders who have no direct economic ties with the firm. Therefore, CSR is placed within the domain of voluntarism (Walden and Schwartz, 1997).

Due to inconsistencies in the definition of CSR, there are various views as to which topics and issues are involved (see Table 1.2). Increasingly, CSR is unfolding into two sides: social and environmental (Watts et al, 1998). Social responsibility encompasses human rights, employee volunteerism and philanthropy, community development, corporate governance, workplace issues and education and training. Environmental responsibility covers the issues of pollution abatement, environmental conservation and restoration. However, some CSR issues do not easily fall into either side; they include the economic impact of business operations, health and safety, relationship with government agencies, supplier and customer relations, environmental and social reporting, transparency and communication.

Table 1.2 Different classifications of CSR issues

Classification by Frederick	Classification by Business for Social Responsibility	Classification by Holme
• Economic growth and efficiency • Education • Employment and training • Civil rights and equal opportunity • Urban renewal and development • Pollution abatement • Conservation and recreation • Culture and the arts • Medical care • Government	• Business ethics • Environment • Human rights • Volunteerism and philanthropy • Community development • Corporate governance • Workplace issues	• Human rights • Employee rights • Environmental protection • Community involvement • Supplier relations • Reporting, disclosure, transparency • Principles and codes • Consumer education, product usage and stewardship • Communication • Corruption

Sources: Adopted from Frederick et al, 'Business and Society: Corporate Strategy, Public Policy, Ethics' (1992); Business for Social Responsibility, 'Introduction to Corporate Social Responsibility, (2002); Holme et al, 'Corporate Social Responsibility: Making Good Business Sense' (2000).

The growth of the CSR concept is linked to important societal changes that include the following (Warhurst and Mitchell, 2000; McWilliams and Siegel, 2001; Business for Social Responsibility, online):

Increased interests in business Due to globalization, liberalization and increased foreign direct investment worldwide business today has become a critical factor in the economic and social development of developed and developing countries as well as countries in transition. Business creates jobs, contributes to national economies in terms of tax payments, large corporations support other small and medium size enterprises. Business also increases inputs in communities where it operates via contributions to education, health, social services and cultural development. Business has become one of the main subjects of local, national and international environmental debates.

Shrinking role of government Developed countries of Europe and North America were set on the way of decreasing the role of the state in Western society by expanding the role of market and privatizing state-owned industrial sectors. The UK and the US governments cut down on social programs and started to promote personal responsibility for welfare provision (Logan, 1998). Liberalization of trade in the developing countries of Asia, Africa and Latin America and transition of planned economies to market economies increased the expansion of private capital worldwide. Today, private business is a powerful economic and social force, which is expected to fill in the vacuum created by the diminishing role of the governments to play an increasingly prominent role in the global economic, social and cultural development.

Increased interest in social and environmental issues Public concerns about social and environmental impacts of business operations are increasing as more evidence is revealed about negative impacts of business operations on the natural environment and human health. The formation of environmental responsibility as a separate issue within CSR can be linked to the growth of the environmental movement in the 1960s, which was concerned with environmental implications of economic growth, including nonrenewable energy resources, global warming, climate change, depletion of the ozone layer, degradation of land and water resources and species extinction (Bartelmus, 1994).

Competitive labor market Highly skilled and professional employees are considered as a competitive advantage to the companies. However, employee attitudes about their employers are becoming a critical issue for business. Employees are interested not only in stable employment, fair wages and safe and comfortable working conditions, but in ethical values pursued by companies and the extent of environmental and social impacts of corporate operations.

Supply chain responsibility Companies are being asked to account for the behavior of their business partners up and down the supply chain. Not only environmental impacts could be traced from raw materials to manufacturing, distribution and retail of products, but also ethical aspects of business transactions between stages in the supply chain such as use of child labor and poor health and safety conditions in the production of materials by partners in the supply chain could be attributed to the final products of companies' operations. Downstream responsibilities of companies operations often concern safe use and disposal of products. Therefore, responsible businesses increasingly require auditing and verification of environmental and social performance of their suppliers, as well being asked to account for implications related to their product use and disposal.

Growing investor pressure Financial institutions are increasingly considering environmental and social risks related to business operations in granting credit. Social and environmental performance of business has become a focus of financial concerns due to successful lawsuits against business corporations on the grounds of adverse social and environmental performance leading to substantial compensation payments and related conflicts between the state, businesses and communities, which complicates business operations and endangers financial investments.

Growing peer pressure Competition on the basis of environmental and social conduct is becoming more important for companies. Benchmarking of corporate social and environmental performance by independent organizations and media has a potential to discourage investors, shareholders and consumers from engaging in certain businesses based on ethical considerations. Business and industrial associations are starting to promote CSR strategies through producing guidelines for corporate responsible behavior or voluntary codes of conduct.

Demands for increased disclosure Accounting is a part of public information given by a company to justify its behavior (Lehman, 1995). Companies increasingly experience demands to demonstrate accountability and transparency of corporate operations in relation to their social and environmental impacts. Demands to disclose information on business operations concerning employees, communities, natural environment, corporate contributions to social and cultural development are often pushed by special interest groups and nongovernmental organizations (NGOs). Corporate stakeholders often expect to be involved in decision-making concerning issues that affect them via consultation and communication, where corporate disclosure plays an important role.

The focus on CSR increased further with the rise of the concept of sustainable development (Shrivastava, 1995), which is built on three pillars: economic growth, ecological balance and social progress (Stigson, 1999). Sustainable development calls for growth to be environmentally and socially sustainable with input from governments, industry and civil society (WCED, 1987). Agenda 21, a comprehensive action plan for sustainable development adopted at the Earth Summit in Rio de Janeiro in June 1992, calls for strengthening the role of business and industry in progress towards sustainable development objectives. Chapter 30 of Agenda 21 states:

> Business and industry, including transnational corporations, play a crucial role in the social and economic development of a country. A stable policy regime enables and encourages business and industry to operate responsibly and efficiently and implement longer-term policies. Increasing prosperity, a major goal of the development process, is contributed primarily by the activities of business and industry (UN, 1993, p. 29).

Business and industry are encouraged to undertake voluntary initiatives, promote and implement self-regulation and encourage greater responsibility in ensuring their activities have minimal impacts on human health and the environment (UN, 1993). Therefore, CSR is considered as a principal vehicle for business and industry to contribute towards sustainable development (Watts et al, 1998; Warhurst, 2001). Agenda 21 stresses the need for business and industry to recognize environmental management as among the highest corporate priorities and as a key determinant to sustainable development. Business and industry are encouraged to undertake environmental reporting, adopt codes of conduct for best environmental practice, incorporate cleaner production policies and take actions for improved environmental awareness (UN, 1993) and are also being encouraged towards more effective management of raw materials, energy and waste, as well as the manufacture of more environmentally friendly products (Brack, 1998).

Likely business benefits of CSR include potential to increase financial performance, reduce costs, enhance business reputation and image, improve productivity and quality, attract employee loyalty, improve relationships with the regulators, and strengthen the access to capital (Business for Social Responsibility, online). Increasingly, companies are responding to changing social and environmental values of society and incorporate the concept of CSR in their mainstream business strategies (see Table 1.3). The nature of CSR strategies differs across businesses as various businesses can have different social, economic and environmental impacts. Implementation of CSR depends upon other factors such as the nation in which the company operates, business sector, company size, corporate culture and the degree of corporate commitment (Kolk, 2000; McWilliams and Siegel, 2001). Sectors that cause significant environmental and social impact, that are widespread and have major economic impacts towards the development of nations are increasingly required to demonstrate corporate social responsibility of their business operations (Taylor et al, 1994).

Corporate Social Responsibility in Mining

This book deals with aspects of CSR in the mining industry, which is one of the most important sectors of the world economy. Mining sector is highly developed in many countries and in some cases, defines the success of national economies (Dorian and Humphreys, 1994). Because economic growth is important for the overall improvement of global standards of living and the provision of conditions for the enhancement of environmental protection (Brack, 1998), the extractive industry has a significant role in contributing towards sustainable development at local, national and global levels (Auty, 1993).

Table 1.3 Typical CSR strategies for business

Mission, Vision and Values Statements Companies incorporate CSR in their business decision-making by placing a prominent place to CSR in their core mission, vision and values documents, which provide insight to companies' values, culture and strategies for achieving these aims.

Corporate Governance Many companies have established corporate social responsibility committees on their Boards of Directors to review strategic plans, assess progress and offer guidance about emerging issues of importance. Some boards that do not have these committees have the full board consider issues of corporate social responsibility.

Executive Management Responsibility Some companies have a senior officer with CSR as a responsibility. The participation of this individual in critical company decisions ensures that CSR considerations are taken into account. In other cases, senior managers collectively consider the interests of key stakeholders before important decisions are made.

Strategic Planning A number of companies are beginning to incorporate CSR into their long-term planning processes, identifying specific goals and measures of progress or requiring CSR impact statements for any major company proposals.

General Accountability In some companies, in addition to the efforts to establish corporate and divisional social responsibility goals, there are similar attempts to address these issues in the job descriptions and performance objectives of as many managers and employees in order to identify how each person can contribute to corporate social responsibility.

Communications, Education and Training Many companies emphasize the awareness of responsible behavior and publicize the importance of corporate social responsibility internally, include it as a subject in management training programs, and provide managers and employees with decision-making processes that help them achieve responsible outcomes.

Employee Recognition and Rewards Most companies understand that employees tend to engage in behavior that is recognized and rewarded and avoid behavior that is penalized. The system of recruiting, hiring, promoting, compensating and publicly honoring employees all can be designed to promote corporate social responsibility.

Social and Environmental Auditing and Reporting A growing number of companies assess, audit and report their social and environmental performance on a regular basis in meeting the expectations of investors, employees, customers, business partners on a range of corporate social responsibility issues.

Use of Influence Some socially responsible companies recognize that they can play a leadership role in influencing the behavior of others, from business partners to industry colleagues to neighboring businesses. They understand that ultimately it is in everyone's best interests to have as many companies as possible honoring the requirements and expectations of corporate social responsibility.

Source: Adapted from Business for Social Responsibility, URL: http://www.bsr.org.

CSR in the mining industry is viewed as a mechanism for maximization of positive and minimization of negative social and environmental impacts of mining, while maintaining profits (Acutt et al, 2001). Although mining is one of the oldest industrial sectors in the world, by no means is it slowing down its international operations. Clearly, the liberalization of investment regimes in many developing countries and countries with economies in transition, has contributed to the expansion of the mining industry in remote areas of the world, thus raising more concerns relating to social and environmental issues (Ascher, 1994; Warhurst and Bridge, 1997). Developing countries and countries in transition have undergone significant changes in their political, economic, legal and social regimes and these changes have been felt throughout whole nations and involve all sectors of the economy, including the extractive industry and the way that society, industry and government approach environmental and social problems (Peterson, 1995; Veber, 1999). The complexity of CSR in the mining industry relates to the fact that mining companies originating from developed countries of the Western Europe, North America and Australia, operate many mining projects in developing countries of Asia, Africa and South America. The significant drivers for CSR may not necessarily exist within the developing countries due to the lack of economic, legal and institutional development.

CSR is particularly relevant to the mining industries, because they cause significant effects upon economic, social and environmental dimensions, and are particularly viewed amongst the most damaging and dangerous industrial sectors (see Table 1.4). The growth of public concern about corporate environmental and social responsibility in mining is often linked with environmental disasters and human rights incidents that took place in the mining, petroleum, chemical and energy industries (Shrivastava, 1995; Warhurst, 2001). These incidents include the nuclear reactor explosion in Chernobyl in 1986, Rhine pollution caused by fire at the Sandoz chemical factory in Switzerland in 1986, the oil spill from the Exxon Valdez in Alaska in 1989, acid mine drainage to the Alamosa River from the Summitville Gold Mine in 1992, toxic leaching from waste rocks of the OK Tedi copper mine to Ok Tedi and Fly Rivers in Papua New Guinea in 1994 (see Table 1.5). Improvement of environmental and social performance in the mining industry is viewed as a matter of business survival (Taylor et al, 1994).

There are several reasons for the mining industry to consider CSR as an important business issue (Walker and Howard, 2002; Rae and Rouse, 2001), they include:

- Public opinion of the mining sector is generally negative. Opinion about natural resource extraction industries is influenced mostly by concerns over environmental and social performance than by performances in areas of product pricing, quality and safety (Rae and Rouse, 2001).
- Pressure groups have consistently targeted the mining sector at local and international levels, challenging the legitimacy of the industry. Various environmental, community and indigenous groups regularly protest against development of mining projects worldwide. Large NGOs have developed

campaigns specifically targeting the mining industry such as Oxfam's Mining Campaign and Friends of the Earth International's Mining Campaign.

- Maintaining 'a license to operate' is a constant challenge. Mining companies are facing resistance from numerous social organizations to expand mining operations. Opposition to mining operations is based on accusations of lack of community engagement, impacts on local agricultural land use, pollution and related health impacts and the influx of work-seeking migrants.
- Corporate social responsibility is especially important for mining companies, as mining companies often operate in remote areas of the world, which are often economically underdeveloped and lack provision of social welfare services. Most importantly, since the mining industry is rapidly expanding in developing countries and countries in transition, where regulatory protection of the environment and state provision of the social welfare may not be effective as in developed countries, mining companies have to demonstrate greater CSR to improve economic and social welfare of communities in which they operate.

Table 1.4 Level of environmental impacts by industries

Level of environmental distraction	Typical industries
Damaging, hazardous and dangerous	Agriculture and Fisheries, Armaments, Automobiles, Chemicals and Plastics, Energy: Coal, Gas and Nuclear, Forestry and Timber, Heavy Engineering, Metals and Mining, Road Transport, Shipping, Tobacco, Waste disposal and Water Supply
Wasteful and polluting	Catering, Computers, Cosmetics, Electrical Equipment, Electronics, Food Manufacture, Health Services, Hotels, Light Engineering, Leisure & Tourism, Packaging and Paper, Property Development, Publishing, Railways, Retailing, Telecommunications and Textiles
'Silent destroyers'	Accounting, Advertising, Banking, Broadcasting, Charities, Civil Services, Education, Financial Services, Insurance, Local Government and Social Services

Source: Adapted from Taylor et al, 'The Environmental Management Handbook' (1994).

Table 1.5 Selected incidents and environmental disasters which contributed towards public concern about CSR

Time periods and changes in corporate strategies	Selected incidents and environmental disasters related to operations of major industrial sectors
1960-80s Growing social awareness	Seveso, Italy, 1976 (toxic chemical release) Amoco Cadiz, 1978 (oil spill-off in France)
1980-95 Attempts to address environmental and social impacts	Bhopal, India, 1984 (toxic chemical release) Chernobyl, Ukraine, 1986 (nuclear reactor explosion) Sandoz, Switzerland, 1986 (fire at the chemical factory) Exxon Valdez, 1989 (oil spill-off in Alaska) Summiville Mine, USA, 1992 (acid drainage to the Alamosa River) Ok Tedi Mine, Papua New Guinea, 1994 (toxic leaching from mine to Ok Tedi and Fly Rivers)
1995-present Networking to prevent future incidents	Shell: Brent Spar, UK and Nigeria, 1995 Grasberg Mine, Indonesia, 1995 Marcopper Mine, Philippines, 1996 Los Frailes Mine, Spain, 1998 Aurul SA Mine, Romania, 2000

Source: Adapted from Hamann, 'Mining Companies' Role in Sustainable Development: the 'Why' and 'How' of Corporate Social Responsibility from a Business Perspective' (2003).

In the context of the mining industries, the notion of CSR goes beyond impact mitigation; apart from acceptance of responsibility for corporate actions and minimization of social and environmental impacts; CSR also aims to enhance social and economic benefits of business operations. Mining companies are beginning to accept their environmental and social responsibility by undertaking a range of initiatives and by adopting policies and procedures for improving their environmental and social performance and contributing towards sustainable development of the communities in which they operate (see Table 1.6).

CSR appeals for mining companies to respond to concerns of its stakeholders, including shareholders, employees, customers, local communities and the general public. For mining companies, local communities are considered as one of the most important stakeholders in the context of CSR, as they are significantly affected by mining operations in terms of economic and social welfare and via mining impacts on the natural environment.

Table 1.6 Examples of initiatives in the mining industry for sustainable development and CSR

Environmental initiatives	Socio-economic initiatives
• Improved environmental planning. • Implementation of mining environmental management tools and strategies. • Conduct of regular environmental audits and reviews and installation of monitoring system. • Pollution abatement and control. • Improved waste management and recycling at mine site. • Implementation of cleaner mine technology, including upgrading and replacement and adoption of green technologies. • Environmental restoration and conservation.	• Consultation with local communities. • Adoption of open information policy. • Communication with all groups, not just the authorities. • Assistance to the community as an entitlement, not as a gift from the company. • Local development funds with support from the highest corporate and government levels. • Contributions to the development fund from the company. • Training and support provided to local communities for negotiations. • Efforts made to diversify the local economy and lessen the dependence on the company. • Social and cultural baseline studies, as well as social impact assessments.

Source: Adapted from Bartelmus, 'Environment, Growth and Development the Concepts and Strategies of Sustainability' (1994); Acutt et al, 'Towards Evidence of the Costs and Benefits of Tri-sector Partnerships' (2001).

Evaluation of Corporate Social Responsibility in Mining

There are several criteria to measure CSR, which primarily include expert evaluation, assessment of corporate environmental and social performance, and content analysis of Annual Reports and other corporate documents (Balabanis et al, 1998). The Corporate Citizenship Company assesses CSR through the assessment of specific interactions between business and its stakeholders such as shareholders, employee, consumers, business partners, government, community and the environment. Another approach to the assessment of CSR is an analysis of business practices within certain areas of CSR such as accountability and transparency, corporate governance and ethics, social impacts, business relationships, product quality and stewardship, etc (Lambert and Elix, 2001). Table 1.7 shows two different sets of assessment criteria that analyze CSR of companies in considerable detail. As impacts of businesses and consequently the approaches to CSR vary across businesses, the criteria for assessment of CSR need to be adjusted in order to make them relevant to specific industry or sector.

Table 1.7 Selected criteria for assessment of CSR

Assessment criteria by Corporate Citizenship Company	Assessment criteria by Lambert and Elix
Shareholders	• Accountability and transparency
• Return of investment	• Corporate governance and ethics
• Corporate governance	• Financial returns and performance
Employee	• Business relationships
• Salary and benefits	• Products and services quality and stewardship
• Health and safety	• Regulatory compliance
• Training and staff development	• Management
• Diversity	• Market position
• Communications	• Employment practices
Consumers	• Employee relations
• Price/value	• Workplace safety and issues
• Quality of product	• Environmental management
• Advertising policy	• Environmental performance
Business partners	• Social impact
• Jobs sustained	• Community relations and involvement
• Payment of bills	• Community investment
• Technology transfer	• Selection and monitoring of suppliers
Government	• Charitable gifts
• Tax contribution	• Commercial initiatives in the community
• Local economic impact	• Performance in other countries
• Transfer pricing policies	
Community	
• Charity contributions	
• Community investment	
• Commercial sponsorship	
Environment	
• Sustainable raw materials	
• Emissions to water and air	
• Energy efficiency	
• Waste management	
• Reduced packaging	
• Consumer education	

Source: Adapted from Corporate Citizenship Company, URL: http://www.corporate-citizenship.co.uk; Lambert and Elix, 'Assessing Corporate Social Performance: a Brief Overview of the State of Play' (2001).

Corporate environmental and social performance measurement could be used by external stakeholders for regulatory, control, influence and risk minimization purposes and by the firm itself for purposes of goal-setting, control, monitoring and improvement of corporate processes and benchmarking with competitors (Thoresen, 1999). Corporate environmental and social performance assessments are often framed within the concept of sustainable development or product life cycle and use environmental efficiency as a tool for assessment. Azapagic and Perdan (2000) propose to assess corporate performance using indicators for sustainable development for the industry, a mixture of quantitative and qualitative indicators, that could be applied to three dimensions of industrial sustainability: economic, social and environmental. Ranganathan (1999) proposes a framework to measure corporate performance using a set of key environmental and social indicators at a firm level. Both assessment frameworks are presented in Table 1.8.

Corporate social reporting is an important strategy for corporate social responsibility. Analysis of corporate social disclosure is one of the popular tools for the assessment of CSR of businesses. Corporate social disclosure has grown significantly over the last 20 years and can be described as an area of accounting research that covers both voluntary and mandatory disclosures made by companies regarding issues that are important to the community at large and have more than just an economic nature (Gray et al, 1995; Tilling, 2001). Social disclosure includes disclosures relating to the interaction between an organization and its physical and social environment inclusive of disclosures relating to human resources, community involvement, the natural environment, energy, and product safety. Environmental disclosure can be defined as being part of wider corporate social disclosures (Deegan and Rankin, 1996; Tilt, 2001). Corporate social disclosure has several roles (Tilling, 2001):

- Assessing the social and environmental impacts of corporate activities;
- Measuring effectiveness of corporate social and environmental programs;
- Reporting upon a company's discharging of its social and environmental responsibilities;
- External and internal information systems allowing comprehensive assessment of all corporate resources and social, environmental and economic impacts.

Several theories that explain a company's motivation to disclose social and environmental information including:

- Regulations and standards (Azzone et al, 1996);
- Legitimacy theory (Patten, 1991);
- Political economy theory (Cormier and Gordon, 2001);
- Stakeholder theory (Roberts, 1991);
- Role of public and external pressure (Walden and Schwartz, 1997; Neu et al, 1998);
- Information costs and benefits (Cormier and Gordon, 2001).

Table 1.8 Indicators for assessment of corporate environmental and social performance

Indicators for sustainable development for industry by Azapagic and Perdan

Environmental indicators	Economic indicators	Social indicators
Environmental impacts: Resource use; Global warming; Ozone depletion; Acidification; Eutrophication; Photochemical smog; Human toxicity; Ecotoxicity; Solid waste	*Financial indicators*: Value added; Contribution to GDP; Expenditure on environmental protection; Environmental liabilities; Ethical investment	*Ethical indicators*: Preservation of cultural values including Stakeholder inclusion; Involvement in community projects; International standards of conduct including Business dealings; Child labor; Fair prices; Collaboration with corrupt regimes; and Intergenerational equity
Environmental efficiency: Material and energy intensity; Material recycling; Product durability; Service intensity	*Human-capital indicators*: Employment contribution; Staff turnover; Expenditure on health and safety; Investment in staff development	*Welfare indicators*: Income distribution; Work satisfaction; Satisfaction of social needs.
Voluntary actions: Environmental management systems; Environmental improvements above the compliance levels; Assessment of suppliers		

Key elements of corporate environmental and social performance by Ranganathan

Environmental performance	Social performance
• *Material use* Quantities and types of materials used and resource inputs.	• *Employment practices* Provision of safe working environment; financial and job security; non discrimination practices; professional development.
• *Energy consumption* Quantities and types of energy used or generated.	• *Community relations* Contribution to community development; job creation; tax payment; philanthropy; employee volunteerism.
• *Non-product Output* Quantities and types of waste created before recycling, treatment or disposal.	• *Ethical sourcing* Engagement in fair trading practices (no child or forced labor, safe working conditions, fair wages) with suppliers, distributors and partners.
• *Pollution release* Quantities and types of pollutant release to air, water and land.	• *Social impact of product* Contribution of products and services to social welfare, equity and the meeting of basic human needs.

Source: Adapted from Azapagic Perdan, 'Indicators of Sustainable Development for Industry: a General Framework' (2000); Ranganathan, 'Signs of Sustainability: Measuring Corporate Environmental and Social Performance' (1999).

Several groups of users are interested in corporate social and environmental disclosures, including: academia, community interest groups, employees, NGOs, general public, government, financial community, local community, regulators and policy makers, shareholders, trade and industry (Tilt, 1994; Azzone et al, 1997). There are also a variety of media for social and environmental disclosure (Tilt, 1994; Halme and Huse, 1997; Williams and Ho Wern Pei, 1999; Line et al, 2002; Yuen and Yip, 2002), which include:

- Advertisements or articles published detailing companies' activities;
- Annual Reports;
- Booklets or leaflets produced to address the social activities of the company;
- CD Reports;
- Community Reports;
- Environmental Reports;
- Labeling of products to promote environmental and other concerns;
- Press releases;
- Supplements to the Annual Reports or produced at interim dates;
- Video tapes;
- Websites.

Annual Reports are the most publicized and visible documents produced by the companies (Halme and Huse, 1997; Neu et al, 1998; Santema and Van De Rijt, 2001). Annual Reports possess a certain degree of credibility in comparison with other types of corporate media because it goes through the auditing process of its financial information (Neu et al, 1998). Annual Reports are viewed as an important source of social and environmental information (Harte and Owen, 1991; Tilt, 1994; Neu et al, 1998), because they are produced regularly by major companies and required by legislation making comparison relatively easy (Tilt, 2001). Even though the level of environmental disclosure has increased over the years, companies still provide relatively little detailed environmental information in their Annual Reports and it is often qualitative and not quantitative (Harte and Owen, 1991; Adams et al, 1995; Bullough and Johnson, 1995).

Increasing numbers of companies publish reports that specifically describe environmental concerns and activities (Davis-Walling and Batterman, 1997). Environmental Reports represent a tool that companies use to explain and illustrate their environmental policy, their principal problems, their performance and information on environmental consequences of the companies' activities (Azzone et al, 1997). Environmental Reports have moved from simply demonstrating a company's commitment to the environment, to a position now of communicating environmental performance. Environmental Reports show a company's openness towards stakeholders and the importance of strategic environmental management (Azzone et al, 1997). Environmental Reports are still evolving; companies produce reports covering both environmental and social issues such as Social and Environmental Reports and increasingly try to target wider issues via producing Sustainability Reports and Corporate Social Responsibility Reports.

Companies are increasingly using the World Wide Web to disseminate environmental and social information and in some cases exceeding corporate social disclosure in Annual Reports (Herbst, 1998; Williams and Ho Wern Pei, 1999). In 2000, 65 per cent of companies from Fortune magazine's Global 500 list used the Internet to report on environmental or social issues, either on a global or domestic scale (Line et al, 2002). Companies not only post their environmental reports on the Web, but also place site-specific reports, updated news relating to environment, other consumer and product related material on environment and papers (Herbst, 1998). Companies are successfully combining web-based reporting with hard copy reports, and some no longer produce hard copies at all (Line at el, 2002). Websites are an attractive mechanism for disseminating information due to several factors: accessibility, low costs of allocating information compared to Annual Reports, frequent update of information, advantage of including graphics and photographs, feedback from information users, ability to overcome space limits compared to Annual Reports, etc (Walton and Alabaster, 1996; Williams and Ho Wern Pei, 1999; Wheeler and Elkington, 2001).

Research clearly demonstrates a growth of environmental and social disclosures by companies (Harte and Owen, 1991; Gray et al, 1995; Tilt, 2001). Not only the quantity, but also the quality of disclosures has also changed, i.e. by the end of the 1980s the emphasis shifted from employee information to environmental information (Niskala and Pretes, 1995). However, the level of corporate social and environmental disclosures is dependent upon several factors and conditions. Corporate social and environmental disclosure varies from country to country due to differences in accounting regulations, governmental actions, national culture, economy, existence of pressure groups, and severity of environmental problems (Harte and Owen, 1991; Meek et al, 1995; Williams and Ho Wern Pei, 1999).

It has been established that companies operating in environmentally sensitive industries such as mineral extraction, oil, gas, chemicals industry and forestry tend to disclose more environmental and social information that companies in other industrial sector (Patten, 1991; Adams et al, 1995; Neu et al, 1998; Tilt and Symes 1999; Nisanen and Nieminen, 2001). In addition, large companies disclose more environmental information than small companies (Meek et al, 1995; Niskala and Pretes, 1995; Deegan and Rankin, 1996) and the level of environmental and social disclosures is often influenced by the pressure placed by external publics on the companies (Tilt, 1994; Neu et al, 1998).

Several specific studies have been undertaken on the disclosure of social and environmental information within the extractive sector. Warhurst (1998) analyzed social information reported by 50 mining companies rated by market capitalization by the Financial Times. The study incorporates analysis of several sources of information: Annual Reports, Environmental Reports, Health and Safety Reports, other printed material and websites. The study concluded that an increasing number of mining companies publish Environmental Reports but few publish Social or Community Reports. Information is most commonly generic and does not provide details for specific sites for issues of interest. However, some companies including Rio Tinto, BHP, WMC, Inco, Placer Dome and Newmont have demonstrated a shift from defensive reporting strategies towards a more pro-active engagement with environmental and social issues.

Frost (2000) conducted a study on environmental reporting practices of 60 Australian extractive companies based on content analysis of Annual Reports. The study revealed that some companies are not disclosing any information on their environmental impacts and that in general, it is the larger companies who have higher level of environmental disclosure. Moreover, companies that report higher profits disclose environmental information to a greater extent than less profitable companies, and companies that receive greater media coverage disclose more environmental information.

Different assessment criteria have been developed to evaluate environmental reports of mining companies. For example, WWF in Australia assessed the quality of environmental reporting by Australian mining companies, judging the reports according to the following criteria (Rae, 1998; WWF Australia, 2000):

1. External verification (assessment of the report by external advisor or independent auditor for credibility of a report).
2. Environmental and social issues (relevancy and complexity of addressed issues).
3. Environmental policy.
4. Environmental management progress.
5. Data (relevancy, credibility and consistency of data, reference of data to overall company and individual site levels).
6. Compliance (obtaining of licenses, spill and accidents).
7. Stakeholder participation.
8. Targets.
9. Format (printed and electronic format of a report), dissemination and feedback.

Environmental Reporting Australian Roundtable held in Canberra in 1997 conducted a workshop identifying essential criteria for an environmental report (Dibly and Newson, 1998). The exercise proved that the most important criteria are: independent environmental audit results, environmental compliance to legislation and codes, environmental incidents and emergency response, quantification of environmental impacts, environmental objectives and targets aimed at improvement, organizational policy on the environment, pursuits towards sustainable development, purpose of environmental report and determining the significance of environmental impacts, i.e. risk assessment, and capital and operating environmental expenditure.

Summary

- Corporate social responsibility is a broad concept that has been developing since the 1960s. Development of the CSR concept is based on several management theories and organizational principles, including stakeholder theory, legitimacy theory, corporate citizenship, etc.

- The growth of public concern about CSR in the extractive industry is linked to adverse environmental and social impacts of mining operations and especially to environmental disasters that took place in mining and petroleum industries.
- Assessment of CSR includes expert evaluation; assessment of corporate environmental and social performance; and content analysis of corporate documents.
- Analysis of corporate social disclosure often involves content analysis of corporate documents and analysis of disclosures on the interaction between the firm and its physical and social environment inclusive of disclosures on human resources, community involvement, natural environment and safety.

Chapter 2

Mining Towards Sustainable Development

World Mining Industry

Mining is the world's second oldest and most important industry after agriculture (Down and Stocks, 1977). Mining is currently the fifth largest industry in the world and it plays a crucial role in world economic development, and the trade of mineral commodities represents a substantial part of international trade (Madeley, 1999; Pastizzi-Ferencic, 1992). Mining impacts upon national economies in terms of employment generation, especially new mining projects which can create large, but short-term demand for employment, capital and raw materials inputs, and by the production of new materials for use in the manufacture of semi-finished and finished products. Exploitation of mineral resources significantly contributes to overall economic development and has economic impacts at the local, regional and national level, providing revenue generation, business development, technological improvements and export earnings (Dorian and Humphreys, 1994; Miller, 1999). Many developing countries are heavily reliant upon the mineral sector as a driving force for economic development (see Table 2.1) and transition economies also often have large mineral sectors (Dias and Begg, 1994; Zank, 1995).

In the 1980s-90s, fiscal crises in developing countries and in those of transition economies have encouraged many countries to pass legislation to promote foreign investment. Since 1989, over 75 countries have liberalized their investment regimes for mining. Moreover, privatization of formerly state-owned enterprises including natural resources, are being opened up to private and foreign investments (Warhurst and Bridge, 1997). Privatization of the mineral sector in developing countries has been accompanied by several important issues, which are largely relevant to transition economies as well, including property rights, royalties, policy reform and privatization transaction issues, environmental protection and employment (Zank, 1995). In response to economic and political reforms, the international mining industry is expanding in areas that were formerly closed for mineral exploitation for legal, political and economic reasons and new mining projects progress into remote areas of the world (Madeley, 1999). Moreover, many international companies are restructuring their operations to capture strategic advantage through mergers, acquisitions and alliances (Warhurst and Bridge, 1997).

Table 2.1 List of developing countries and their reliance on fuels, minerals and metals exports, %

Country	Fuels, minerals and metals share of merchandise exports 1965	1985	1990
Angola	6	97	82
Australia	13	44	34
Bolivia	93	82	69
Brazil	9	-	16
Canada	28	-	19
Chile	89	64	57
China	15	25	10
Colombia	18	-	32
Congo D.R.	4	89	89
Ghana	12	30	35
India	10	25	8
Indonesia	43	75	48
Liberia	72	65	65
Mali	1	-	0
Mexico	22	64	43
Papua New Guinea	0	51	61
Peru	45	70	55
Philippines	11	-	12
Sierra Leone	2524	39	38
South Africa	24	39	14
US	8	-	6
Venezuela	97	94	87

Source: Adapted from World Bank Report, 'World Development Report 1992: Development and the Environment' (1992).

In some countries, governments participate in mining projects in order to receive a greater share in benefits and participate in management of mineral resource exploitation. Nationalism is viewed as one of the reasons for government participation in mining projects. For example, a growing expansion of foreign capital in mining projects in developing countries raises public concerns of foreign domination in national economies (Padmore, 1992).

Dorian and Humphreys (1994) categories the effects of mining into two categories:

1. *Primary effects* occur directly as a result of mining operations. The primary impacts and benefits of mining include revenues and profits, foreign exchange, employment, regional development and infrastructure development.
2. *Secondary effects* arise from the linkages between the mining and other industries. Secondary effects are not necessarily planned for and are

commonly a by-product of the original objectives. For example, in transitional economies, specific secondary impacts are the training of employees and the transfer of technology. There are some negative secondary effects such as effects on the environment, socio-cultural effects and negative economic impacts.

Warhurst (1998) categorizes the effects of the mining industry on development into three broad spheres: economic, social and biophysical. The economic sphere includes the effects on relative economic benefits, the distribution of natural resource based commodity rents, and effects on local and remote communities. The social sphere includes socio-political effects on the rights of individuals and groups, human health and working conditions, as well as socio-cultural effects. The biophysical sphere includes effects on the health of ecosystems, on bio-diversity conservation, on air, water and marine resources, minerals, forests and soils.

Environmental Concerns in Mining

Mining is sophisticated process which involves extraction of minerals from the earth's crust and which generally causes irreversible damage to the natural environment (Hore-Lacy, 1992). Mining consists of a several operational phases and every stage bears its own specific characteristics, each having different impacts on the natural environment (Balkau, 1993). The stages of mining include:

- *Exploration* – includes surveys, field surveys, drilling and exploratory excavation.
- *Project development* – includes site development with construction of roads and buildings, underground work on access tunnels, erection of treatment plants, overburden placing, preparation of disposal areas, construction of service infrastructure such as power lines or generating plants, railways, water supplies and sewage and laboratories.
- *Mine operation* – operations may vary from underground mining to surface mining from open pits or placer deposits, hydraulic mining in or near riverbeds. Newer processes may also include heap leaching of tailings' dumps, bio leaching of surface heaps or deposits and solution mining of buried deposits.
- *Beneficiation* – this is on-site processing, which may include communication to reduce particle size, flotation using chemicals, gravity separation or magnetic, electrical and optical sorting, ore leaching with a variety of chemical solutions.
- *Mine closure* – the final stage of mining operations, mine closure and rehabilitation intend to mitigate environmental impacts caused.

Concern over the effects of mining operations on the environment has long been recognized.[1] The extent and scale of mining impacts on the environment is often large and vastly spread (Down and Stocks, 1977; Sengupta, 1993; Hester and

Harrison, 1994; Ramani, 1995). Table 2.2 presents a comprehensive list of major environmental issues associated with mining activity. In addition, other impacts of mining include track traffic, vibration and air blasts, sedimentation and erosion.

Table 2.2 Major environmental issues in mining

- Destruction of habitat and bio-diversity at the mine site
- Ecosystem/habitat/bio-diversity protection in adjacent land
- Landscape/visual impact/loss of land use
- Site stabilization and rehabilitation
- Mine waste/tailings disposal
- Sudden failure of tailings facilities
- Abandoned equipment, solid waste, sewage
- Air emission
- Dust
- Climate change
- Energy consumption
- Changes in river regimes
- Effluent discharges and acid drainage
- Groundwater alteration or contamination
- Hazardous wastes and chemical residues
- Hazardous chemicals handling, safety, workplace exposure
- Noise
- Radiation
- Workplace health and safety
- The impact of metals toxicology on the marketing of metals
- Cultural and archaeological values
- Public health and urban settlement issues around mines

Source: Adapted from Balkau and Parsons, 'Emerging environmental issues for mining in the PECC region' (1999).

The extent of the environmental hazards and the costs of eliminating them can be specific to certain sites and dependant upon several factors including size of operation, geographical location, climate, method of mining, local geology and mineral characteristics. As mineralogy and geochemistry of different mineral deposits differ from each other, therefore the pollution potential consequently varies (Down and Stocks, 1977; Warhurst and Noronha, 2000). There are several mining methods available such as open pit, open cast, underground mining, dredging and marine mining, however, surface mining is the most popular and the most damaging as well (Down and Stocks, 1977; Sengupta, 1993). Surface mining operations include the removal of vegetation, top soils, rocks and other strata that lie above mineral or fuel deposits, which gradually change the physical environment, influence climate, air and water quality. Surface mining creates large volumes of overburden, waste rock, acid mine drainage, airborne dust and other contaminants. These by-products are deposited on land, and they penetrate waterways and pollute the air. In some instances, mining activities can have an adverse effect on human health, as well as wildlife (Sengupta, 1993; Hester and Harrison, 1994; Eggert, 1996). Adverse occupational health effects associated with mining include injuries and accidents, respiratory conditions, i.e. tuberculosis, pneumoconiosis, chronic heavy metal exposure and chemical exposure (Ramani, 1995; Maclean and Warhurst, 2000).

The existence of important mining regions in industrialized countries has shown that problems such as land degradation, acid mine drainage and tailings stability, are not simply confined to developing countries, but are issues that face the industry as a whole (Warhurst and Insor, 1996). Therefore, mining is viewed as a cause for substantial changes to the natural environment of mining that have long-term implications at local, regional and global scales (Miller, 1999). As a result, the mining industry is currently undergoing a transformation, as the industry faces new challenges. Mining companies have begun to restructure their operations responding to liberalization of investment regimes, new production technologies and the growth of environmental concern (Crowson, 1997; Warhurst, 1998).

Social Concerns in Mining

Mining operations not only affect the natural environment, but also influence societies in which they operate directly and indirectly via economic and environmental effects. Socio-cultural impacts from mining vary in scope, intensity and importance relative to the scale of mining (Clark, 1992). Social impacts are those that affect the interaction between people or groups and cultural impacts are those that affect the basic beliefs and values of the people or groups (Clark, 1992). As shown in Table 2.3 social impacts of mining are categorized into primary and secondary (Acutt et al, 2001). Primary impacts include issues directly linked to the project and are typically addressed through mitigation. Secondary impacts include wider social and economic issues, which are more difficult to relate directly to a specific project or to the responsibility of a particular operating company. Clark (1992) categorizes social and cultural issues arising from mining development into several groups:

- Disruption of the physical environment and environmental degradation.
- Disruption of societal organization and cultural values. For example, the employment of men at mine sites, while bringing an opportunity to participate in cash economy, increases the burden on agriculture and on women who remain in the villages. In addition, sacred sites can be affected by mining operations as well (Madeley, 1999).
- Inequitable revenue distribution. Socio-economic impacts of mining also include wage disparities between mine employees and local community and low compensation (Stevenson, 1996; Madeley, 1999; Pressend et al, 2001).

- High inflation or even hyperinflation.
- Increased and unresolved issues on land access/loss of compensation.
- Adjustment problems of relocation and migration. For example, during the last hundred years, mining has meant that probably 100 million people, most of them in developing countries, have been removed from the land where they lived and farmed (Madeley, 1999).
- Increased costs of government of providing services.
- Loss of control by local people over major decisions.

Table 2.3 Examples of social impacts of mining

Primary social issues	• Disturbance to agricultural land and livelihoods • Increased road traffic • Encroachment on spiritual and burial sites • Cultural disturbance • Displacement of families or entire village
Secondary social issues	• The spread of disease • Changes in migration patterns • The change of world views and cultural values • Altering economic, livelihood and purchasing system • Economic dependency • Inflation • Competition for land resources • Shifting marriage patterns • Health care practices

Source: Adapted from Acutt, 'Towards Evidence of the Costs and Benefits of Tri-sector Partnerships' (2001).

Mineral projects also cause various social impacts through the development of infrastructure, roads, railways, hydroelectric engineering and town building for workers (Madeley, 1999). The extent of socio-cultural impacts of mining depends upon the level of development already present in the area of proposed mineral development, which can range from a very primitive level of development as in Papua New Guinea, to very advanced levels of development as in Canada (Clark and Clark, 1999). Today, the mining industry is moving into remote areas in the search of minerals, where new development lacks established industry and little or no infrastructure (Khanna, 1999). Therefore, mining often becomes the leading sector in the area of its operations, providing employment and infrastructure development (Ali and Behrendt, 2001). Mining companies have a tremendous influence on the areas of their operations and may result in mine-dependent societies, which makes the mining companies, in part, responsible for the well-being of the communities (Khanna, 1999).

Socio-cultural effects of mining include effects on traditional lifestyle of communities. In many cases, mining activity is conducted in areas of indigenous people's livelihood. Through associated social and environmental impacts, mining companies significantly influence indigenous people and their communities, posing a threat to their way of life and culture and causing disputes in relation to property rights (Ali and Behrendt, 2001). Therefore, land and access issues arise which relate to social and cultural impacts of mining, particularly in Australia and Canada (Clark and Clark, 1999). These impacts on the local community via public health issues including HIV and AIDS, which is increasingly becoming a major threat to workers, especially in Africa (Pressend et al, 2001). The negative impacts of mining project developments on the physical health of individuals and communities have been the source of extensive research. Premature death and illness due to environmental health risks account for one fifth of the total burden of disease in the developing world (Lvovsky, 1999). The connections between health and the environment have been established by research in social, economic and cultural factors in the etiology of disease, but there is still clearly a need for further research (Maclean and Warhurst, 2000).

There are several factors that define socio-cultural impacts (Clark and Clark, 1999), including:

1. Social and cultural impacts are largely irreversible.
2. The largest changes are to traditional values.
3. The transition to cash economy is the largest catalyst to change.
4. Impacts arise because the process of change is rapid but the capacity of change is slow.
5. Greatest conflicts arise between the tribalists and developmentalists.
6. Socio-cultural impacts have local, regional and national implications, which are largely not understood by government.

Socio-cultural changes brought by mining operations within local and indigenous communities occur during the life cycle of mining project development. The research of Clark and Clark (1999) has shown the following:

1. The early stages of exploration and development result in the most conflict, in terms of total number of areas of conflict, between the development and the indigenous people.
2. The mineral development phase is normally one of reconciliation between the development and the people and is the time when the greatest disruption occurs within the social-cultural system of the indigenous people.
3. Long-term mineral development results in periodically renewed conflicts between the development and the indigenous peoples, particularly the younger generations who may feel disenfranchised from the benefits of the project – largely reflecting the breakdown of the cultural peer system.
4. Exhaustion and abandonment of a mineral development venture is the most traumatic phase, leading to major social conflict within the area and a complete reassessment of social and cultural values.

Often communities that are close to mineral resource exploitation projects expect to benefit, or be impacted by the project in a various ways (Davy, 2001). Several expectations are formed in relation to mining project development, including the development of infrastructure and the promotion of other businesses. Currently, due to privatization of mining operations in developing countries, almost all new mining projects are financed by private capital. Therefore, the expectations of the communities in the role of corporations and nongovernmental organizations (NGOs) in provision of social infrastructure and services have increased (Davy, 1999). Furthermore, there is concern that after mine closure, the mining sites may result in barren landscapes and this concern in also related to the lifetime of mineral projects, which may last for only a few decades. Therefore, mining companies are expected to encourage other businesses to invest in the region (Ali and Behrendt, 2001).

Companies are beginning to realize that public perception and expectations are challenging corporate strategy, whereas in the past, commercial rationale was sufficient to justify a business decision (Khanna, 1999). In response to public pressure, companies have found that they have to move away from the traditional use of public relations towards community consultation (Khanna, 1999). It is becoming widely accepted that local communities are a crucial part of this consultation process and in many cases they are demanding compensation, which covers infrastructure, training for mine and non-mine employment, and social services including education and health (Crowson, 1998; McMahon and Strongman, 1999).

Contribution of the Extractive Industry to Sustainable Development

Discussions related to the link between development and the environment started in the 1970s. In 1972, the Club of Rome published *The Limits to Growth*, a report that concluded, 'if the present growth trends in world population, pollution, food production, and resource depletion continue unchanged, the limits to growth on this planet will be reached within the next one hundred years' (Meadows et al, 1972). The following developments on the subject such as the Conference on the Human Environment held in Stockholm in 1972, leading to the establishment of United Nations Environment Program (UNEP), also contributed to the debates on development and the environment. In 1974, UNEP and the United Nations Conference on Trade and Development (UNCTAD) convened the Cocoyoc seminar in Mexico on Patterns of Resource Use, Environment and Development, which already used the term sustainable development, which was used in relation to the discussions on the relationship between economic growth and the natural resource base upon which it depends.

In 1983, the General Assembly of the United Nations established the World Commission on Environment and Development (WCED). During 1984-1987, the Commission conducted meetings and public hearings in ten countries, sponsored studies and reports which further developed the discussion on sustainable development. The Brundtland Report, *Our Common*, the report written by the

World Commission on Environment and Development (WCED), defines sustainable development as:

> ...development that meets the needs of the present generations without compromising the ability of future generations to meet their own needs (WCED, 1987, p. 43).

The most notable conference, which was held on the topic of environment-development relations, was the 1992 United Nations Conference on Environment and Development (UNCED), known as the Earth Summit, which took place in Rio de Janeiro. Three important documents were adopted at the Earth Summit by consensus: Rio Declaration, Statement of Principles on Forests and Agenda 21, plan for achieving sustainable development in the 21st century. To ensure the follow-up of UNCED, the Commission on Sustainable Development was established in December 1992 to monitor and report on implementation of the Earth Summit agreements at local, national and regional levels. Consequently, the idea of sustainable development was accepted by international and intergovernmental organizations and has become official policy of the United Nations and the World Bank, and for many countries throughout the world.

The notion of sustainable development lies in progress within economic, social, as well as environmental dimensions. The principles of sustainable development are described as social progress that recognizes the needs of everyone; effective protection of the environment; prudent use of natural resources; and maintenance of high and stable levels of economic growth and employment. The objectives of sustainable development include: reviving growth; changing the quality of growth; meeting essential needs for jobs, food, energy, water, sanitation; conserving, enhancing the resource base, reorienting technology and managing risks; merging environment and economics in decision making (Bartelmus, 1994). However, for many countries, which are eager to raise their standard of living and increase economic growth, sustainable development implies a low rate of economic growth via decreasing their energy and mineral resource exploitation (Miller, 1999).

The term sustainable development is used to reflect a growing concern about the interaction between economic activity and the quality of environment. If renewable resources are extracted at a rate greater than their regenerative capacity, their utilization cannot be considered sustainable. In relation to mineral resources, the depletion of natural resources is a major concern in the debate about sustainable development (Cowell et al, 1999; Miller, 1999). Sánchez (1998) argues that the depletion of mineral resources should be compensated by 'the generation of new wealth, which, in form of useful lasting capital, can benefit present and future generations'. Hore-Lacy (1992) agues that if we properly take into account the needs of future generations and environmental effects, there is no reason for us to restrain the use of minerals. Although the resources of the Earth are finite, the limits of the supply of resources are far away from true and have no practical meaning and therefore mineral deletion is not an issue for foreseeable future. Minerals are regarded as nonrenewable resource, but compared with renewable materials such as wood and food; the amount actually in the earth's crust is vast. There are arguments that unlike fossil fuels, many metals and minerals can be recycled without losing their chemical or physical properties. Therefore, this means that the issue of depletion and the classification of metals as nonrenewable resources are two concepts that are increasingly being challenged (Crowson, 1998; Sánchez, 1998). Moreover, the discovery of new mineral deposits and advances in technology now make possible the recovery of minerals from previously unprofitable deposits and even from waste stockpiles and add to existing stocks of mineral resources (Sánchez, 1998).

Sustainable development demands that the environmental impacts of natural resource exploitation, extraction, transformation, consumption and recycling and final disposal should be minimized. This means that land that has been used for mining operations must be rehabilitated to a state that allows successive uses to benefit from it. The same applies to tailings and waste rock disposal areas, as well as for industrial sites (Sánchez, 1998). Mining in terms of sustainable development can be interpreted as a transparent process which makes sure that appropriate revenues generated by exploration of nonrenewable resource are suitably invested to ensure the future development of long-term sustainable livelihoods for the members of the affected community (Epps, 1996). The mining industry is being increasingly pressured to minimize social and cultural disruptions by taking into account views and values of local communities and the general public. This is particularly important for mineral exploration and extraction, which often takes place in the ancestral land of autochthons (Sánchez, 1998).

A study recently conducted by PricewaterhouseCoopers of 32 world-class mining and mineral organizations revealed that 18 of the studied organizations use the term of sustainable development for various internal and external uses, and have an understanding of sustainable development which primarily focuses on local operational impacts, particularly on employees and local communities (PricewaterhouseCoopers and MMSD, 2001) (see Table 2.4). There are several key factors that drive mining companies to address sustainable development, including: enhancement of shareholders value; survival of the business in the long term; improvement of risk management and improvement of relationships with local communities. The WBCSD views corporate social responsibility as the third pillar of sustainable development, along with economic growth and ecological balance, and therefore, a key component for a sustainable future (Watts et al, 1998). Many mining companies consider sustainable development in their corporate strategies and therefore develop environmental, health and safety, and ethics policies that relate to the area of sustainable development (PricewaterhouseCoopers and MMSD, 2001).

Table 2.4 Priority of issues on understanding sustainable development by mining and minerals industry

- Impact on people in local communities
- Economic impact on local communities
- Impact on the lives of employees
- Interaction with local communities (e.g. consultation processes)
- Environmental impact in the local area
- Considering the long term effects of corporate decisions on the environment, society and economy affected by corporate operations
- Interaction with government and regulators
- Economic impact in countries of operation
- Contribution of global environmental impacts (e.g. climate change)
- Economic, environmental and social impact of products throughout their life-cycles
- Educating our employees and local communities on the value to society that mining provides
- Environmental, economic and social impacts of suppliers and contractors

Source: PricewaterhouseCoopers and MMSD, 'Mining and Minerals Sustainability Survey 2001' (2001).

International Influences upon the Extractive Industry

Since environmental problems are transnational, activities affecting the natural resource sector in one geographical area will have effects globally. There are number of bodies which influence the mining industry for better environmental and social practice (Khanna, 1999):

- Environmental pressure groups, NGOs such as Minewatch, Greenpeace, Friends of the Earth and the Mineral Policy Centre, which have generally a negative attitude to mining.
- International organizations such as the World Bank, UNEP provide the mining industry with guidelines.
- National governments, which are responsible for establishing legislation and ensuring compliance.
- Mining associations, which assist their members to improve their environmental, social and economic performance.
- International media, whose coverage of mining disasters can be intense.

UNEP has a number of activities specifically addressing the mining industry, including activities of its units, particularly, the Production and Consumption Unit, the Water Unit and UNEP programs, such as the Cleaner Production, Awareness and Preparedness for Emergencies at Local Level (APELL) and Environmental

Technology Assessment have also been integrated into mining activities. The objective of UNEP's mining activities is to improve environmental performance of the mining industry world wide through (UNEP, 1999):

- influencing international and national government polices and company practices on environmental management, regulations and practice;
- monitoring, interpreting and communicating trends in mining and the environment;
- seeking, with partners, cost-effective solutions to major environmental problems in mining.

In recent years, national governments have adopted several international environmental conventions under the auspices of the United Nations. In addition, regional bodies such as European Union concluded other agreements. Balkau and Parsons (1999) outline the most important conventions and protocols that have affected and will continue to affect the mining industry (see Table 2.5).

Established in 1991, the International Council on Metals and the Environment (ICME) brings together major nonferrous and precious metals mining and primary metal companies on a worldwide basis. It was set up with support of the Mining Association of Canada and with financial support of eleven founding members. ICME represents approximately 60 per cent of the Western world's production of copper, lead, nickel, silver and zinc. ICME acts as an industry representative on environment and health issues at international level and it develops and promotes criteria for responsible policies and practices in relation to environment and health (Bomsel et al, 1996).

The Global Mining Initiative brings together many of the world's largest mining and minerals companies under the auspices of the WBCSD, which aims to ensure that the industry which is essential to the well-being of a changing world is responsive to global needs and challenges.[2] The principal action of the Global Mining Initiative has been to launch an independent study, the Mining, Minerals and Sustainable Development project, which aimed to promote industry's awareness of new socio-environmental challenges with specific respect to sustainable development (MJ, 2002).

Voluntary initiatives have become important tools in encouraging improvement in performance of the mining industry in a range of areas from environmental performance to environmental reporting. International and national mining associations are working towards promotion of sustainable practices in the mining industry and introduce guidelines, codes of practice and environmental charters, which are voluntarily adopted by their members. This trend is seen as industry self-regulation along with increased governmental control (Dias and Begg, 1994). The development of codes for the mining industry tries to provide leadership in improving performance and transparency (Wells, 2001). Environmental codes are voluntary and set principles for companies to follow in order to improve their environmental performance. Environmental codes are common in a small number of countries, predominantly in Australia, Canada, Brazil and the USA. There are a number of minerals industry codes such as: Minerals Council of Australia – Code for Environmental Management; Canadian Institute of Mining, Metallurgy and Petroleum – Environmental Code of Conduct; Mining Association of Canada – Policy and Environmental Management Charter; and International Council on Metals and the Environment has adopted Environmental Code as well (Atkinson, 1999; Balkau and Parsons, 1999; Cottrell and Rankin, 2000).

Table 2.5 List of important conventions and protocol affecting the mining industry

Instrument	Impact on the mining industry
World Heritage Convention, 1972	Affects permission to mine if there is a potential to impact natural or cultural values.
Montreal Protocol, 1987 (Vienna Convention)	Fire protection technology affected. Changed refrigeration practices, especially in the deep gold mining industry in South Africa.
Basel Convention, 1989	Likely to impact trading of some metals and is of greater concern to the smelting industry.
Biodiversity Convention, 1992	Affects access to land, because it addresses endangered ecosystems and indigenous peoples' rights.
Kyoto Protocol, 1997 (Framework Convention on Climate Change)	Likely to lead to energy price changes, controls on the release of carbon dioxide, methane and other greenhouse gases, and may affect coal markets.

Source: Adapted from Balkau and Parsons, 'Emerging Environmental Issues for Mining in the PECC Region' (1999).

Governmental Regulations for Environmental Protection in Mining

Between 1985 and 1995, over 90 nations introduced or commenced working on new or major revisions to mining legislation. Reforms to the mineral regulatory system require changes in mining law and other laws dealing with matters such as land, taxation, imports and exports, waters, foreign investment, foreign exchange, labor practices, safety and environment (Otto, 1997). Determining the environmental policy for sustainable development of the natural resources sector is a national priority in a number of developed and developing countries. There are typical environmental policies in the natural resources sector, including pollution abatement, pollution control, waste reduction at sources, waste disposal, transport and trade in hazardous waste, recycling, conservation, technology upgrading or replacement, green technologies and environmental restoration (Dias and Begg, 1994). National governments use a variety of instruments for the implementation of environmental policies in the private sector (Dias and Begg, 1994), which are divided into three groups:

1. *Command and control regulations* – aim to protect the environment by setting standards and procedures to restrict the amount and types of pollution allowable, such as emission standards, quality standards, and process and product standards. These regulations may be in a form of statute law or may be incorporated in contract, agreement or license.
2. *Economic incentives and disincentives* are used in some circumstances when command and control mechanisms are inappropriate. These instruments include subsidies for environmentally friendly investments, tradable permits, and numerous forms of emission charges, fiscal disincentives and fines.
3. *Environmental impact assessment* (EIA) is used as precondition for issuing licenses or permits to new industrial activities. In 1988, EIAs became obligatory for certain major projects in European Community member states. The EIA process aims to provide decision-makers with an indication of the environmental implications of major projects before a decision is made to initiate a project. Data and details on likely impacts are complied and measures proposed to avoid, reduce or remedy significant adverse effects.

Australia, Canada and USA have several common elements in their environmental regulations related to the mining sector, which include: reserving some land areas from mining activities (national parks and wildlife management areas), requiring environmental impact assessment prior to governmental approval of projects, setting limits for discharges of waste and contaminants to water, requiring monitoring as part of permit, using enforcement options for violation of permit conditions, requiring reclamation of surface area upon cessation of mining operations. In industrial countries, especially in Australia, Europe, Japan and North America, a regulatory framework for safeguarding the quality of land, water and air affected by mining and mineral processing is becoming increasingly complex (Acquah, 1994). This growing concern about environmental degradation occurred during a period of rapid liberalization in developing countries, which is related to promotion of foreign investment (Warhurst and Insor, 1996). In developing countries, environmental regulations designed specifically for mining and mineral processing have been until recently uncommon; however, a few developing countries (e.g. Chile and Brazil) have adopted extensive regulatory frameworks (Acquah, 1994). The concerns in relation to degradation of natural resources in developing countries is focused on management of natural resources for sustainable and productive growth, and fair and equal allocation of benefits from natural resource exploitation within the country (Newcombe, 1999). Developing countries' governments and those of economies in transition have come to realize that their attitudes must shift from one of development versus environment to one of development and environment (Dias and Begg, 1994).

After the breakdown of the former Soviet Union, the transition to more democratic political systems, and the emergence of market economies in Central and Eastern Europe, nearly all governments in the region have adopted more stringent environmental regulations based on US and Western Europe standards. Several governments in Central and Eastern Europe, including the Czech Republic, Hungary, Slovakia, Poland and Bulgaria, have created special Environmental Protection Funds to collect public revenues from air, waste and water charges, pollution fines, grants and special taxes and allocate them to solve environmental problems. Similar Funds have been established in Russia, Belarus, Ukraine and Kazakhstan. However, few governments in the region have adequate financial resources to clean up the environmental damage that has occurred over the past half century or to enforce new environmental regulations effectively (Rondinelli and Vastag, 1998).

The link between natural resource exploitation and environmental protection is increasingly viewed as a national priority for many governments of both developed and developing countries (Ascher, 1994; Dias and Begg, 1994; Balkau and Parsons, 1999), and this link is clearly reflected in the mining sector (Eggert, 1996; Warhurst and Insor, 1996). In many developed, developing and transition nations, efforts to reform mineral regulatory systems have been slow. Often this slow rate of progress is rooted in the lack of a political consensus regarding key issues. Divisions of authority between central, regional and local government, between competing government ministries and departments, and pressures brought to bear on politicians by special interest groups can all act to hinder the introduction of new legislation and administrative procedure (Otto, 1997). Although the industry is increasingly coming to accept its share of responsibility towards the environment, there are limits to what companies can and will do, particularly in developing countries or economies in transition, where there is all too often an absence of appropriate government regulations (Poulin and Sinding, 1993; Bomsel et al, 1996). Moreover, environmental regulations alone are unlikely to address environmental problems (Warhurst and Bridge, 1997; Warhurst, 1999) and the industry is called upon to undertake initiatives beyond regulatory requirement in order to solve environmental problems arising from mining operations.

The Role of the Mining Industry

The mining industry today is very much concerned about the possible impacts of environmental and social issues on its business activity and therefore those issues have become key agenda items (Buckley, 1992; Epps, 1996; Sinding, 1999). Not only the enforcement of regulations, public pressure on the mining industry and ethical considerations (Welford, 1998; Khanna, 1999), but also the realization that improvement of environmental performance can reduce costs and enhance the image and reputation of business (Azapagic and Perdan, 2000), drive mining companies to challenge their strategies with regard to environment. In the developed world, the industry's response to environmental issues and sustainable development underwent three major stages (Azapagic and Perdan, 2000):

1. Reactive approach (1970-1980s), when the industry developed its environmental performance forced by regulations and used end-of-pipe solutions as an option for solving environmental problems.
2. Proactive approach (1980-1990s), when companies realized that pollution prevention, cleaner production, reduction of waste at source and more efficient use of materials bring more benefits in terms of environmental performance, reduction of costs and increase of profits.
3. New approach, which companies are starting to integrate environmental performance into business strategies and development, communicate externally through environmental reporting.

There are a number of ways in which the mining industry can avoid undue environmental impacts through systematic application of preventive technologies, processes and procedures (Balkau, 1993), including:

- *Mine development and operation* A number of mining techniques are in widespread use depending on the site-specific characteristics, and the environmental and other constraints that may be imposed.
- *Beneficiation and processing* The processes chosen depend on the ore characteristics and location; however there is a research into cleaner and safer options.
- *Dust control* Dust can be controlled on a mining site through careful design and selection of excavation, transport and crushing equipment.
- *Noise reduction* Noise should be one of the criteria for choice of mobile transport equipment and machinery. Traffic movement should be planned to avoid as much as possible steep grades and noise sensitive areas.
- *Recycling and reuse* Greater potential exists for reuse of wastewater and processing effluents.
- *Effluent treatment* Mine drainage and effluents will often need treatment prior to discharge.
- *Disposal of processing effluents* Effluents from mines and processing operations are usually discharged after suitable treatment to water bodies near sites. Disposal to tailing dams may be feasible for some effluents if the volumes are within the capacity of the dam.
- *Waste disposal* More options often exist for managing waste disposal.
- *Accidents and emergencies* All mines should make provision for dealing with unexpected incidents. Collapse of mines or tailings structures is an ever-present risk, and worst-event scenarios can help the management plan for such eventualities.

Table 2.6 Responses of the mining industry to mitigation of impacts on environment and society

Environment	Society
• Research and development – industry solves or minimizes environmental problems through the development of better technologies. • Management reorganization – Many mining companies are putting in practice and auditing environmental management systems, publishing reports on their environmental performance and preparing statement of environmental policy. • Environmental management standards – industry is increasingly become assigned to standards for environmental management, such as ISO 14001. • Environmental reporting – informing stakeholders on environmental performance and activities. • Voluntary agreements – some mining companies are going beyond regulatory standards for emissions and environmental quality and enter in voluntary agreement aiming to reduce pollution and improve environmental performance.	• Employment practices – provision of a safe working environment; financial and job security; equal opportunities; and professional development. • Involving stakeholders – industry has developed various ways of communicating with governments, industry, environmentalists, communities, indigenous people, employees and independent organizations. • Community development – contribution to community development, including job creation; taxes paid/tax breaks received; philanthropy; and employee volunteerism. • Tri-sector partnership – cooperation for sustainable development involving government, industry and civil society organizations.

In addition, to achieve cleaner mining, mining companies should reduce the consumption of energy and materials and promote the use of cleaner technologies (Seymoar et al, 1998). All these actions have an important management implication, and mining companies have developed several responses (see Table 2.6) in order to mitigate the impacts on the environment, employees and local communities (PricewaterhouseCoopers and MMSD, 2001), which are regarded as the most important industry challenges for sustainable development.

Warhurst and Bridge (1997) argue that competition and environmental imperatives have increased the incentive for mining companies to invest in innovation and assimilation of new technologies. These technologies enable the profitable extraction and processing of lower grade and more complex ores, reduce production costs and improve processes for pollution prevention, process control, remediation and re-use. Liberalization of investment regimes and foreign investment in the mining sector, including joint ventures and other forms of strategic alliances between multinational mining companies and domestic producers, may provide an opportunity for the transfer of the skills and technological resources in order to improve environmental performance in developing countries (Warhurst and Bridge, 1997).

Voluntary agreements between industry and the government increased in the 1990s. Industry is agreeing to take measures that go beyond existing regulatory requirements in exchange for commitments by governments to provide various types of regulatory 'relief' such as simplified permitting process or reduced frequency of reporting and inspections. However, voluntary agreements usually involve larger companies.

Responses of the Extractive Industry to Environmental and Social Concerns

Due to the rise of environmental and social concerns and increased pressure from international organizations and national governments on the mining industry, mining companies are developing strategies to improve their environmental and social performance. This section therefore focuses on several responses developed by the mining industry that are important for mitigation of impacts on the natural environment, on health and safety of the workforce and on local communities. The mitigation of these impacts is closely related to the concept of sustainable development and its implications to the mining industry. The mining industry understands the concept of sustainable development in relation to their operation primarily focusing on impacts to local communities, employees and the local environment (see Table 2.4). These particular responses support the concept of social and environmental responsibilities of corporations that concentrate on areas such as environment (embracing environmental protection, pollution abatement, conservation and environmental reporting), workplace issues, community involvement, and philanthropy. This section refers to the implications that mitigation of environmental and social impacts has on corporate strategies regarding reorganization of environmental management, environmental reporting, health and safety, human resources, stakeholder involvement, community development and tri-sector partnerships.

Management Reorganization

Krut and Gleckman (1998) define environmental management as 'the management of institutions and people that impact the environment'. Using the definition of Organization for Economic Cooperation and Development (OECD), environmental management covers issues such as energy efficiency, waste minimization, cleaner production and eco-efficiency (OECD, 2000). The concept of environmental management incorporates several common principles (OECD, 2000), including:

- The simultaneous achievement of economic and environmental benefits and the integration of environmental management into mainstream management practice.
- The application of an integrated, preventive environmental strategy to production processes, and to products throughout their life cycle.
- The minimization of risks to human health, and the sustainable use of natural resources.

Environmental Management Systems (EMS) were developed to improve environmental performance of corporations (Kolk, 2000). EMS are the collection of procedures, activities, organizational structures and engineering controls employed by an organization to direct and regulate its activities in order to ensure improvement of environmental performance and minimization of environmental damages (Welford, 1998). EMS are defined by the British Standards Institution (cited in Welford, 1998, p. 38) as: 'the organizational structure, responsibilities, practices, procedures, processes and resources for determining and implementing environmental policy'. There are several basic stages for development of environmental management systems (Dias and Begg, 1994; Watkins, 1998; Welford, 1998), including:

- Policy statements and commitments.
- Plans and programs.
- Integration of plans and programs.
- Measurement, audits and review of environmental management performance against policy, plans and programs.
- Education and training.
- Publication of environmental performance information.

According to Brophy (1998), environmental policy is a foundation upon which an organization should base its interactions, decision-making and future policies in relation to environmental concerns. Environmental policy generally outlines a company's commitments to environmental management and is developed before any environmental management activity and should be revised once changes in effects and performance are identified. Typically, an environmental policy includes the company's values, statements, specific objectives and targets with regard to environmental issues (Welford, 1998) and covers several issues, including:

- Attitude towards environment.
- Overall environmental goals.
- Commitment to audit and review environmental performance.
- Commitment to meet or surpass environmental standards.
- Conservation of natural resources.
- Minimization of environmental effects.
- Commitment to provide healthy workplace.

- Commitment to liaise with local community and society at large on environmental issues.
- Commitment to train and educate personnel on environment, health and safety issues.
- Special commitments to areas such as energy, waste, land and water management.

Having EMS in place would mean that action plans are drawn up to mitigate or avoid the impacts. A key step towards mitigation of environmental impacts is seen in EIA, which should be managed from the outset of the mining project (Balkau, 1993). The EIA is a decision-making tool, used by both the project proponents and regulating agencies and it identifies the consequences of implementing the mining operations on the biogeophysical environment, human health and welfare (Sinding, 1999). Plans and programs of the mining companies should be based upon predictions of the environmental impact assessment and should be followed by regular monitoring and auditing of environmental performance (Warhurst and Noronha, 2000). The EMS requires establishment of organizational structure (Warhurst and Noronha, 2000). There must be assigned personnel with environmental expertise who implement plans and programs and carry out regular monitoring and report to top management (Balkau, 1993). Therefore, an appropriate communication between the site staff and the management should be established and adequate finances should be allocated to carry on the environmental activities (Balkau, 1993; Warhurst and Noronha, 2000).

Mining companies should carry out regular auditing of environmental performance (Balkau, 1993). The audit monitors a company's practice to see that it meets environmental policy objectives, as well as compliance with international, regional and national legislation. The audit provides a basis for drafting future environmental policies and taking steps to improve environmental performance. Environmental audits can alert companies to areas in which it does not meet current or proposed legislative requirements, and may avoid unnecessary expenditure on litigation costs, fines and damages. Audits also can encourage more efficient techniques and waste reduction measures. Information from the audits can be used by lenders in judging the viability of a project and to raise the public profile of the company (Dias and Begg, 1994). According to Welford (1998) key objectives of environmental audits are to:

1. verify compliance;
2. identify problems;
3. formulate environmental policy;
4. measure environmental impact;
5. measure environmental performance;
6. confirm effectiveness of environmental management system;
7. provide a database for corrective action and future plans;
8. develop company's environmental strategy;
9. communicate.

Education and training is an important component of EMS as an obvious internal change is required at mines to increase the level of awareness of environmental issues, potential roles of cleaner technologies and strategies in the industry. Awareness training must begin with the managerial staff and engineers on sites, which will motivate these individuals to work towards minimizing environmental impacts and extending knowledge of pollution issues to other employees. To aid in the training process, mines may require assistance from consultants, or may need to hire people with appropriate knowledge and skills (Hilson, 2000).

The first formally adopted standard for EMS was developed in 1992 by the British Standards Institution (BSI), and BS 7750 became operational for company certification in 1995 (Krut and Gleckman, 1998). There are other EMS standards such as ISO 14001, adopted by the International Organization for Standardization, and the European Union's EMAS (the Eco-Management and Audit Scheme). ISO 14001 is an international, voluntary EMS standard, which is a part of ISO 14000 series that covers standards in the field of environmental management tools and systems. Companies use ISO 14001 as a standard for internal auditing purposes, for self-declaration or for third party certification (Kolk, 2000). ISO 14001 defines environmental management system to include several elements: environmental policy, assessment of environmental aspects and legal and voluntary obligations, a management system, series of periodic internal audits and reports to top management, and a public declaration that ISO 14001 is being implemented. ISO certification is rapidly growing in developed countries, such as Japan, Germany, UK, Sweden, and USA and in some developing countries such as Thailand, India and Malaysia (Kolk, 2000).

The certification of environmental management systems possesses several important implications for the companies. Among advantages of certification of EMS are: more information enables better control and increased consistency of business practices; reduced operating costs (waste, energy, water and effluent reduction, recycling and process changes) and quality mark, which improves communication with regulators and the public. However, among disadvantages may be such issues as implementation costs, time and effort spent on certification and quality of certification standards is lower than internal management system (Kolk, 2000).

The definition used in ISO 14031, 'environmental performance is the company's achievement in managing any interaction between the company's activities, products or services and the environment' (Rikhardson, 1999). Generally, data on environmental performance include data on several areas, such as: raw material consumption, use of chemical substances, water consumption, energy consumption, emission to air, emission to water, solid waste, noise and odor (Ranganathan, 1999). There are different sets of indicators for measuring corporate environmental performance that are based upon the notion of sustainable development, and include indicators of material use, energy consumption, non-product output and pollutant release (Ranganathan, 1999).

Companies in Central and Eastern Europe also increasingly incorporate environmental concerns in their business management and show improvement in their environmental performance. There are also cases of successful integration of environmental management systems in countries of transition economies (Buckley et al, 1999). However, in general countries of transition economies are slower in the integration of environmental management in their business management practices.

Environmental Reporting

In the 1990s, companies in North America and Europe produced corporate Environmental Reports, and at least 1,000 companies worldwide now produce stand-alone Environmental Reports (White and Zinkl, 1999). Mining companies make information available to the general public via Annual Reports, Health and Safety Reports, Social Reports, other printed material and corporate websites (Warhurst, 1998). The first companies to release Environmental Reports were those in the petrochemical sector, starting in the 1980s, while mining sector started to produce their own Environmental Reports only in the 1990s (Scott, 2000).

The reasons for corporate environmental and social reporting are rooted in Agenda 21, adopted by the UNCED. Principle ten of the Agenda 21 promotes the disclosure and dissemination of information by industry on environmental performance. It recognizes the need for a mean to evaluate these actions, using indicators and markers to measure progress towards the goals of Agenda 21, and to provide a solid basis for decision-making at all levels (Warhurst, 1998). Corporate reporting reflects corporate attitude to environmental issues and there are several reasons why companies make this information public (Warhurst and Bridge, 1997; Khanna, 1999):

- An emerging issue is that financial analysis is now beginning to look closely at a company's environmental performance and future liabilities may well affect a company's share price performance on the stock exchange.
- Reports are an effective tool within an organization, helping to integrate environmental management with other management and accounting systems, and putting the message of corporate responsibility across to employees.
- Industry facing up to environmental challenges and legislation.
- Growing public demand for adequate information on corporate performance and responsibility.
- Environmental reporting and continual assessment and improvement, is a requirement of industry standards such as ISO 14001, whereby companies have to account for their actions to stakeholders.
- Environmental reporting helps to improve technologies and leads to more responsible environmental management of mineral resources.

A number of concerns have been expressed about corporate environmental reporting: first, that published data may be unreliable, second, many companies are selective about the material they include in their reports, and third, data is not comparable either within a report, between reports of different years, or between reports from different companies even within the same sector (Azzone et al, 1996; Hopkinson and Whitaker, 1999). Therefore, many argue that the solution to these discrepancies in environmental reporting practice may be found in standardization of reporting format and performance measures in order to achieve an effective environmental performance measurement and evaluation and inclusion of a third party statement (Hopkinson and Whitaker, 1999; Scott, 2000).

Specific challenges for environmental reporting include development of generally accepted and recognized standards. For example, the US-based Coalition for Environmentally Responsible Economies has established a Global Reporting Initiative (GRI) which aims 'to bring together the numerous initiatives on corporate environmental reporting that have development independently around the world, and to help them to shape them into one set of coherent, consistent global standards' (Cottrell and Rankin, 1998). GRI in agreement with the Mining, Minerals and Sustainable Development Program developed a mining sector supplement to the core GRI Sustainability Reporting Guidelines (Global Reporting Initiative, online).

In most countries environmental reporting is voluntary and there are no definite rules concerning its form and content (Azzone et al, 1996). Niskanen and Nieminen (2001) argue that because of the absence of regulation, companies manage themselves in relation to the objectivity of environmental information released. One way to increase the credibility of information is via third party verification (Scott, 2000; Line et al, 2002). In addition, companies are expected not only disclose their own activities in respect to environmental and social issues, but also those of their subsidiaries (Tilt, 1994), due to the assumption that global economic actors have a global environmental and social impact and that they have a responsibility to disclosure those impacts to stakeholders (Line et al, 2002). Global environmental reporting is increasing among global companies as they recognize that their national Environmental Reports cannot demonstrate the entire picture of their environmental management (Krut and Moretz, 2000).

There is a close link between performance and corporate reporting, although this link is not always accepted. Neu et al (1998) suggest that environmental disclosures are not necessarily false; however they will always be only a partial representation of environmental performance. Environmental disclosures aim to manage public impressions of a company's activities and are selective in order to highlight specific organizational operations, such as positive actions, re-formulating actions for important publics and ignoring actions raised by less powerful publics. Researchers also note that companies do not generally disclose negative information (Guthrie and Parker, 1990). In addition, a study undertaken by Deegan and Rankin (1996) of Australian companies, which were successfully prosecuted by environmental authorities, showed that even though companies were publicly and successfully prosecuted it does not affect their environmental disclosure policies in Annual Reports.

Companies need to be transparent about the effects they have on communities, including reporting on the repair of any damage done to the environment. According to Guerin (2000) there are several critical issues that need to be addressed in an environmental report, which include:

- account for the triple bottom line;
- spotlight real issues, impacts and priorities;
- develop specific, measurable, attainable and relevant targets, and have them verified;
- integrate annual and environmental reporting with each other;
- focus on financial market users, NGOs and community groups;
- quantify any savings and account for costs;
- make use of the Internet but do not go paperless;
- engage and re-engage stakeholders;
- communicate;
- use feedback analysis to determine reporting effectiveness and to uncover strengths and weaknesses in the company's environmental performance.

Health and Safety

Generally, there are three key trends concerning integration of health and safety with other management systems: firstly, with quality management systems, secondly, with environmental management systems and thirdly, integration of quality, safety and environment into one system (Honkasalo, 2000; Herrero et al, 2002). According to Ranganathan (1999) and Wheeler and Sillanpää (1997) health and safety policies refer to the general employee/human resources domain. Balkau and Parsons (1999) include workplace health and safety as one of the environmental issues for mining operations. Moreover, Clinch (2001) argues that there is a significant link between environmental and health and safety policies and that in some cases, the effects of policy in one area will have negative effects on the other. Therefore, an integrated approach may improve the effectiveness and efficiency of policies in both areas (Clinch, 2001). Honkasalo (2000) argues that the integration of occupational health and safety matters into environmental management systems can help companies to avoid duplicated measures and find solutions that handle both environmental and health risks in an optimal way. Vassie et al (2000) states that in some cases the introduction of quality management systems leads to an improvement in health and safety performance and that the principles of quality management apply equally to health and safety management. Herrero et al (2002) states that quality principles are readily applied to safety, such as: objective of zero accidents; analysis of incidents; documenting the politics of safety, procedures and the instructions; safety committees; participation of the workers; statistical analysis; all accidents and injuries could be prevented.

According to Hine et al (1999), there are key safety principles such as: prevention of injuries; employee involvement; responsible safety management; work safety as a condition of employment; risk minimization; safety improvement; training; and finally, incorporation of safety with productivity. Companies commonly have several key systems for managing health and safety, including formulation of safety policy, conducting risk assessment, establishing an accident reporting system, written safety working procedures, incentive schemes, safety audit system, training programs and health surveillance (Vassie et al, 2000). According to Wheeler and Sillanpää (1997), a cycle of continuous improvement of effective management of occupational safety and health includes:

- leadership and commitment;
- management review;
- assessment of policy and strategies objectives;
- defining organizational structure and responsibilities;
- identifying hazards and assessing risks requirements and opportunities;
- planning programs, allocating resources, setting performance indicators and targets, production of manuals and initiating training;
- implementation of programs;
- monitoring performance against targets and indicators;
- conducting internal audits;
- verification and certification;
- publishing occupational health and safety statistics.

According to Warhurst (1998), health and safety practices are an integral part of corporate social responsibility of mining companies and they should involve ensuring that a health and safety system is in place, and regular health and safety auditing and reporting. In addition, occupational health and safety standards have been developed in many countries. For example, the British Standard, Guide to Occupational Health and Safety Management Systems (BS 8800, 1996) illustrates the possibilities for building up an occupational safety management system as a management system on its own (Honkasalo, 2000). The international standards of ISO 9000 have helped many organizations to implement quality systems and regulations based on ISO 9000 have been created to guide companies in developing systems for the management and prevention of worker risks. There is also the American Association of Industrial Hygiene, Occupational Health and Safety Management System (OHSMS), which is a guide for safety management and worker health (Herrero et al, 2002). Some companies have derived their safety management systems from environmental standards such as ISO 14001, which would involve policy, planning, implementation and operation, measurement and evaluation, review and improvement (Santos-Reyes and Beard, 2002). In 1999, British Standard Institute has published OHSAS 18001 (Occupational Health and Safety Assessment Series) specification and accompanying OHSAS 18002, guidelines for implementation. The standard is compatible with ISO 9001 and ISO 14001.

Employee/Human Resources Policies

Apart from occupational health and safety, issues of human resources cover firstly, a general human resource policy which includes remuneration, career development, diversity and representation, and secondly, a code of ethics for individual rights and responsibilities (Wheeler and Sillanpää, 1997). Within the employee/human resources area, companies address several issues (Mathews, 1993; Warhurst, 1998; Business Impact, online), including:

- human resources policy;
- officer for human resources;
- recruitment practices;
- training, education and professional development;
- rewarding employees;
- redundancy;
- employee well-being.

Important areas of recruitment concern equal opportunities, non-discrimination and representation. Regarding recruitment policies of mining companies, important indicators of corporate social responsibility would be willingness of mining companies to employ local community members and indigenous people (Warhurst, 1998).

Many companies promote career development of their personnel, and establish mentoring programs, promote and reward skills development, and encourage volunteering (Business impact, online). In the case of the mining industry, companies may conduct education of employees regarding local community culture and values (Warhurst, 1998). Concerning the rewards to employees, companies offer performance related pay, such as profit sharing schemes, use of shares to create incentives and establishing company pension plans. In cases of redundancy, companies should involve forward planning to avoid layoffs, communication of the company's intentions, and a search for creative alternatives. However, if redundancy is inevitable, companies should provide assistance such as finance or career counseling to help employees find new jobs (Mathews, 1993; Business Impact, online).

There are several options available to improve the health and well-being of employees, including actions to promote health, safety and well-being, offering health plans extended to families, offering company housing provision, health screening and occupational health monitoring of all workers connected with site operations (e.g. miners, truck drivers and suppliers), encouraging health club membership, and encouraging rehabilitation programs to people who have problems with smoking, drinking and drugs (Mathews, 1993; Warhurst, 1998; Business Impact, online).

Stakeholder Involvement and Community Development

Companies today are aware of the power of stakeholders to influence opinion in relation to legitimacy of their business and that stakeholders attitudes and opinions could threaten their well-being, license to operate and hence survival in terms of restrictions on sources of raw materials or methods of operations (Grafé-Buckens and Hinton, 1998). The company's stakeholders are a variety of individuals and/or groups that are affected by the companies operations. Wheeler and Sillanpää (1997) give a comprehensive list of potential stakeholders with their classification (see Table 2.7).

To engage stakeholders in the process of considering environmental impacts and implications of businesses, companies can adopt environmental stakeholder initiatives which can be defined as all activities undertaken by a company to increase the communication with stakeholders in environmental issues relating to the company and its business practice (Grafé-Buckens and Hinton, 1998). Companies have adopted various mechanisms for engaging their stakeholders; Grafé-Buckens and Hinton (1998) have categorized those into three major ways:

1. Provision of information to stakeholders of the company's activities, which covers an area of corporate environmental reporting.
2. Consultation with stakeholders, which incorporates one way process of information flow from stakeholder to the company, such as feedback forms in Environmental Reports.
3. Stakeholder participation, where the company is engaged in active dialogue with its stakeholders.

Table 2.7 List of stakeholders

Primary social stakeholders
- employees and managers
- investors
- customers
- suppliers and business partners
- local communities

Primary non-social stakeholders
- natural environment
- non-human species
- future generations

Secondary social stakeholders
- government and civil society
- social and third world pressure groups and unions
- media and commentators
- trade bodies
- competitors

Secondary non-social stakeholders
- environmental pressure groups
- animal welfare pressure groups

Source: Adapted from Wheeler and Sillanpää, 'The Stakeholder Corporation: a Blueprint for Maximizing Stakeholder Value' (1997).

As the mining industry has a close relation to the life of the local communities in which it operates, via benefits and impacts, the local community is viewed as a primary stakeholder for the mining industry. Mitigation of the social and economic impacts of mining operations on local communities is an important pillar of sustainable development in the mining industry, which allows local communities to continue functioning and contribute to sustainable livelihood (Miller, 1999). It is important to differentiate categories of community involved in, or affected by, mining operations (IIED, 2002):

- Occupational communities – households or families who derive all or most of their income from mining.
- Residential communities – households or families who live within the geographical area affected by mining. They may live in close proximity or many miles away, such as on a river polluted by mine. These communities fall into two types: those in existence before the mine was built and those that have developed as a result of the mining operations.
- Indigenous communities – households or families with an ancient and cultural attachment to the land where mining occurs or has an impact.

Indeed, the management of community issues is becoming an important factor in determining the contribution of mining operations to the sustainable development agenda. Research conducted by Grafé-Buckens and Hinton (1998) identified reasons for company's interest in stakeholder engagement:

- Companies are responding to growing public awareness and concern for environmental issues, demand for increased corporate accountability and pressure and influence of stakeholders.
- Companies' willingness to engage a particular stakeholder group depends upon their power and influence to change company's public image and ability to operate.
- Companies are motivated to satisfy stakeholder demands for information about their business and to gain acceptance and approval of current and new projects.

McPhail and Davy (1998) have developed several recommendations for embracing social concerns in the decision-making of companies within the mining, oil and gas sectors, including:

- *Adopt a policy on social issues and develop capacity.* Policies form the basis for implementing management, measurement and reporting systems, supported by senior managerial commitment.
- *Identify stakeholders and acknowledge the legitimacy of their perspectives.* Acknowledging all stakeholders' perspectives is very important. For example, failure to acknowledge the legitimacy of customary laws in parallel with

national or regional laws pertaining to natural resource exploitation will almost certainly result in conflict with the owners or occupiers of land.

- *Identify social risks and opportunities.* Managing social risks and preventing impacts is preferable to solving them retroactively.
- *Assess social and environmental impacts thoroughly and integrate where appropriate.* An analysis of the social and environmental impacts of project alternatives should form the basis for avoiding or reducing impacts to acceptable levels.
- *Recognize public involvement as integral to protect sustainability.* Stakeholder involvement encompasses information exchange, consultation, participation, and joint decision-making.
- *Delineate responsibilities for social provisions.* Investment in social and community provisions is an integral part of doing business and is viewed as part of the license to operate from the local communities.
- *Aim for social equity in revenue distribution, compensation, and other social investment.* Public and private sectors should ensure that some project revenues accrue to key stakeholders in an equitable manner. For example, a proportion of government income from royalties, taxes, or equity might be allocated for local development activities to ensure that those most affected enjoy some of the benefits.
- *Develop partnerships in support of sustainable development.* Companies should actively pursue partnership approaches for delivery of social programs in support of sustainable development.
- *Develop mechanisms for long-term representation of stakeholders and conflict resolution.* Public involvement should continue throughout the lifetime of the project.
- *Evaluate the effectiveness of social investment.* The simplest measure is the cash value of a company's contributions including grants, donations, training, and enterprise development activities. However, new indicators should be developed considering aspirations of local communities.

According to International Institute for Environment & Development (IIED, 2002), in order to help communities work towards sustainable development, companies adopt various approaches, which address the following issues:

- Revenue distribution and use.
- Projects, Funds and Foundations.
- Supporting small local businesses.
- Employment and skills development.
- Conflict and dispute resolution.
- Community health initiatives.
- Mine closure.
- Community participation in decision-making.
- Improved social impact assessment.

It is becoming widely accepted that the local community is a part of the consultation process of environmental effects, and also of social, economic and cultural benefits and costs (McMahon and Strongman, 1999). According to Wood (1998), a company can adopt several actions for community involvement:

- Appointment of officer to act as a contact with community and the media.
- Research into community attitudes in order to understand community expectations.
- Organization of site visits and public displays in order to attract community support.
- Arrangement of public forums for various groups of people.
- Explanation and advertisement of companies intentions and operations.
- Consultation with local communities.

Strongman (1998) suggests several ways mining companies can integrate community concerns in their operations (see Table 2.8). Local communities often demand compensation, which covers infrastructure, training for mining and non-mining employment, and social services including education and health (McMahon and Strongman, 1999).

Table 2.8 Community actions of mining companies towards sustainable development

Mining stage	Community actions
Exploration stage	• identify and inform communities • overcome mismatch of interests
Project design stage	• build respect and trust • consult with affected communities • minimize negative impacts
Operational stage	• forge effective community partnerships • set up joint monitoring arrangements • support training and capacity building • outsource jobs to the community
Closure Stage	• physical reclamation • help community prepare for mine closure

Source: Adapted from Strongman, 'Mining and the Community from Enclave to Sustainable Development' (1998).

Companies are increasingly required to perform positive impacts on the communities in which they operate in order to meet the needs of company's stakeholders. Many mining companies have developed social investment programs in the communities affected by or near to their projects (Carter, 2000), including initiatives such as:

- Directly, through general management or specialist community department.
- Participating in alliances with other businesses that have community involvement activities.
- Giving money or organizing employee volunteer programs with NGOs or local organizations that do community development work.
- Contributing to government programs that promote community development in the locality or region.
- Establishing and facilitating community development initiatives through corporate foundations.

One of the most important groups of stakeholders is indigenous peoples' community, especially in the age of globalization and expansion of the mining industry in areas of historical livelihood of indigenous peoples. The way mining companies manage their relationships with the local and indigenous communities is a critical factor in the success of their activities. Companies have come to realize that close relations with the local community is important to enhance community capacity to deliver social services, to promote growth of the local economy and to allow the traditional cultures and lifestyles of indigenous people to flourish, particularly after mine closure (ICME, 1999).

There are a number of reasons why mining companies pay attention to relations with indigenous people, including: increasing contact with indigenous communities; national legislation on acknowledgement rights of indigenous people and titles to traditional territories; realization of corporate risks that can arise from confrontation with indigenous people; increasing acceptance of shareholders expectations and recognition of vulnerability of indigenous people to social, cultural and economic impacts from mining projects (Hill, 1999).

Mining companies have adopted various strategies in order to improve relations with indigenous people, including community consultation. Community consultation is used to learn about the needs, concerns and aspirations of the indigenous communities in order to ensure community participation and improving living standards and to ensure that local communities understand the consequences brought by mining in the area (Hill, 1999). The early and ongoing practice of community consultation creates trust and keeps expectations of communities realistic. Mining companies can approach the consultation directly with affected communities, through local indigenous organizations, or using independent professionals in order to conduct workshops and discussion groups. In addition, negotiated agreements often incorporate consultation and joint decision-making (Hill, 1999).

Some mining companies formulate their relations with indigenous communities through negotiated agreements, such as the Cominco-NANA Agreement (Red Dog), the Raglan Agreement between Falconbridge and Makivik Corporation, the Musselwhite Agreements between Placer Dome, First Nations groups and government shareholders, and WMC's Agreements with the Bla'an cultural communities (Tampakan). These agreements cover several elements such as (Hill, 1999):

- Preferential employment opportunities for local indigenous people.
- Education and training commitments.
- Financial compensation.
- Preferential business opportunities.
- Environmental commitments.

Common business approaches to community development, or social investment, include charity giving to areas such as education and youth development or through a company-funded Foundation, Trust Fund or local NGO, specifically for that purpose (Crowson, 1998; McMahon and Strongman, 1999; Business Partners for Development Natural Resources Cluster, online). Recently, some companies have shifted from company-led approaches to community development, to programs that are managed through partnerships with NGOs and community groups (Business Partners for Development Natural Resources Cluster, online). Government, industry and third sector address community development related to mining in various ways, and sometimes these activities include some sort of cooperation between the sectors. The third sector is somewhere between the private and public sectors, it includes voluntary, nonprofit, charitable, quasi-governmental, nongovernmental and tax-exempt organizations (Schwartz, 2000).

Government, Industry and NGOs

There is now a growing consensus that for the management of social and environmental issues to be truly effective, it should involve constructive input from each of three main groups of participants – government, private enterprise and society (Acquah, 1994; Davy, 1999). Recommendations for sustainable development developed during United Nations Summits in Rio de Janeiro in 1992 and Johannesburg in 2002 state that sustainable development could be achieved when businesses, governments and civil society not only make their individual contributions, but also exercise a high degree of cooperation in the areas affecting the sustainability.

One of the concepts that captures this cooperation for sustainable development, and particularly relevant for the extractive industries, is tri-sector partnership, which refers to a voluntary partnerships, which promote community development involving the company, civil society organization and local and/or central government (Warner, 2000a, 2000b). In the context of the extractive industries, the tri-sector partnership could be used in a variety of ways for the benefits of society, environment and economy (Warhurst, 2001):

- Political and social risk management, relationship management including the resolution of local disputes.
- Integration of business strategies with regional development plans, rural poverty adjustment schemes and local education programs.
- Promotion of education, new skills and local procurement.
- Public consultation, cooperation and involvement with local organizations.

- Closure planning, decommissioning, site remediation, viable future land-use options and employment options.
- Environmental management after closure, voluntary codes of practice.
- Funding social investment projects aimed at local community development.
- Management of foundations as vehicles for social investment.

Corporate social responsibility employing mechanisms of tri-sector partnerships can minimize the negative impacts and risks of social issues related to mining project development (Acutt et al, 2001). Tri-sector partnerships enable mining companies in the process of trust building that is lies as a foundation principle of corporate social responsibility, and they enable greater agreement on delivery of specific initiatives for the benefit of society and the environment. Moreover, tri-sector partnerships lead to joint, coordinated actions for community development based on the use of complementary resources and distribution of responsibilities between businesses, governments and civil society organizations (Warner, 2000b).

The success of tri-sector partnerships depends upon the strengths, competencies and resources, which businesses, governments and civil society can bring together. Mining companies could contribute their expertise in supply chain and project management, engineering and infrastructure and logistical expertise to the partnerships. Governments can provide strategic co-ordination and expertise in public services. Civil society organization could contribute their capacity to mobilize community participation (Warner, 2000a; MEM, 2002). Tri-sector partnership could help building trust and confidence in relationships between the industry, government and civil society organizations, by allocating roles and responsibilities between the partners and defining expectations from each other.

Davy (2001) suggests that partnership approach to social investment offers a number of valuable benefits. Firstly, bilateral 'corporate-community' partnership helps to moderate increasing community expectations of unconditional belief that might lead to conflicts and potential disruption to operations. Secondly, partnerships between the company and civil society that include government agencies (true tri-sector partnerships) have helped to improve the long-term physical and political sustainability of social investments. Thirdly, the partnerships have helped to define local priorities for social investment, assign responsibilities, apportion costs, establish accountabilities and resolve conflicts. Fourthly, partnerships improve understanding and trusts. Fifthly, the partnership may guarantee the success of social investment and may help to protect both local and global corporate reputation. In addition, the benefits of tri-sector partnership to companies include potential cost-saving via levering resources, skills sharing, reduced dependency, networking and collaboration (Warner, 2000b; Acutt et al, 2001).

Cooperation between government, industry and NGOs for sustainable development can be grouped into several categories:

1. Cooperation between the government and industry via negotiated agreement, when the government delivers initiatives.
2. Cooperation between the government and industry, when the government assigns nongovernmental organization for delivering initiatives.
3. Companies establish nonprofit organizations through which they direct initiatives.
4. Cooperation between the industry and already established nongovernmental organizations, when nongovernmental organizations deliver initiatives.

Mining projects bring various economic benefits, and the governments derive income from the mining projects through royalties, taxes or equity. Although this money goes to various end uses, sometimes a part of these revenues are directed towards a special goal to benefit the communities affected by project development. These investments include improved infrastructure, services, education, or access to funds for local enterprise initiatives. In other cases, a government may offer investment within a project to those directly affected by the extraction development (McPhail and Davy, 1998). In some instances, the companies have interest in social investment of these state-earned project revenues and consequently, companies and governments resolve issues related to allocation of project revenues through project negotiations and concession agreements (Pershing, 1999).

A number of countries especially those which economies are based on oil revenues have already created mechanisms for ensuring sustainable development of their nations. The reason for such behavior is more clearly explained by the Norwegian Ministry of Finance in its 1998 budget. It suggests that while petroleum revenues provide countries with sizeable surpluses both on the current account of the balance of payments and in central government finances as well as opportunities to maintain growth and employment, single sector economies are also more vulnerable to internal and external instabilities and cost pressures (Pershing, 1999). Petroleum revenues could not be considered as income in a usual sense since it involves a depletion of petroleum wealth. Another important factor is that revenues from the petroleum sector show greater variations over time than other revenues, partly as a result of fluctuations in the price of crude oil. The dependence on petroleum revenues implies that without careful planning, declines in oil prices would require substantial alterations in economic policy (Pershing, 1999).

Examples of successful long-term management initiatives designed to provide the benefits of sales of national natural resource bases in the future exist in Kuwait (Kuwait Investment Authority), Norway (Norwegian Government Petroleum Fund), and Oman (General Reserve Fund). Similar initiatives have also been established in Iran, Iraq, Saudi Arabia and Venezuela. In Kuwait, the Governmental Department of Finance originally managed the surplus revenues from oil sales. In 1960, the General Reserve was created, consisting of all the State's investments (Pershing, 1999). In 1976, the Government of Kuwait created a Reserve Fund for Future Generations in order to ensure the prosperity of its citizens after the oil era, by allocating an initial US$ seven billion into the Fund and setting a mechanism for annual contributions of ten per cent from Government revenues (Kuwait Information Office, online). Presently, assets from the Reserve Fund for Future Generations (now managed by the Kuwait Investment Authority) are invested in international stocks, foreign bonds, major currencies, and various economic projects, under the supervision of economic and financial experts in Kuwait and advised by international financial institutions. The Fund plays a major role in implementing economic and social policies for local development, and regional and international cooperation, investing in a diverse portfolio (Pershing, 1999).

In Norway, the Government established a Norwegian Government Petroleum Fund in 1996 as an integrated part of the central Government fiscal budget and is set in order to reflect Government savings. Money is allocated to the Fund only when the budget, including petroleum revenues, shows a surplus. The Fund allocates its assets in low risk bonds, with currency distribution matching Norway's import weights. Thus, the Fund's income is formed from net petroleum revenues and any returns on investments (Pershing, 1999). Oman achieved remarkable economic prosperity after the oil shock in the 1970s, supported by a steady increase in petroleum output and exports. The Government of Oman established a State General Reserve Fund in order to create savings for the future generations (IMF, 1998). In Alberta, Canada, the Government created the Alberta Heritage Savings Trust Fund in 1976, when resource revenue was high and the economy was booming. They decided that a portion of Alberta's oil and gas revenues should be set-aside for future generations (Government of Alberta, online).

In the 1980s and 1990s there has been a significant increase in nongovernmental organizations active in development. The number of NGOs registered in the OECD countries has grown from 1,600 in 1980 to 2,970 in 1993, and the total spending of these NGOs has risen from US$ 2.8 billion to US$ 5.7 billion in current prices (Edwards and Hulme, 1998). Recently, and particularly after Cold War, development policy and aid transfer have come to be dominated by 'New Policy Agenda', which is based upon two sets of beliefs derived from neo-liberal economics and liberal democratic theory. The first, set of beliefs reflects that NGOs are a preferred channel for service-provision in deliberate substitution for the state, which minimizes their direct role in the economy and supports NGOs in providing welfare services to those who cannot be reached through markets. The second set of beliefs sees NGOs as vehicles for democratization and an essential element for attaining civil society. NGOs are supposed to act as a counter-weight to state power – protecting human rights, opening up channels of communication and participation, providing training grounds for activists and promoting pluralism as a result of these developments. Governments have therefore been prepared to channel increasing amounts of official aid to, and through, NGOs (Edwards and Hulme, 1998).

In some cases, the government involves third sector participation, where the government maintains control of the policy-making process and uses the third sector organizations as service providers (Schwartz, 2000). The nature of government-third sector collaboration can be seen as running on a continuum ranging from that of principal-agent relationship to that of communion type partnership. Principal-agent relationship implies that the government maintains control of the policy making process, while the NGO is used for provision of services (Schwartz, 2000). Examples of successful cooperation between the government and NGOs with appropriate funding from the petroleum industry are present in the Province of Alberta in Canada (Alberta Heritage Savings Trust Fund), and the State of Alaska in the USA (Alaska Permanent Fund Corporation). In 1976, the Alaska Permanent Fund was established with the idea of generating earnings from oil development for both current and future generations, pursuing an idea to transform a nonrenewable resource, oil, into a renewable resource, money, that would not only last, but also increase over time. The state of Alaska has received 900 million US dollars from the sale of oil leases in 1969. The Alaska Constitution has prohibited any dedicated funds, and in 1976 the legislation was implemented to set aside a Permanent Fund of at least 25 per cent of the state's oil royalties and related income (Rasmuson, 1993). It took four years after the Constitutional authorization before the Legislature passed a bill in 1980 creating the Alaska Permanent Fund Corporation, providing for independent management by a Board of Trustees, a cautious list of authorized investments – principally government bonds – and a dividend program for the public from the Fund's earnings (Rasmuson, 1993).

Mining companies have developed various approaches to social issues, including the creation of corporate foundations, collaboration with international organizations, community foundations, local NGOs through grant making and via sponsorship and donations (McMahon and Strongman, 1999; Miller, 1999). Some mining companies have established different types of foundations and trusts, endowed with the funds from the company's profits. These foundations take different forms and are engaged in a range of activities, for example, BHP has established the BHP Trust that is a grant-giving nonprofit organization, targeted to Australian community and gives funds for NGOs.

Summary

- Mining is an important industrial sector, which has significant effects upon the development in three major spheres: economic, environmental and social. In terms of economic impact, the mining industry contributes to business development, national revenues, employment and export earnings. The mining industry causes irreversible damage to the natural environment, including destruction of habitat and bio-diversity, air and water emission, land degradation, etc. The mining industry influences communities in which it operates directly through employment and disturbance of cultural sites, and through economic and environmental effects.
- International debate including national governments, international organizations, environmental NGOs, mining associations and international media on sustainable development, which includes progress within economic,

social as well as environmental dimensions, pressure the mining industry to minimize social and environmental disruption caused by its operations. Mining in terms of sustainable development can be interpreted as a transparent process which makes sure that appropriate revenues generated by exploitation of nonrenewable resources are suitably invested to ensure the future development of long-term sustainable livelihoods for the members of affected community (Epps, 1996).

- The mining industry's response to environmental and social issues and sustainable development underwent three major stages: reactive approach in the 1970s and 1980s, proactive approach in the 1980s and 1990s and finally, new approach, which integrates environmental performance with business strategies and development. The mining industry is very much concerned about the possible impacts of environmental and social issues on its business activity and therefore those issues have become key agenda items in their operations.
- Improved mitigation of environmental and social impacts as a response of the mining industry to an increased pressure from stakeholders. Key strategies for sustainable development and CSR within the mining industry include management reorganization, adoption of environmental reporting, improvement of health and safety practices, establishment of sound employee policies, involvement of stakeholders and participation in community development.

Notes

[1] As far back as the 16th century, Georgius Agricola, whose writings are considered to be the first scientific and technical texts on mining and metallurgy, also referred to the negative effects of mining on the environment. Georgius Agricola stated that: 'The strongest argument of the detractors [of mining] is that the fields are devastated by mining operations…' (cited in Down and Stocks, 1977, p.7). By the beginning of the 20th century many industries had started reclamation works, but it was very seldom that these were successful. Other problems, such as pollution abatement, were seldom tackled before World War II, and it must be mentioned that many problems did not exist, or were in a minor form until the age of modern technology (Down and Stocks, 1977).

[2] Global Mining Initiative, URL: http://www.globalmining.com.

Chapter 3

Sectoral Specifics of CSR in Mining

Introduction

The level of CSR differs from one mining company to another. CSR is a response of mining companies to the environmental and social concerns of their stakeholders. Therefore, the extent to which a mining company is engaged in the CSR practices depends on the scope of environmental and social concerns and strength of stakeholder demands. The variety and intensity of environmental and social concerns are primarily connected to the severity of environmental and social impacts of mining operations, depending on type of minerals mined, mining methods, beneficiation processes, and chemicals used and released. The difference in responses to CSR concerns by different mining sectors does not stop there. Different mining sectors have a ranging degree of environmental and social awareness, which is often linked to the history of environmental stewardship and past environmental incidents related to the mining of specific metals and minerals. The scale of mining operations and geographical distribution of mineral deposits contribute towards the intensification of environmental and social concerns. Most importantly, the structure, ownership and leadership in the mining sector determine the formation and delivery of CSR strategies.

This chapter explores and compares CSR strategies within two mining sectors: gold and diamond mining. The comparison of these two mining sectors is conducted across structure, ownership, geographical distribution, history of environmental and social concerns and range of CSR strategies. Although these mining sectors are different in the sense that gold is a metal and diamond is a mineral, which leads to variations in mining methods and environmental impacts, gold and diamonds both belong to the precious metals and minerals group and primarily serve the same market, i.e. jewelry industry. Notwithstanding the fact that gold and diamonds are rare minerals, these are truly international industries, which are developed in many countries around the world and are under the influence of large mining corporations.

World Gold Mining

Gold is mined in many countries around the world; major gold-producing countries include South Africa, USA, Australia, Canada, China, Russia and Indonesia (see Table 3.1).[1] In the 1970s and 1980s, gold mining was mainly under the control of private capital in market economies, while in centrally planned economies it was under state control (Ericsson and Tegen, 1992). However, since the 1990s, during the transition of centrally planned economies to market economies, gold production in the transition economies was widely privatized.

Table 3.1 World's top 20 gold producing countries by volume, tons

Country	1995	1996	1997	1998	1999	2000
South Africa	522	495	489	474	450	428
US	319	329	351	364	342	355
Australia	254	289	311	313	300	296
China	133	145	157	161	156	162
Canada	150	164	169	164	158	155
Russia	142	130	137	127	138	155
Indonesia	74	92	101	139	152	139
Peru	57	65	75	89	128	133
Uzbekistan	64	71	82	81	86	86
Papua New Guinea	55	53	50	63	65	76
Ghana	53	51	56	73	78	74
Brazil	67	64	59	55	54	53
Chile	49	56	53	47	48	50
Philippines	29	31	34	35	33	35
Mali	-	-	17	22	25	30
Mexico	20	25	26	26	23	27
Kyrgyzstan	-	-	17	22	20	22
Zimbabwe	26	27	24	27	30	21
Colombia	24	23	21	22	21	21
Rest of the world	234	163	128	160	226	231
Total	2,272	2,273	2,357	2,464	2,568	2,573

Source: Adapted from Murray (1997, 1998a, 1998b, 1999); MJ (2001a).

In recent years, worldwide demand for gold has been about 4,000 tons per year. Most of this, c. 85 per cent, is used for jewelry and about 12 per cent is used for industrial applications, such as electronics and dentistry, and the rest is bought by investors (Young, 2000). The annual supply of gold also equals around 4,000 tons, of which 2,500 tons are mined, 400 tons come from official sector sales, 1,000 tons is old gold scrap and 100 tons come from disinvestments (Gold Fields Mineral Services, 1997, 1999; Murray, 1998b; MJ, 2001a).

Historically, gold has been the principal medium of international monetary exchange, but its role has significantly changed in recent years (US EPA, 1994). The main consumers of gold for jewelry fabrication are Italy, India, China, US, Taiwan and Saudi Arabia. Much of the demand for gold jewelry comes from Asia and the Middle East, where it is often purchased as an investment (Young, 2000). Governments in the industrialized world hold significant reserves of gold. Countries such as US, Germany, France, Italy and Switzerland possess almost half of the world's total gold reserves, which were estimated by the World Gold Council (2001) at around 32,915 tons in 2001. The International Monetary Fund, European Central Bank and other institutions also possess significant gold reserves. Governments now hold enough gold to supply current demand for more than eight years and investors hold an estimated 24,000 tons of gold. These figures show that there is enough high purity gold in storage to cover current demand for 14 years, without a single ounce of mine production (Young, 2000).

Gold is traded on several stock exchanges. The most important physical markets for gold are, in time sequence: the Tokyo Commodity Exchange for Industry (TOCOM), the loco London Market in Hong Kong (loco – delivery in), the Zurich Gold Pool, the London Gold Market and the Commodity Exchange (Comex) in New York (Laulajainen, 1990). Gold markets and gold prices are influenced by several factors including the central banks' selling of gold reserves, fluctuations in interest rates, oil prices, consumer demand, financial crisis, and tensions in some large gold-producing countries (Cai et al, 2001).

In 2000, the top ten gold mining companies (World Gold Analyst, 2000) extracted 31.4 million ounces of gold, which comprises around 30 per cent of the total world gold production (see Table 3.2).[2]

Table 3.2 Gold production and turnover of top ten gold mining companies in 1998-2000

Company	Gold production, million ounces			Turnover, million US dollars		
	1998	1999	2000	1998	1999	2000
AngloGold	5.8	5.6	5.5	2,296	2,324	2,299
Newmont Gold	4.1	4.2	4.9	1,454	1,399	1,555
Gold Fields	2.8	3.1	3.9	972	947	1,114
Barrick Gold	3.2	3.7	3.7	1,298	1,432	1,320
Placer Dome	2.9	3.1	3.0	1,145	1,162	1,413
Rio Tinto	2.8	2.9	2.7	7,112	7,197	7,875
Homestake Mining	2.1	2.1	2.2	798	748	667
Freeport-McMoRan C&G	2.2	2.4	1.9	1,757	1,887	1,869
Normandy Mining	1.4	1.7	1.9	1,484	1,356	1,324
Ashanti Goldfields	1.5	1.6	1.7	600	582	582
Total	28.9	30.4	31.4	-	-	-
Total world production, tons	2,510	2,550	2,550	-	-	-
Share of top ten producers in world production, %	41	37	38	-	-	-

Source: Adapted from Amey 'Gold, Minerals Yearbook 2000' (2001) and companies' Annual Reports for 1998, 1999, 2000.

Amongst the list of top ten mining companies there are mining corporations, which specialize in mining of more than one metal or mineral, such as Rio Tinto, Placer Dome, Freeport-McMoRan Copper and Gold, and Normandy Mining. The turnovers of those companies, which cover all mining operations, not only gold mining, are not directly comparable (see Table 3.2).

Most top gold mining companies were registered under the legislation of developed countries such as Australia, Canada, UK and USA, but conduct gold mining operation in many developed countries of Asia, Africa and Latin America (see Table 3.3). Leading gold mining companies are entering into partnership for the development of large mines around the world. Large gold mining companies are absorbing small and medium sized companies and the largest companies merge as a consolidation phase takes place in the gold mining industry. The trend of consolidation in order to maintain gold reserves is likely to continue, and it is probable that future gold market will be under the control of ten major mining companies (Amey, 2001).

Table 3.3 Geography of operations of top ten gold mining companies in 2000

Company	Corporate headquarters	Locations of operations in 2000
AngloGold	South Africa	Argentina, Australia, Brazil, Mali, Tanzania, Namibia, South Africa, and USA
Ashanti Goldfields	Ghana	Ghana, Guinea, Tanzania and Zimbabwe
Barrick Gold	Canada	Canada, Peru, USA and Chile (mine closure)
Freeport-McMoRan C&G	USA	Indonesia
Gold Fields	South Africa	Ghana and South Africa
Homestake Mining	USA	Argentina, Australia, Canada, Chile and USA
Newmont Mining Corporation	USA	Indonesia, Mexico, Peru and USA
Normandy Mining Ltd.	Australia	Australia, Brazil, Canada, Chile, Cote D'Ivoire, New Zealand and Turkey
Placer Dome	Canada	Australia, Canada, Chile, Papua New Guinea, South Africa, USA and Venezuela
Rio Tinto	UK	Argentina, Australia, Brazil, Chile, Indonesia, Papua New Guinea, USA and Zimbabwe

World Diamond Mining

Diamonds are produced in more than 22 countries of which the most important producers, in order of value of production are: Botswana, South Africa, Russia, Australia and Namibia (Jennings, 1995).[3] Angola, Australia, Botswana, Canada, Democratic Republic of Congo, Namibia, Russia and South Africa are the largest rough diamond producers, which account for 96 per cent of the world's diamond production in carats and 90 per cent in value (see Table 3.4 and Table 3.5).

Table 3.4 Natural rough diamond production in 1995-1999, million carats

Country	1995	1996	1997	1998	1999
Australia	40.8	42.0	40.2	40.8	29.7
D.R. Congo*	20.0	21.9	17.0	25.8	22.0
Botswana	16.8	17.7	20.1	19.8	21.3
Russia	16.9	19.2	19.4	19.8	16.2
South Africa	9.1	10.2	9.8	10.9	9.7
Angola	1.9	3.8	3.3	3.0	3.6
Canada	-	-	-	0.3	2.5
Namibia	1.3	1.4	1.5	1.5	1.6
Sierra Leone	0.3	0.2	0.4	0.3	0.6
Ghana	0.8	0.7	0.9	0.7	0.5
Guinea	0.5	0.5	0.5	0.3	0.5
CAR	0. 6	0.6	0.5	0.5	0.4
Tanzania	-	0.1	0.1	0.1	0.2
Liberia	-	-	0.1	-	0.2
Zimbabwe	0.2	0.5	0.6	0.1	-
South America	2.0	1.7	2.1	1.6	1.3
Others	0.5	0.7	0.5	0.3	0.8
World Total	111.7	121.2	117.0	125.8	111.1

* Democratic Republic of Congo, formerly Zaire

Source: Adapted from MJ, 'Diamonds: De Beers' Perspective' (1999); Rombouts, 'The Diamond-mining Scene' (2000).

Also, small-scale diamond mining exists in Brazil, Central African Republic, China, Ghana, Guinea, Indonesia, India, Ivory Coast, Liberia, Sierra-Leone, Tanzania and Venezuela. In many cases, it is difficult to obtain reliable official statistics on foreign trade of rough diamonds because of the secrecy connected to security and price maintenance. In many cases it is possible only to make estimates of production and sales. Grades of diamond deposits range from 3.5 carats per 100 metric tons to 600 carats per 100 metric tons (Jennings, 1995) and annual diamond production from major mines range from 0.7 million carats to 40 million carats (see Table 3.6).

For the purpose of trade regulation, raw diamonds are classified into two large groups: jewelry or gem quality diamonds, which are suitable for cutting (from one millimeter in diameter) and industrial diamonds, which include brown and black diamonds, and are not recognized as precious stones. The latter are used in industry. Industrial diamonds are used for the manufacture of instruments used for cutting, grinding or tooling (Balazik, 1999). Natural diamonds are used in metallurgy and car production, radio electronics, optics, mining, geological exploration, medical equipment, etc. (Teslenko, 1997). However, the importance of natural diamonds for industrial use has been challenged by the expansion of synthetic materials which are produced in the USA, Russia, Ireland and South Africa (Balazik, 1999).

Table 3.5 Diamond production in value, million US dollars

Country	1993	1995	1996	1997	1998	1999
Botswana	1,443	1,764	1,890	1,650	1,577	1,613
Russia	1,074	1,400	1,200	1,305	1,284	1,523
South Africa	1,081	1,050	1,020	983	1,058	985
D.R. Congo*	255	450	400	897	661	725
Angola	140	750	1,100	806	692	544
Australia	386	360	400	322	400	437
Canada	-	-	-	-	33	422
Namibia	338	422	460	410	389	414
Sierra Leone	-	-	-	120	-	138
Guinea	-	-	-	117	-	119
CAR	-	-	-	109	-	92
Venezuela	-	-	-	34	-	59
Tanzania	-	-	-	19	14	37
Liberia	-	-	-	30	-	32
Ivory Coast	-	-	-	12	-	24
Ghana	-	-	-	16	-	19
China	-	-	-	17	-	15
Brazil	-	-	-	131	-	-
Other	824	-	-	17		-
World Total	5,401	7,286	7,570	6,995	6,728	7,253

* Democratic Republic of Congo, formerly Zaire

Source: Adapted from Jennings (1995); MJ (1998); Picton (1996, 1997); Rombouts (1999, 2000).

Jewelry quality diamonds are very rare minerals. World mining of jewelry quality diamonds is small and diamond ores differ from each other by their content (Gush, 1994). Production of jewelry quality diamonds accounts for 15-17 per cent of the total diamond production (Abramov, 1994) and equals to 55,000 carats or 11 tons of jewelry diamonds per year (Balazik, 1999). Taking into account that the price of jewelry diamonds is considerably higher than of industrial diamonds, miners receive various profits depending on the deposit. Because of the nature of rough diamonds and their valuation, sharp differences in diamond prices occur depending on their quality. It is therefore better to refer to rough diamond production according to its value (Gush, 1994; Jennings, 1995). The value of diamond production differs from mine to mine dependent upon quality and price of the mined diamonds (see Table 3.7).

Table 3.6 Production from world's major diamond mines, thousand carats

Mine	Country	1996	1997	1998	1999
Argyle	Australia	40,000	40,200	40,900	29,700
Jwaneng	Botswana	11,200	12,472	12,688	11,400
Udachny	Russia	-	10,200	10,200	9,300
Opara	Botswana	-	6,746	6,300	9,070
Jubilee	Russia	-	2,700	4,200	5,200
Mbuyi-Maji	D.R. Congo	6,500	6,800	6,500	5,000
Venetia	South Africa	4,300	4,336	4,495	3,739
Ekati	Canada	-	-	250	2,510
Finsch	South Africa	2,100	2,226	2,165	1,737
Sytykan	Russia	-	500	700	1,700
Premier	South Africa	1,600	1,452	1,392	1,574
Marsfontein	South Africa	-	-	532	882
Lethlakane	Botswana	-	893	785	878
Namaqualand	South Africa	700	796	768	736
Namdeb onshore	Namibia	-	793	705	716
Kimberley	South Africa	600	612	683	563
Namdeb offshore	Namibia	471	481	497	514
Koffiefontein	South Africa	137	136	158	127

Source: Adapted from Picton (1997); MJ (1997, 1998); Rombouts (1999, 2000).

The value of a diamond lies in its beauty and rarity; therefore it is always difficult to place a precise value on such qualities (Gush, 1994). Moreover, diamond prices have little or no relation to the cost of their extraction (Bergenstock and Maskulka, 2001). Jewelry quality diamonds have a number of features as a trade product: firstly, the demand is relatively stable, and secondly, diamonds are long lasting. The current diamond market is one of the most stable markets of consumer goods (Green, 1993). High and stable prices for diamonds are explained by two reasons: firstly, diamonds are relatively rare minerals and are considered to have high aesthetic value; and secondly, the regulation of diamond supply to the diamond market, in which De Beers has been engaged for the most part of the 20th century (Green, 1993).

The main feature of the current market of rough diamonds is its monopolization. De Beers controls a large share of the market (see Table 3.8). Through the Central Selling Organization (CSO), De Beers Groups controls 85 per cent of jewelry diamonds in the world's market. This organization buys diamonds on the basis of agreements with diamond mining producers and then resells them to the consumers, i.e. diamond cutters. The total rough diamond supply is controlled by the CSO and comes from three sources: De Beers/Centenary-owned mines; outside suppliers contracted to the CSO (cartel members); and open market purchases via buying offices in Africa, Antwerp and Tel-Aviv (rough output purchases from countries that have not signed an agreement with De Beers) (Bergenstock and Maskulka, 2001). Jewelry diamonds are cut in several large centers for production of polished diamonds, which are located in New York, Antwerp, London, India and Israel (Tichotsky, 2000). Newly created centers have been formed in China, Malaysia, Thailand and Vietnam (Kommersant-daily, 1993). Certain specialization has occurred, dependent upon the weight of diamonds processed. For example, large diamonds are traditionally polished in the US, medium size stones are polished in Belgium, Israel, Netherlands, Russia and South Africa, and small stones are cut in India (Teslenko, 1997).

Table 3.7 Production from world's largest diamond mines in value, million US dollars

Mine	Country	1997	1998	1999
Jwaneng	Botswana	1,022	1,141	1,140
Udachny	Russia	969	969	930
Argyle	Australia	322	398	426
Ekati	Canada	-	33	422
Opara	Botswana	405	346	363
Venetia	South Africa	390	405	355
Jubilee	Russia	175	252	338
Other mines	Russia	65	63	255
Namdeb onshore	Namibia	258	229	233
Namaqualand	South Africa	147	138	132
Premier	South Africa	116	111	126
Namdeb offshore	Namibia	108	109	113
Lethlekane	Botswana	103	86	110
Marsfontein	South Africa	-	80	100
Finsch	South Africa	143	119	96
Mbuyi-Maji	D.R. Congo	75	65	75
Kimberley	South Africa	67	81	70
Koffiefontein	South Africa	20	25	20

Source: Adapted from MJ (1998); Rombouts (1999, 2000).

Table 3.8 World's top rough diamond producers in 1999

No.	Company	Diamond production, US dollars	Diamond production, thousand carats
1	Debswana (De Beers and Government of Botswana)	1,612,550	21,348
2	ALROSA	1,523,000	16,200
3	De Beers	799,835	8,476
4	Argyle Diamond Mines Pty Ltd (Ashton/Rio Tinto)	426,000	29,700
5	Ekati Diamond Mine (BHP/Dia Met)	421,680	2,510
6	Namdeb (De Beers and Government of Namibia)	345,780	1,230
7	Marsfontain Mine (De Beers/Southern Era)	99,666	882
8	Catoca Mine	92,160	1,440
9	MIBA	75,000	5,000
10	Luzamba	51,800	185
11	Namco	41,374	274
12	Baken Mine (Transhex Mining)	36,800	160
13	Merlin (Ashton/Rio Tinto)	10,752	84
14	Ocean Diamond Mining	7,200	48
15	Bellsbank-Ardo (Rex mining)	6,300	35
	World's total	7,253,020	1,110,58

Source: Adapted from Rombouts, 'The Diamond-mining Scene' (2000).

Environmental and Social Concerns in Gold and Diamond Mining

Gold and diamond mining companies originate from both developed and developing countries. The important feature of both industries is a strong presence of western-based companies, operating in developing countries. Gold and diamond mining have slightly different environmental impacts. However, social concerns are formed by other factors such as presence of indigenous groups, level of economic development in the mining area, stakeholder groups involved, etc.

Among environmental and social challenges that the gold mining industry face today are:

- concerns about the use of cyanide in production processes;
- increasing environmental and social regulation;
- the need to achieve global environmental and social standards in developing countries;

- tailings storage facilities design and closure;
- emission controls and restrictions;
- improvement in mining technology negatively impacting upon employment levels.

The gold industry is concerned with land access processes at both the exploration and mining stages and the need for stakeholder agreement and enhancement of its global reputation when seeking new ventures (Mitchell, 1999; Cottrell and Rankin, 2000; Hilson and Murck, 2001). In addition, small-scale gold mining sometimes involves illegal mining, and raises concerns of environmental degradation and negative impacts on the health of miners in developing countries including Brazil, French Guiana, Ghana and Guyana (Avocat et al, 2001; Heemskerk, 2001).

Some of the proactive gold mining companies have confronted the existing challenges in order to improve resource efficiency, stakeholder liaison to prevent future confrontations, identification of new markets, and changes to corporate culture (Cottrell and Rankin, 2000). For example, Homestake's McLaughlin Gold Mine in California successfully operates under the world's strictest environmental regime. The McLaughlin site is notable for its innovative process-design criteria, fail-safe tailings and waste disposal systems, extensive, ongoing rehabilitation and environmental monitoring systems. These environmental management initiatives have not resulted in substantial extra costs, and indeed many have improved the efficiency of the mine and made the overall operation more economical (Warhurst, 1999).

A good example of a successful adoption of a tri-sector partnership for community development is Minera Las Cristinas, CA (MINCA) in Venezuela, a gold mining project with 70 per cent ownership by Placer Dome Inc. The principal community development effort was oriented towards partnership in improvement of the health care. MINCA along with interested parties, including USA-based NGOs, government health agencies at national, regional and local levels, and community leaders through signing a Memorandum of Understanding, organized a provision of health services to mine worker, their families and the general community (BPD Natural Resources Cluster, online).

A good example of integrating indigenous people in decision-making within the gold industry comes from the Musselwhite gold mine in Ontario, Canada, which is operated by Placer Dome Inc. First, Placer Dome has developed a sustainability policy which identifies the company's basic principles, including corporate commitment, public responsibility, social progress, environmental stewardship and economic benefits. Following negotiations with stakeholders, complex issues were formalized into three agreements: firstly, Ontario-Windigo-Shibogama Planning Agreement, secondly, Musselwhite General Agreement and thirdly, General Compensation Agreement. The Musselwhite General Agreement is signed by several parties: Placer Dome, First Nations (i.e. indigenous communities), and the federal and provincial governments, and addresses several areas of concerns including: environment, culture and heritage, employment, business and other economic opportunities and First Nations administration. The lessons from the Musselwhite experience show that such agreements must be accepted as fair by all signatories, agreements must provide for mechanisms to deal with unforeseen problems as they occur, and agreements have to be deliverable (ICME, 1999).

Environmental impacts in diamond mining vary depending on the mining method. The environmental impact of diamond mining is similar to those existent in open pit mines, the problems stem from waste disposal, leaching and ground water pollution (Bew, 1995). A major diamond mine may process 150 million tons of kimberlite and therefore produce significant volumes of waste and barren rock, as most kimberlite fields are barren (Jennings, 1995). In small-scale diamond mining, for example in Sierra-Leone, mining activities are severe and devastating for land resources. After an area is mined the land is left exposed and degraded without reclamation and therefore it is unsuitable for agriculture and other use. Other impacts of African diamond mining include soil erosion, exposure to flooding, removal of vegetation, surface drainage, water pollution and health hazards (Sipkins, 1995). Moreover, in Angola, diamond mining activities are accompanied by river re-routing and water pollution that threatens the health of the local population via pollution of drinking water and the related spread of dysentery, cholera and the increase in the number of mosquitoes carrying malaria (Morris, 1996). In the case of marine mining, the environmental impacts include reduction of biota (Gibbons and Wickens, 1999), a disturbance of sediments, seabed smothering, potential oxygen depletion, and toxicity effects due to the suspension of heavy metals and/or pesticides (Tarras-Wahlberg and O'Toole, 2000).

The history of diamond mining shows significant disrespect for human rights, for example, apartheid in South Africa was often used to benefit the mining companies. Moreover, diamond mines often become a subject of disputes and even military conflicts. For example, the civil war in Sierra Leone is largely a resource war between rebels, who control much of the diamond mining in the east of the country, and the democratic government. Similar conflicts are related to the diamond mines in the Democratic Republic of Congo and Angola (Ali and Behrendt, 2001). However, human rights violations by mining companies have diminished over recent years (Ali and Behrendt, 2001).

Land and access issues are the greatest concern of the mining companies around the world, and they are directly related to social and cultural issues surrounding unresolved land claims particularly in Australia and Canada (Clark and Clark, 1999). One of the important issues in Australian diamond mining is the land rights of Aboriginal people. The issue of native title is connected to the treatment of cultural values of Aboriginal communities by mineral development on Aboriginal territories. An interesting example of legal arguments over land rights is the Mabo case that led to a statutory recognition of native title in the area of diamond mining development in East Kimberley, Australia. Rio Tinto used this concept for the development of its Aboriginal and Torres Strait Islander Policy, which targeted on developing good relationships with communities, neighboring the Argyle diamond mine (Harvey, 2000).

Indigenous knowledge has also become a necessary component of environmental assessment. The trend stems from the pursuit of political and property rights by indigenous people around the world and the recognition of these rights by nation states. An interesting example of community involvement comes from the Northwest Territories in Canada, where prior to the proposed diamond mine, BHP Diamonds Inc. had developed a holistic approach in involving indigenous people and their knowledge in environmental impact assessment. Indigenous people were accepted as integral players in the identification, assessment, and mitigation and monitoring of environmental and socio-economic impacts (Stevenson, 1996).

CSR Strategies of Gold Mining Companies

Environmental Strategies of Gold Mining Companies

Corporate strategies towards improvement of environment and community involvement can be analyzed with the use of corporate social disclosure revealed in Annual Reports, Environmental and Social Reports and corporate websites. Mining companies focus on several major areas of CSR strategies in their disclosure such as environmental practices, employee policies, health and safety, community development and corporate philanthropy. The analysis of corporate documents (including Annual Reports, Environmental Reports and corporate websites) produced by ten top gold mining companies in 1998-2000 shows a general increase in the volume of corporate social disclosure.

Larger companies such as Freeport-McMoRan Copper & Gold, Rio Tinto, Placer Dome and Normandy Mining disclose environmental and social information in significant detail, while smaller gold mining companies such as Ashanti Goldfields and Goldfields are the companies who disclose the least environmental and social information (see Table 3.9). The important foundation for the environmental reporting is an establishment of corporate environmental policy. Eight of ten top gold mining companies report having a corporate environmental policy. Usually, corporate environmental policies are adopted by the Board of Directors and are regularly revised. Often companies include the texts of their corporate environmental policies in their Environmental Reports or on corporate websites. In some instances, the companies produce separate publications on their corporate environmental policies (Placer Dome and Rio Tinto). Some companies adopt integrated policies that embrace the company's commitments to environmental, economic, social, health and safety issues. For example, Placer Dome has a sustainability policy, which incorporates the company's environmental, social and economic commitments. Furthermore, the companies adopt corporate environmental policies for their global operations (Placer Dome and Rio Tinto).

Table 3.9 Environmental policy and commitments of gold mining companies in 1998-2000

Disclosure categories	AngloGold	Ashanti Goldfields	Barrick Gold	Freeport-McMoRan C&G	Gold Fields	Homestake Mining	Newmont Mining	Normandy Mining	Placer Dome	Rio Tinto	Total
• Developed environmental policy	✓		✓	✓		✓	✓	✓	✓	✓	8
• Commitment to meet regulatory requirements and standards	✓		✓	✓	✓	✓	✓	✓	✓	✓	9
• Commitment to improve environmental performance	✓			✓	✓	✓		✓	✓	✓	7
• Commitment to principles of sustainable development	✓			✓		✓		✓	✓	✓	6
• Commitment to minimize environmental impacts and effects			✓	✓	✓	✓	✓	✓	✓	✓	8
• Commitment to reclaim and return land to beneficial users			✓	✓		✓	✓	✓	✓	✓	7

Note: ✓-statement is present

Source: Adapted from gold mining companies' Annual Reports, Environmental Reports for 1998-2000 and corporate websites for 2000/2001.

Most commonly, the gold mining companies form their environmental policies around the commitments to: meet regulatory requirements and standards; improve environmental performance; base their activities upon principles of sustainable development; minimize environmental impacts; and reclaim land after operations. All gold mining companies allocate environmental staff, which draw plans for environmental actions and conduct environmental management operations. An important foundation for environmental management is the presence of established environmental management systems. Although seven out of the top ten gold mining companies report having an EMS in place and only four of them report having their EMS certified to ISO standard 14001. Nevertheless, most companies disclose their engagement in environmental monitoring, waste, water and tailings management, rehabilitation and re-vegetation practices, and environmental impact assessments (see Table 3.10).

Table 3.10 Environmental management in gold mining companies in 1998-2000

Disclosure categories	AngloGold	Ashanti Goldfields	Barrick Gold	Freeport-McMoRan C&G	Gold Fields	Homestake Mining	Newmont Mining	Normandy Mining	Placer Dome	Rio Tinto	Total
• Environmental office/department/staff	✓	✓	✓	✓	✓	✓	✓	✓	✓	✓	10
• Environmental action plan	✓	✓	✓	✓	✓	✓		✓	✓	✓	9
• Commitment to industry's code of conduct	✓			✓		✓		✓	✓	✓	6
• Environmental management system	✓		✓	✓		✓		✓	✓	✓	7
• Certification of ISO 14001				✓		✓			✓	✓	4
• Environmental monitoring (water, air quality, waste, noise, etc.)	✓	✓		✓		✓	✓	✓	✓	✓	8
• Tailings management	✓			✓		✓	✓	✓	✓	✓	7
• Overburden management				✓				✓	✓	✓	4
• Waste recycling and management		✓		✓		✓		✓	✓	✓	6
• Water management	✓	✓	✓	✓	✓	✓	✓	✓	✓	✓	10
• Air protection management				✓	✓	✓		✓	✓	✓	6
• Noise reduction								✓			1
• Environmental rehabilitation and reclamation	✓	✓	✓	✓	✓	✓	✓	✓	✓	✓	10
• Re-vegetation practices	✓	✓	✓	✓	✓	✓	✓	✓	✓	✓	10
• Energy efficiency								✓	✓	✓	3
• Protection of wildlife			✓			✓	✓	✓	✓	✓	6
• Conduct ecological risk assessment	✓			✓		✓		✓	✓	✓	6
• Environmental impact assessment	✓	✓	✓	✓		✓	✓	✓	✓	✓	9

Note: ✓-statement is present

Source: Adapted from gold mining companies' Annual Reports, Environmental Reports for 1998-2000 and corporate websites for 2000/2001.

Seven out of the top ten gold mining companies reported to have environmental audits, and six of them reported to have environmental audits conducted by external specialists or organizations. Companies also disclose information on the frequency of environmental audits and include third party verification statements in their Environmental Reports (Homestake Mining, Normandy Mining and Rio Tinto) (see Table 3.11).

Table 3.11 Environmental education, training and research in gold mining companies in 1998-2000

Disclosure categories	Companies										
	AngloGold	Ashanti Goldfields	Barrick Gold	Freeport-McMoRan C&G	Gold Fields	Homestake Mining	Newmont Mining	Normandy Mining	Placer Dome	Rio Tinto	Total
• Education and training of personnel of environmental issues		✓	✓	✓		✓	✓	✓	✓	✓	8
• Conduct environmental research	✓			✓		✓	✓	✓	✓	✓	7
• Cooperation with external organizations on environmental research	✓			✓				✓	✓	✓	5
• Review of environmental performance		✓		✓		✓		✓	✓	✓	6
• Conduct regular environmental audits	✓	✓		✓		✓		✓	✓	✓	7
• Conduct independent environmental audits	✓			✓		✓		✓	✓	✓	6
• Reporting on environmental activities		✓	✓	✓	✓	✓		✓	✓	✓	8

Note: ✓-statement is present

Source: Adapted from gold mining companies' Annual Reports, Environmental Reports for 1998-2000 and corporate websites in 2000/2001.

Gold mining companies mention having certain procedures for environmental reporting. Some companies also report in more detail on site operations, even producing separate publications, such as Environmental Site Reports (Placer Dome and Normandy Mining). Mostly, the information concerning environmental performance is qualitative. Some companies comprehensively disclose environmental performance data, including the number of environmental incidents (Normandy Mining, Placer Dome and Rio Tinto), rates of rehabilitation (Homestake Mining, Normandy Mining and Rio Tinto), air emissions (Homestake Mining, Normandy Mining, Placer Dome and Rio Tinto), waste (Homestake Mining, Freeport-McMoRan C&G and Rio Tinto), and water consumption and discharge (Homestake Mining, Normandy Mining and Rio Tinto). However, the metrics used for the description of environmental performance data often differs and does not provide means for comparison between the companies.

An inventive way of measuring environmental and social performance within mining companies is the use of specially developed integrated performance indicators according to which data is regularly collected and system for improvement of operations is installed. For example, Normandy Mining has introduced internally developed five-star program, which aims to drive continuous improvement in the management and performance of safety, community and environment issues. Via assessing performance against a series of defined levels and setting minimum requirements, Normandy formulates targets beyond minimum requirement to promote continual improvement. Furthermore, to ensure consistency of devised program, the company conducts audit by external professionals across the group's operations (Normandy Mining Limited, 2000b).

Only four out of the top ten gold mining companies produce separate Environmental Reports, two started this practice in 1996 (Homestake Mining and Rio Tinto) and others started in 1997 (Normandy Mining and Placer Dome). The companies give various titles to their regular reports that address social, environmental, and health and safety issues. For example, Rio Tinto entitles its reports as Social and Environment Report (1998, 1999) or Social and Environment Review (2000). Homestake Mining entitles its reports as Environment, Health and Safety Report, while Normandy calls it simply an Environment Report. Placer Dome is the only company, which produces Sustainability Reports (1998, 1999 and 2000). In addition to the overall Environmental Report, which covers the entire operations of the company, some companies produce site-specific reports, for example Normandy has produced Environment Reports for each gold mining operation and Placer Dome produced Sustainability Reports for each of its subsidiaries.

The gold mining companies also produce other publications where they disclose information on environmental, health and safety and community issues. For example, Rio Tinto produced such publications as Mining and the Environment; Global Business, Local Neighbors; Human Rights Guidance; Towards Sustainable Development; Rio Tinto WA Future Fund; Rio Tinto and Aboriginal People; Rio Tinto and Kormilda Collage; Education with Communities; and Rio Tinto Business and Education and Communities program.

Corporate websites are a useful resource for information related to environment, health and safety and community issues. The companies place texts of their corporate environmental, health and safety and community policies, and also place electronic copies (PDF or HTML) of their hard copy Annual Reports and Environmental and Social Reports.

Employee Policy and Health and Safety in Gold Mining Companies

Commitments to health and safety issues are often reflected in corporate health and safety policies (AngloGold, Barrick, Homestake Mining, Normandy Mining, Placer Dome and Rio Tinto). Most commonly, the health and safety commitments of gold mining companies are related to health and safety standards, continuous improvement of performance and aiming towards zero injuries (see Table 3.12).

Most disclosed health and safety themes included information about dedicated health and safety departments, health and safety action programs and employee training on health and safety issues. The gold mining companies are more likely to disclose information on health and safety issues in their Annual Reports and some companies include relevant information in their Environment, Health and Safety Reports.

Mining companies closely cooperate with trade unions for compliance with health, safety and employment regulations (e.g. AngloGold and GoldFields). For example, AngloGold develops plans for effective compliance with employee related legislation (Employment Equity Act) in negotiation with trade unions and associations in the South Africa region. The plan includes AngloGold Employment Equity Philosophy, monitoring of plan operations and analysis of employment policies, procedures and practices. Employment Equity Forum, a consultative body comprising representatives of corporate management and trade unions, is created to monitor plan implementation (AngloGold Limited, 2000).

In relation to health and safety performance data, the gold mining companies focus on lost time injury frequency rate and number of deaths (Ashanti, Gold Fields, Freeport-McMoRan C&G, Homestake Mining, Normandy Mining, Placer Dome and Rio Tinto). The companies that have operations in African countries are more likely to have appropriate programs for HIV and AIDS (AngloGold, Ashanti, Gold Fields, Placer Dome and Rio Tinto). These programs usually aim to combat the spread of the virus through the introduction of education programs, participation in national initiatives for raising awareness (Debswana), assisting affected families and investing in HIV vaccine (AngloGold). Companies also provide anonymous testing and confidentiality for HIV-positive employees, commit to nondiscrimination on the basis of HIV, and provide employees with health and counseling advice and prophylactic antibiotics (AngloGold and De Beers).

Table 3.12 Employee policy, health and safety strategies of gold mining companies in 1998-2000

Disclosure categories	Companies									
	AngloGold	Ashanti Goldfields	Barrick Gold	Freeport-McMoRan C&G	Gold Fields	Homestake Mining	Newmont Mining	Normandy Mining	Placer Dome	Rio Tinto	Total
• Developed employee policy	✓	✓		✓				✓	✓	✓	6
• Employee program/plan	✓		✓	✓	✓			✓	✓	✓	7
• Employee training	✓	✓	✓	✓	✓			✓	✓	✓	8
• Employee health and education systems that are extended to families	✓		✓	✓				✓	✓	✓	6
• Developed health and safety policy	✓		✓			✓		✓	✓	✓	6
• Commitment to health and safety standards	✓	✓			✓	✓		✓	✓	✓	7
• Commitment to improve health and safety performance	✓	✓		✓	✓	✓		✓	✓	✓	8
• Dedicated health and safety department/office	✓	✓	✓	✓	✓	✓	✓	✓	✓	✓	10
• Health and safety action program	✓	✓	✓	✓	✓	✓		✓	✓	✓	9
• Training of personnel on health and safety practices	✓	✓		✓	✓	✓		✓	✓	✓	8
• Occupational health monitoring of all workers connected with site (miners, truck drivers, etc.)	✓				✓	✓		✓	✓	✓	6
• HIV and AIDS policy or program	✓	✓			✓				✓	✓	6
• Health and safety audit	✓	✓		✓		✓		✓	✓	✓	7
• Reporting on health, safety issues	✓	✓				✓		✓	✓	✓	6

Note: ✓-statement is present

Source: Adapted from gold mining companies' Annual Reports, Environmental Reports for 1998-2000 and corporate websites in 2000/2001.

Community Development and Stakeholder Engagement in Gold Mining Companies

Management of social issues is becoming high on the agenda in the operations of gold mining companies. Six companies have mentioned that they have developed community or social policies (see Table 3.13). The companies consider and address community concerns and some companies have developed separate community and social programs (AngloGold, Barrick, Freeport-McMoRan C&G, Normandy Mining, Placer Dome and Rio Tinto). Although many gold mining companies operate in countries with strong presence of indigenous people, none of them disclosed information concerning an indigenous people's officer or department and only one company mentioned having developed an indigenous people's policy (Normandy Mining). In addition, only two companies reported having a dedicated department or office for social and community issues. The companies often contribute to community life via improving infrastructure, building roads and supplying water to communities; investing in education by building and supporting local schools; investing in health services by building or donating to hospitals; and investing in environmental conservation, education and research. The gold mining companies cooperate with local NGOs, government agencies and international organizations on a variety of issues, including employment, land rights, education, research, health and community development.

Community development programs within the gold mining companies include a range of targeted initiatives such as:

- Organization of consultation forums with local communities and other stakeholders concerning employment, environmental impacts, preservation of cultural heritage and support of community organizations (e.g. Placer Dome).
- Signing negotiated agreement with local communities and indigenous people concerning employment in the mining industry, environmental research and conservation (e.g. Placer Dome and Freeport McMoRan C&G).
- Organization of cross-cultural training to improve understanding of community issues by operational staff (e.g. AngloGold).
- Support of non-mine training programs for local community members, which aim to improve self-sufficiency of local communities (e.g. Barrick).
- Support to communities affected by natural disasters such as flooding (e.g. AngloGold).
- Support of local business development. Mining companies assist small-scale miner who will be displaced by large-scale mining (e.g. AngloGold, Placer Dome).

Eight companies disclosed that they are engaged in charity giving, which includes direct donations to charitable organizations and support of various local organizations. Six gold mining companies reported to have Foundations or Trusts for philanthropic purposes, which are studied later in Chapter 6.

Table 3.13 Community development and stakeholder engagement strategies of gold mining companies in 1998-2000

Disclosure categories	Companies: AngloGold	Ashanti Goldfields	Barrick Gold	Freeport-McMoRan C&G	Gold Fields	Homestake Mining	Newmont Mining	Normandy Mining	Placer Dome	Rio Tinto	Total
• Social/community policy			✓	✓			✓	✓	✓	✓	6
• Indigenous people's policy								✓			1
• Department/office for social and community issues				✓						✓	2
• Social/community program	✓		✓	✓	✓			✓	✓	✓	7
• Social/community research	✓			✓				✓	✓	✓	5
• Consult with stakeholders	✓	✓		✓	✓	✓	✓	✓	✓	✓	9
• Address community concerns on a regular basis		✓	✓	✓		✓		✓	✓	✓	7
• Willingness to employ local community members	✓		✓	✓		✓	✓	✓	✓	✓	8
• Education and training of local community members	✓		✓	✓		✓	✓	✓	✓	✓	8
• Commitment to discuss and land rights with community				✓				✓		✓	3
• Contribution to community life (school, clinics, etc.)	✓	✓	✓	✓		✓	✓	✓	✓	✓	9
• Compensation system for affected communities						✓	✓		✓	✓	4
• Reporting on social/community issues				✓				✓	✓	✓	4
• Cooperation with local NGOs for community development	✓		✓	✓	✓	✓	✓	✓	✓	✓	9
• Cooperation with industry associations	✓			✓		✓		✓	✓	✓	7
• Cooperation with international organizations	✓							✓	✓	✓	4
• Cooperation with government on social programs	✓		✓	✓	✓	✓	✓	✓	✓	✓	9
• Charitable giving	✓	✓	✓	✓			✓	✓	✓	✓	8
• Establishment of philanthropic foundation	✓			✓	✓		✓	✓		✓	6

Source: Adapted from gold mining companies' Annual Reports, Environmental Reports for 1998-2000 and corporate websites for 2000/2001.

CSR Strategies of Diamond Mining Companies

Environmental Strategies of Diamond Mining Companies

Diamond mining companies disclose information concerning their CSR strategies and practices in their Annual Reports, Environmental and Social Reports and corporate websites. Corporate documents of the top seven diamond mining companies were analyzed for the period of 1998-2000.[4] The selected diamond mining companies operate in countries in Africa, Asia and Australia (see Table 3.14 and Table 3.15). It is difficult to identify the total diamond output of these companies and compare it with the world's diamond production because De Beers is a shareholder of Debswana and Namdeb, and therefore, the total annual production of De Beers includes a portion of output from these enterprises.

Table 3.14 Geography of operations of the top diamond mining companies in 2000

Company	Corporate headquarters	Operations locations
Argyle Diamonds	Australia	Australia
De Beers	South Africa	Botswana, Namibia, South Africa, Tanzania
Debswana	Botswana	Botswana
Ekati	Canada	Canada
Namdeb	Namibia	Namibia
Rex Mining	Belgium	South Africa
Southern Era	Canada	Angola, South Africa

Table 3.15 Diamond production, thousand carats

Company	1998	1999	2000
De Beers	31,200	32,300	36,500
Argyle Diamonds	40,840	42,800	26,500
Debswana	19,773	21,348	24,651
Ekati	-	663	1,417
Namdeb	1,280	1,290	1,320
Southern Era	282	353	436
Rex Mining	35	24	31

As in the case with gold mining companies, large diamond mining companies such as De Beers, Argyle Diamonds and Ekati, followed by Debswana and Namdeb are more prominent in their social disclosure than smaller companies such as Rex Mining and Southern Era. Most diamond mining companies established corporate environmental policies, which include commitments to meet regulatory requirements, to minimize environmental impacts and conduct environmental rehabilitation after mine closure (see Table 3.16).

Table 3.16 Environmental policy and commitments of diamond mining companies in 1998-2000

Disclosure categories	Argyle Diamonds	De Beers	Debswana	Ekati	Namdeb	Rex Mining	SouthernEra	Total
• Developed environmental policy	✓	✓	✓	✓	✓		✓	6
• Commitment to meet regulatory requirements and standards	✓	✓	✓	✓	✓		✓	6
• Commitment to improve environmental performance	✓	✓		✓				3
• Commitment to principles of sustainable development	✓		✓	✓	✓			4
• Commitment to minimize environmental impacts and effects	✓	✓	✓	✓	✓		✓	6
• Commitment to reclaim and return land to beneficial users	✓			✓	✓		✓	4

Note: ✓-statement is present

Source: Adapted from diamond mining companies' Annual Reports for 1998-2000 and corporate websites in 2000/2001.

Six diamond mining companies reported to have a corporate environmental policy. In the case of Ekati, which is a subsidiary of BHP, the diamond mine has been operating under BHP Environmental Policy in 1998-1999 and in 2000, BHP adopted an integrated Health, Safety, Environment and Community Policy, which was then applied to the Ekati Diamond Mine. Agryle has also an integrated environment, health and safety policy.

The diamond mining companies include texts of their environmental policies in their Environmental Reports and on their corporate websites. In 2000, Argyle and Ekati worked under comprehensive environmental policies, which inter alia included the following important commitments to the environment:

- to comply with regulatory requirements and standards (Argyle, Ekati),
- to minimize environmental impacts resulting from operations, products and services (Argyle and Ekati),
- to aim for continuous improvement of performance (Argyle and Ekati),
- to ensure management systems for control and monitoring of environmental impacts (Argyle and Ekati),
- to conduct research (Ekati),
- to conserve resources and materials (Argyle and Ekati),

- to protect the environment (Argyle and Ekati),
- to consult with stakeholders (Argyle and Ekati),
- to communicate with government and the community on environmental issues (Ekati),
- to report on environmental performance (Argyle).

Other important environmental disclosures include allocation of environmental staff, development of environmental action plans, establishment of environmental management system, conduct of environmental monitoring, protection of wildlife, environmental impact assessment, environmental education, research and regular environmental audits. Amongst the studied diamond mining companies, only Rio Tinto for Argyle is assigned to the industry environmental code of conduct. The least popular environmental disclosures include: tailings management, overburden management, energy efficiency, conduct of risk assessment, cooperation with external organizations on environmental research and conduct of independent environmental audits (see Table 3.17).

An important part of environmental strategies of the diamond mining companies is participation in environmental conservation and protection of wildlife. For example, De Beers has reported in its Safety, Health and Environment Review 2000:

> At all mining operations areas are set aside as nature reserves to protect indigenous wildlife, increase environmental awareness through education and provide outdoor recreation (De Beers, 2000b).

Mining companies invest in wildlife conservation via:

- Establishment of nature reserves, such as De Beers Venetia Limpopo Nature Reserve, which is a component of proposed Vhembe Dongola National Park and transfrontier Peace Park with Botswana and Zimbabwe. The park aims to contribute to the regional economy and benefit local communities through eco-tourism.
- Forming partnerships with environmental protection agencies and environmental NGOs for protecting animals. Argyle formed a partnership in 2000 with Department of Conservation and Land Management to care for wallabies and supported Kununurra Wildlife Carers' Association, which cares for injured animals in East Kimberley region.
- Establishment of animal farms. De Beers allocated some 27,000 ha for establishment of commercial beef production.

Table 3.17 Environmental management in diamond mining companies in 1998-2000

Disclosure categories	Companies							
	Argyle Diamonds	De Beers	Debswana	Ekati	Namdeb	Rex Mining	SouthernEra	Total
• Environmental office/department/staff	✓	✓	✓	✓	✓			5
• Environmental action plan	✓	✓	✓	✓	✓			5
• Commitment to industry's code of conduct	✓							1
• Environmental management system	✓	✓	✓	✓	✓			5
• Certification of ISO 14001	✓	✓	✓		✓			4
• Environmental monitoring (water, air quality, waste, noise, etc.)	✓	✓	✓	✓	✓			5
• Tailings management				✓				1
• Overburden management	✓			✓				2
• Waste recycling and management	✓	✓	✓	✓				4
• Water management	✓	✓	✓	✓				4
• Air protection management	✓		✓	✓				3
• Noise reduction								0
• Environmental rehabilitation and reclamation	✓	✓	✓	✓	✓		✓	6
• Re-vegetation practices	✓	✓		✓			✓	4
• Energy efficiency	✓		✓					2
• Protection of wildlife	✓	✓	✓	✓	✓			5
• Conduct ecological risk assessment				✓				1
• Environmental impact assessment	✓	✓	✓	✓	✓			5
• Education and training of personnel of environmental issues	✓	✓	✓	✓	✓			5
• Conduct environmental research	✓	✓	✓	✓	✓			5
• Cooperation with external organizations on environmental research				✓	✓			2
• Review of environmental performance	✓	✓	✓	✓				4
• Conduct regular environmental audits	✓	✓	✓	✓	✓			5
• Conduct independent environmental audits	✓		✓	✓				2
• Reporting on environmental activities	✓	✓	✓	✓	✓			5

Note: ✓-statement is present

Source: Adapted from diamond mining companies' Annual Reports for 1998-2000 and corporate websites in 2000/2001.

The diamond mining companies report on their environmental performance with a different degree of detail, revealing the number of environmental incidents (Argyle) and their character (Ekati), rate of rehabilitation (Argyle and Ekati), air emissions (Argyle, Ekati), water consumption (Argyle and Ekati) and water discharge (Ekati), consumption of raw materials (Ekati), waste generation (Ekati), license limits and compliance (Ekati), precipitation survey (Ekati) and environmental expenditure (Ekati). Only three companies out of the top seven diamond mining companies produced stand-alone Environmental Reports, these are Argyle, De Beers and Ekati. Ekati started producing stand-alone Environmental Reports since 1998, Argyle since 1999 and De Beers since 2000. Environmental Report for Ekati Diamond Mine is produced in two formats, a summary of the Environmental Report and a full CD-version Environmental Report. Companies also put their Environmental Reports in PDF format on their corporate websites, which are exact copies of the hard-copy reports, some fail to compile a library of previous Environmental Reports and only put the most recent report on their websites. Mostly, the companies place qualitative information in relation to environment, health and safety and community issues on their corporate websites in HTML format.

Employee Policy and Health and Safety in Diamond Mining Companies

In the area of employee health and safety, the most frequent themes to be disclosed are: employee plans and programs, employee training and education, training of employees for health and safety practices, and health and safety audits (see Table 3.18).

Only three diamond mining companies reported to have an employee policy and finally only two companies stated that they had procedures for reporting on health and safety issues. The diamond mining companies provide education and training for their employees in the areas of mining operations (Argyle, Namdeb), literacy (Ekati), local culture (Argyle and Ekati), and health and safety (Argyle, De Beers, Debswana and Namdeb). The companies provide a range of benefits for their employees, including healthcare and education assistance programs (Ekati), which in some cases are extended to family members (Argyle, Debswana and Namdeb), retirement and pension plans (Argyle, De Beers, Debswana, Ekati, Rex Mining), life insurance and medical travel assistance (Ekati).

Companies express the significance of health and safety of their employees (Ekati) and express a commitment to maintain health and safety standards (Argyle, De Beers, Debswana and Ekati) and improve health and safety performance (Argyle, De Beers, Ekati and Rex Mining). Also, the companies set particular goals such as working free of injury and adverse health impacts (Argyle and Ekati). Mining companies adopt zero injury tolerance approach, whereby companies are setting the highest standard to ensure health and safety of their employees. For example, Argyle state in its 1999 Annual Social and Environmental Report:

> We believe that all workplace injuries, illnesses and environmental incidents are preventable and consider rapid progression toward our goal of zero losses as being imperative to maintaining a sustainable business (Argyle Diamonds, 1999, p.8).

In order to achieve ultimate goal of zero injury, mining companies analyze safety incidents, and take appropriate actions to preventive reoccurrence. Moreover, companies undertake risk assessment to record, identify and address significant health and safety risks. Most commonly, the companies report on health and safety performance using lost time injury frequency rate and number of fatal injuries (Argyle, De Beers, Debswana, Ekati and Namdeb).

Table 3.18 Employee policy, health and safety strategies of diamond mining companies in 1998-2000

Disclosure categories	Companies							
	Argyle Diamonds	De Beers	Debswana	Ekati	Namdeb	Rex Mining	SouthernEra	Total
• Developed employee policy	✓		✓		✓			3
• Employee program/plan	✓	✓	✓	✓	✓			5
• Employee education and training	✓	✓	✓	✓	✓	✓		6
• Employee health and education systems that are extended to families	✓		✓		✓			3
• Developed health and safety policy	✓	✓	✓	✓				4
• Commitment to health and safety standards	✓	✓	✓	✓				4
• Commitment to improve health and safety performance	✓	✓		✓		✓		4
• Dedicated health and safety department/office		✓	✓	✓	✓			4
• Health and safety action program	✓	✓	✓	✓				4
• Training of personnel on health and safety practices	✓	✓	✓		✓			4
• Occupational health monitoring of all workers connected with site (miners, truck drivers, etc.)	✓	✓	✓					3
• HIV and AIDS policy or program		✓	✓		✓			3
• Health and safety audit	✓	✓			✓	✓		4
• Reporting on health, safety issues	✓	✓	✓					3

Note: ✓-statement is present

Source: Adapted from diamond mining companies' Annual Reports for 1998-2000 and corporate websites in 2000/2001.

Community Development and Stakeholder Engagement in Diamond Mining Companies

Community issues are very erratically addressed in corporate social disclosure of the top diamond mining companies (see Table 3.19). Only two companies have Community Relations Policies (Argyle and Ekati). In 1998-1999, Ekati operated under the BHP Community Relations Policy and in 2000, under the BHP Health, Safety, Environment and Community Policy. In 2000, Argyle and Ekati operated under Community Relations Policies with several important commitments to:

- consult with stakeholders (Argyle and Ekati),
- cooperate with Aboriginals for their long-term benefit (Argyle),
- recognize and respect the traditional rights of indigenous peoples (Argyle and Ekati),
- implement cross-cultural awareness programs (Argyle),
- commission social and community studies (Argyle),
- respect cultural heritage (Argyle and Ekati),
- communicate and consult with traditional land owners on land management (Argyle),
- implement cross-cultural training programs (Argyle),
- comply with regulatory requirements and standards (Ekati),
- produce a public Community Relations Plan and produce Social and Environmental Report (Argyle),
- promote employment from the local areas through recruitment and training programs (Argyle),
- conduct business based on mutual trust (Argyle and Ekati),
- help local communities meet social needs through development and use of skills and technologies (Ekati),
- engage employees, business partners, contractors, suppliers and visitors in a shared responsibility (Ekati).

Amongst the community involvement section, the most frequent disclosures are: social/community programs, willingness to employ local community members, educating and training local community members, contribution to community life, cooperation with local NGOs and governmental agencies in social and environmental programs, and charity giving. However, none of the companies reported to have an indigenous peoples' policy, although Argyle and Ekati follow Community Relations Policies, which incorporate some indigenous peoples' issues. None of the diamond mining companies reported to have an indigenous peoples' officer or department or to have developed a compensation system for affected communities and cooperation with international organizations. Amongst the least disclosed community related issues are: development of a social and community policy, a dedicated department for social and community issues, conducting social and community research, considering and addressing community concerns on a regular basis, commitment to discuss and negotiate with community over land rights, procedures for reporting on social and community issues and cooperation with industry associations.

Table 3.19 Community development and stakeholder engagement strategies of diamond mining companies in 1998-2000

Disclosure categories	Companies							
	Argyle Diamonds	De Beers	Debswana	Ekati	Namdeb	Rex Mining	SouthernEra	Total
• Development social/community policy	✓			✓				2
• Department/office for social and community issues	✓							1
• Indigenous people officer/department								0
• Social/community program	✓	✓	✓	✓				4
• Conduct social/community research	✓			✓				2
• Consult with stakeholders	✓	✓		✓				3
• Consider and address community concerns on a regular basis	✓			✓				2
• Willingness to employ local community members	✓	✓	✓	✓	✓			5
• Education and training of local community members	✓	✓	✓	✓	✓			5
• Commitment to discuss and negotiate with community over land rights	✓			✓				2
• Contribution to community life (school, clinics, etc.)	✓	✓	✓	✓	✓			5
• Compensation system for affected communities								0
• Reporting on social/community issues	✓							1
• Cooperation with local NGOs	✓	✓	✓	✓	✓			5
• Cooperation with industry associations	✓	✓						2
• Cooperation with international organizations								0
• Cooperation with and contribution to government programs	✓	✓	✓	✓	✓			5
• Charitable giving	✓	✓	✓	✓	✓			5
• Have philanthropic foundation, fund or trust		✓	✓		✓			3

Note: ✓-statement is present

Source: Adapted from diamond mining companies' Annual Reports for 1998-2000 and corporate websites in 2000/2001.

The companies have developed different approaches of integrating community concerns in their decision-making, via organization of meetings and forums with indigenous and local communities (Argyle, De Beers and Ekati), site visits for community residents (De Beers and Ekati) and conducting discussions (Ekati). Companies discuss various issues with the community, including land use (Argyle and Ekati), potential employment (Argyle, Debswana, Ekati and Namdeb), and impacts and benefits of mining operations (Argyle and Ekati). In addition, the companies are conducting studies of traditional ecological knowledge and incorporate it in their plans for the monitoring and mitigation of impacts (Ekati) and conduct cross-cultural awareness training to learn about Aboriginal cultures (Argyle). The companies undertake commitments via agreement with local governments and set specific targets for employing local people and indigenous people (Argyle and Ekati) and express general willingness to hire local and indigenous people (Debswana and Namdeb). The companies provide training for local community members for the employment in mining and other operations (Argyle, De Beers Debswana, Ekati and Namdeb). Also, the companies establish scholarships to members of local and indigenous communities (Argyle and Namdeb).

Companies contribute to community life via charity giving and conducting community programs or through their philanthropic Foundations and Trusts. The community initiatives include fuel purchase programs for some communities (Ekati), support of local arts (Argyle), and support of schools, youth organizations, sporting clubs and community events (Argyle). The strongest method for community engagement in diamond mining companies is cooperation with local NGOs. For example, in case of Ekati Diamond Mine, voluntary and confidential Impact and Benefits Agreements were signed with four Aboriginal groups: the Dogrib Treaty 11 Council and Akaitcho Treaty in 1996, with North Slave Metis Alliance in July 1998, and with the Inuit Kugluktuk and the Kitikmeot Inuit Association in December 1998. However, companies did not report to have any cooperation with international organizations.

The cooperation of diamond mining companies with the Governments focuses on social, economic and environmental issues (Argyle, De Beers, Debswana, Ekati and Namdeb). In the case of Ekati Diamond Mine, BHP Billiton Diamonds Inc. signed the Socio-Economic Agreement with the Government of the North West Territories of Canada, which includes commitments of BHP Billiton Diamonds Inc. to employ local community members, to support and benefit communities and to create business opportunities. In addition, an Environmental Agreement was negotiated with the Governments of Canada and the North West Territories to address environmental concerns, which are not covered by the legislation. As part of this agreement, a nongovernmental agency – Independent Environmental Monitoring Agency has been formed.

Corporate philanthropy is a well-established practice within the CSR sphere and diamond mining companies are no exception. The diamond mining companies engage in charity giving and mainly make donations for courses such as education, and health and safety (Ekati). Companies encourage employees to be involved in communities and contribute their time to community programs (Ekati). Three diamond mining companies reported to have philanthropic Foundations and Trusts, which are studied separately in Chapter 6.

Comparison of CSR Strategies in Gold and Diamond Mining

Within the studied top ten gold mining companies there are three groups of reporters. First group of companies are the leaders in social and environmental reporting (Rio Tinto, Normandy Mining and Placer Dome). Second group of companies (Freeport McMoRan C&G, Homestake Mining and AngloGold) are moderate reporters. Third group (Barrick Gold, Ashanti Goldfields, Newmont Mining and GoldFields) disclose the least amount of information on social and environmental issues. Likewise, top seven diamond mining companies fall into three groups of reporters. First group of companies (Argyle Diamonds, Ekati and Debswana) are the leaders in social and environmental reporting. Second group of companies (De Beers and Namdeb) are moderate reporters. The third group of companies (Southern Era and Rex Mining) hardly discloses any information on their social and environmental strategies and operations. It is notable that first group of reporters amongst gold and diamond mining companies include large mining companies, and those which specialize on mining of more than one mineral. Second and third group of reporters are companies which mostly specialize on the mining of only one mineral.

Top gold and diamond mining companies including AngloGold, Argyle, Ekati, Homestake Mining, Normandy Mining, Placer Dome and Rio Tinto are increasingly concerned about sustainable development in relation to their mining operations. They commit to the principles of sustainable and specifically emphasize their sustainability approach within their corporate environmental policies. An important feature of CSR is a voluntary approach to the environmental and social concerns, which goes beyond legislative requirements. In the case of gold and diamond mining companies it is manifested in their willingness to perform better than the prevailing regulatory requirements and standards (Placer Dome and Rio Tinto). Mining companies have now begun to view environmental, social and employee issues in unity, and gold and diamond mining companies often adopt integrated policies which embrace environmental, economic, social, health and safety issues (e.g. Sustainability Policy of Placer Dome and Health, Safety, Environment and Community Policy of BHP for Ekati).

A growing number of the global top gold and diamond mining companies both in developed (Homestake Mining, Placer Dome and Rio Tinto) and developing countries (Freeport-McMoRan Copper & Gold, De Beers, Debswana and Namdeb) have adopted formal EMS or are certified to EMS standards, namely ISO 14001. Many gold and diamond mining companies now routinely conduct environmental audits, environmental monitoring and environmental impact assessment. Environmental reporting as a way to establish the transparency of mining operations is becoming an established practice with some mining companies, however, these are large mining companies, which have a diverse mining interest and are based in many countries around the world.

Generally, disclosure of environment, health and safety and community related information has grown within the period of 1998-2000, particularly in relation to the disclosure of more detailed information about their environmental policies, commitment and operations. Comparing the analyzed sources of information (Annual Reports, Environmental Reports and corporate websites), more detailed environmental and social disclosure is found in the Environmental Reports than in the Annual Reports. Often, corporate websites contain more environmental and social disclosure than the Environmental Reports.

Almost all top gold and diamond companies disclose environmental information in other media, including Annual Reports and corporate websites, especially Ekati, Freeport-McMoRan, Homestake Mining, Normandy Mining, Placer Dome and Rio Tinto who are extensively using their corporate websites to disclose environmental information. Some companies produce other publications devoted to environment, health and safety and community issues, which are also available on corporate websites in PDF format. In the case of Ekati there are other publications such as Community Partnership Program, Environmental Agreement, Ekati Diamond Mine Facts, Ekati Annual Environmental Plan and Environmental Assessment Executive Summary, Sable Pigeon and Bear Tooth Kimberlitic Pipes, and De Beers produced a publication on De Beers Fund.

Community engagement is becoming a vital issue of CSR amongst mining companies in both gold and diamond mining sectors. The majority of mining companies express willingness to consult stakeholders in community related issues, devise community programs involving local communities, and cooperate with local NGOs and governments. The most important issues that mining companies address in their community engagement practices are education, employment, charitable contributions and traditional values. One of the important issues in community relations is the negotiation concerning land rights of local communities. For example, three out of the global top ten gold mining companies (Freeport-McMoRan Copper & Gold, Normandy Mining and Rio Tinto) and two out of the global top seven diamond mining companies (Argyle and Ekati) are willing to negotiate land issues with local communities. Gold and diamond mining companies are concerned about the possible effects of their operations upon the local communities. Some companies declare willingness to provide employment and train local communities. Some gold mining companies (Homestake Mining, Newmont Mining, Placer Dome and Rio Tinto) have developed systems of the compensation to affected communities.

Several points were established when comparing CSR strategies of gold and diamond mining companies:

- Companies based in developed countries such as US, Australia and Canada tend to disclose in significant detail their environmental and social strategies, while companies based in developing countries such as Namibia, Botswana and Ghana have lesser degree of corporate social disclosure.
- Larger companies in general have a higher degree of development of CSR strategies, including environmental stewardship, environmental management

systems, employee management, health and safety, community development and reporting.

- However, it is the more horizontally integrated companies such as Rio Tinto and BHP that are engaged in mining of different minerals except from gold and diamond, have more developed CSR strategies. These companies also tend to develop more integrated policies for environment, employees, health and safety and community issues.
- Companies that are assigned to Industry Codes of Conducts and international initiatives for sustainable development have more sophisticated CSR programs and reporting systems in place.
- Gold mining companies tend to disclose more environmental and social information in their corporate documents than diamond mining companies and have more integrated approach to CSR strategies.

Companies that are incorporated under the legislation of developed countries in Europe, North America and Australia are more proficient in the development of CSR strategies. These companies tend to have highly developed approaches to environmental management, health and safety, employee management and community development. These companies tend to apply their CSR policies to all their mining operations, including those in developing countries. Therefore, the operations of western mining companies in developing countries may enhance business practices in relation to CSR and sustainable development.

Larger and more integrated companies which have mining operations in different countries and are engaged in different minerals mining such as Rio Tinto and BHP are more exposed to stakeholder demands and allocate more resources to the development of CSR strategies. Companies engaged in mining of one mineral such as Barrick Gold and GoldFields Mining are less innovative in their CSR strategies. Some mining companies are assigned to voluntary environmental codes of conduct, which reflect a high regard to environmental management within the global precious metals and minerals mining industry. For example, AngloGold, Normandy Mining, Placer Dome and Rio Tinto are assigned to the Australian Minerals Industry Code for Environmental Management, Rio Tinto is also assigned to the Global Sullivan Principles; and Rio Tinto and Freeport-McMoRan Copper & Gold have joined Global Mining Initiative.

Gold mining companies in general perform better in relation to disclosing their CSR strategies and practices. Industry leadership and peer pressure play the important roles in the development of CSR. It is the lack of leadership and industry cooperation for sustainable development and CSR within the diamond mining industry, and its relatively small geographical base that hampers the development of CSR strategies. While within the gold mining industry, six out of ten top gold mining companies sign up to industry codes of conduct for environmental management, only one diamond mining companies does so. Seven top gold mining companies reported participation in industry associations and only two diamond mining industries reported so. High degree of industry cooperation between mining companies for sustainable development, and participation in the implementation of

Industry Codes of Conduct and Global Mining Initiative strengthens the approach to CSR strategies amongst gold mining companies.

In the diamond mining industry, De Beers controls the majority of diamond interests around the world and influences the corporate governance practices within the diamond industry. Companies that are formed with the interest of De Beers such as Namdeb and Debswana could benefit from increased CSR performance in De Beers as many mining companies adopt their corporate policies across all mining operations around the world. The lack of cooperation in the diamond mining industry for sustainable development reflects the low level of CSR practices disclosed in their corporate documents. Presence of government interest in formation and operation of diamond mining companies is a specific feature of the present diamond mining industry. Governments in developing countries often own a share in development of diamond mines on their territories (Namdeb and Debswana). This may affect the extent of pressure from the national governments to improve environmental and social performance of mining operations, since governments have direct financial interest in the mining ventures.

Summary

- Corporate social and environmental disclosure has grown over the last decade within the mining industry. Social and environmental disclosure has developed from inclusion of environmental, health and safety issues to greater incorporation of issues relating to community relations and employee policy. Recent trend amongst the mining companies is a disclosure of economic impact of their operations. Corporate social and environmental disclosure is becoming more integrated and reflects greater understanding of concept sustainable development and corporate social responsibility of mining industry.
- The leading media for social and environmental disclosure are stand-alone Environmental Reports and corporate websites. Stand-alone reports have evolved from Environmental Reports in the beginning of 1990s to Environment and Community Reports and recently to Sustainability Reports. Corporate websites are becoming a useful resource for finding information on companies' social and environmental strategies and performances. Overall, corporate websites include more social and environmental information than Annual Reports and Environmental Reports, because companies post on the web their environmental policies, updates and other publications relating to social and environmental operations.
- The content analysis of corporate documents (Annual Reports, Environmental Reports and websites) of mining companies shows that top gold mining companies overall have greater degree of disclosure with regard to social and environmental issues then top diamond mining companies. The leading reporters on social and environmental issues are large, diversified companies, while companies concentrating on mining of only one mineral tend to disclose less

social and environmental information. Overall top gold mining companies have greater degree of social and environmental disclosure than top diamond mining companies, because the gold mining industry is more liberalized and is more responsive to social and environmental concerns due to larger geographical base and influence from the large diversified companies.

Notes

[1] Gold is rare and its abundance in the Earth's crust is on average about 0.004 parts per million. Gold is a soft yellow, highly ductile and malleable metal. Usually gold is alloyed to give in more strength and in these alloys the term carat is used to express the amount of gold present, 24 carats being pure gold. A troy ounce, short for oz, is the adopted metrics of gold output in international practice (1 troy ounce equals to 31.1035 grams in the metric system). Gold was first discovered circa 3,000 BC (Puddephatt, 1978; US EPA, 1994). Gold is usually found in placer and ore deposits. Placer or alluvial gold consists of the weathering of auriferrous rocks, which have been washed into riverbeds or similar situations (Puddephatt, 1978). Gold mining operations consist of three major steps: extraction, beneficiation, and processing. Extraction involves removing the ore material from a deposit. Mining methods for gold ore extraction are selected on the basis of maximum ore recovery, efficiency, economy and the character of the ore body and involve either surface or underground techniques (US EPA, 1994). Four main techniques, which vary with mining operations and depend on the characteristics of ore and economic considerations, are used in the beneficiation of gold ore: cyanidation, flotation, amalgamation, and gravity concentration (US EPA, 1994).

[2] Since the analysis was conducted for 1998-2000, many mining companies which were under study have changed. For example, in 2004, AngloGold and Ashanti Goldfields have merged forming AngloGold Ashanti. Before that AngloGold has released AngloGold towards Sustainability – a Social Investment Report for 2001/2002. On 20 February 2002, Newmont Mining acquired Normandy Mining Ltd. In this study these two companies are reviewed separately up to 2000. In addition, Barrick Gold merged with Homestake Mining Limited on 14 December 2001. Nevertheless, these companies are reviewed separately as the given in the list of top ten gold mining companies for 2000. In 2002, Barrick Gold has started to produce annual Environment, Health, Safety and Social Responsibility Report. Freeport-McMoRan C&G started to publish separate report Economic Social and Environmental Report since 2001. In 2003, for the first time, Gold Fields issues Sustainable Development Report. In 2001, Newmont has produced Environmental Report for Newmont Australia titled 'Now and Beyond' and a series of site specific Environmental Reports. Placer Dome and Rio Tinto continued producing their respective Sustainability Report and Environment and Social Review in the years following the study.

[3] The first diamonds were found in India in 300 BC (Janse and Sheahan, 1995). Contemporary diamond production started in the 19th century after the discovery of diamond deposits in South Africa. Diamonds are unique minerals according to their physical, chemical and geological characteristics. Diamonds are composed of carbon and are the hardest natural substance, as it is No. 10 on Mohs' scale of hardness (Barnes, 1995). Diamond has a specific gravity of 3.52 (Parsons and Kraus, 1969). The weight of diamonds is measured in carats (ct), one ct equals 0.2 grams. Raw diamonds differ significantly by their weight, shape, surface, colour and internal defects (Parsons and Kraus, 1969; Teslenko, 1997). Diamonds are found in a kimberlite, which is a host rock for diamonds in primary deposits, and in secondary alluvial deposits (Janse and Sheahan, 1995). There are 4,000 to

5,000 kimberlites discovered and their number is increasing year by year, of which probably 500 are diamond-bearing; and less than 60 are under exploitation, of which there have been only 15 major producers. Only several major (>3 million carats per annum) discoveries have been made since 1950. These are Argyle in Australia, Opara and Jwaneng in Botswana, Jubilee, Udachnaya and Mir in Russia, and Venetia in South Africa, and Ekati in Canada. Secondary deposits appear in river systems, for example in Angola, West Africa and other parts of the world. There are alluvial deposits in marine situations such as Namibia and the western coast of South Africa (Gush, 1994). The total number of world wide alluvial occurrences is estimated at 4,000 (Janse and Sheahan, 1995; Jennings, 1995). The mining of diamonds is relatively straightforward and not more difficult than the mining of other minerals. Primary deposits or kimberlite pipes are commonly mined by an open pit method or by underground extraction methods towards the end of their economic life (Gush, 1994). The deepest known planned pit is Udachnaya (600m), but Jwaneng and Opara may exceed this. The average life for these mines is c. 30 years (Jennings, 1995). Marine deposits are mined by vast earth-moving operations on the onshore terraces and by very sophisticated mining and extraction ships under the sea (Gush, 1994). Operations in marine diamond mining can extract diamonds from a water depth of 30 to 150 metres (Tarras-Wahlberg and O'Toole, 2000). The recovery of diamonds is based on their physical features. As diamonds are heavy, so a heavy media separation mechanism is used for an effective recovery of diamonds. Also grease belts are used to recover diamonds from the ore, as diamonds are attracted to grease. Another method for recovery of diamonds is X-ray sorting, because some diamonds fluoresce (Gush, 1994).

[4] The diamond mining companies included in this review were selected as defined by their raw diamond production in 1999 (Rombouts, 2000), only those companies which provided three media sources were analysed. ALROSA was excluded from the analysis, because it is explored separately as a case study. Corporate social disclosure of these diamond mining companies was analysed through content analysis of their Annual Reports for 1998-2000, Environmental and Social Reports for 1998-2000 and corporate websites for 2001 or 2002.

Chapter 4

Geographical Specifics of CSR in Mining

Introduction

Chapter four concentrates on geographical specifics that affect the development of CSR within a mining company, they include local political, economic, environmental, social, legal and cultural factors. These factors determine the way mining companies operate and the nature of environmental and social concerns of companies' stakeholders in relation to mining operations. Companies adopt corporate social responsibility as a reaction to stakeholder concerns. Stakeholder concerns and pressures for improvement of corporate social responsibility also develop within circumstances, which depend on the industrial sector, as it was discussed in Chapter 3. Pressures for corporate social responsibility within a specific geographical area, where a mining project operates, are linked to a variety of factors, including:

- Political agenda for protection of natural environment.
- Level of development and implementation of environmental regulations.
- Peculiarities of the local environment, such as ability of the natural ecosystems for self-regeneration. The history of other industrial impacts upon the natural ecosystems.
- Level of protection of the natural environment in the vicinity of the mining operations, such as presence of natural reserve parks near the mine.
- Variety of communities affected by the mining operations. These communities may be indigenous, rural, urban, workman and local.
- Links between the communities and the natural environment (land, forests, water resources, etc), which are affected by the mining operations. Livelihood of certain communities may economically, culturally or spiritually depend upon the natural environments where the mining operations are taking place.
- The property rights for land, mineral and other natural resources.
- Economic development of the area, where the mining company operates, which may vary from developed to completely underdeveloped.

It is true of some developing countries and countries in transition, where the level of institutional development, legal system and economic development is low, that there is a greater danger that mining companies would under-perform in terms of corporate social responsibility. Countries in transition have a strong base for institutional structure, legal system and economic development; however they endure transition from one system to another, which entails economic, political and legal restructuring. For example, the Russian Federation is undergoing a transition from a planned economy to a market economy and from a communist political system to a democratic system. Mining is very well developed within the Russian Federation. The Russian mining sector is very diverse and has been significantly influenced by economic and political transition. The extractive industry in the Russian Federation has faced different issues after the disintegration of the former Soviet Union, for example, political changes have brought center-periphery resource disputes where territorial constituents of the Russian Federation have increasingly claimed more control over mineral resources and their economic benefits (Bond, 1994). Moreover, privatization of the industrial sector has changed the face of the ownership of mineral resource exploitation, transforming from state ownership to private ownership.

This Chapter presents territorial specifics within which CSR strategies of mining companies develop using a case study of gold and diamond mining in the Republic of Sakha (Yakutia), a part of the Russian Federation. The Republic of Sakha is chosen as a case study, because it has highly developed mining sector with gold mining and diamond mining that are the leading extractive industries, which a range of mine related issues. Firstly, the mining industry in the Republic of Sakha has caused significant adverse impacts, including environmental pollution, harmful effects of infrastructure developments, nuclear explosions, social impacts and health problems involving indigenous people. Secondly, privatization and economic changes in the Russian Federation have led to the transformation of the mining industry involving structural, economic and financial implications. Thirdly, the rise of sovereignty has lead to center-periphery disputes relating to sharing of economic benefits from mineral resource exploitation. Fourthly, innovative approaches have helped towards solving environmental, social and economic problems related to the mining industry.

Corporate social responsibility of gold and diamond mining companies in the Republic of Sakha develops within a contextual framework of three main levels: national level (Russian Federation), regional level (Republic of Sakha) and local level (gold and diamond mining areas). Therefore, this Chapter consists of three parts. The first part describes the contextual setting for the development of corporate social responsibility in the Russian Federation as a whole, outlining the national mining industry, the particulars of gold mining and diamond mining industries, environmental misuse in the Russian Federation associated with mining operations and the current development of corporate social responsibility. The second part describes the regional level for the development of corporate social responsibility, i.e. within the Republic of Sakha, outlining the regional mining industry, particularly referring to gold mining and diamond mining in the Republic of Sakha and regional environmental impacts of mining. The third part refers to the local contextual level for development of corporate social responsibility, it describes the extent of environmental impacts caused by gold mining and diamond mining industries to the local environment, thus setting the issues that are to be addressed by the case study of gold and diamond mining companies examined later in Chapters 5 and 6.

Mining in the Russian Federation

The Russian Federation is abundant in mineral resources and their exploitation is a key sector of the Russian economy.[1] The Russian Federation is one of the world's largest mineral producers, contributing nearly 15 per cent to the world's mineral production, which has a great impact on the international mineral market (Anosov, 1999; Dorian and Humphreys, 1994). Each year, the Russian Federation extracts around ten per cent of the world's production of oil, 24 per cent of gas, 20 per cent of nickel and cobalt, five to seven per cent of coal and iron ores, and a large volume of other non-ferrous and rare earth metals (Goskomekologia, 2000).

Since 1991, the Russian Federation has opted for a path of economic, social and political reforms, market liberalization and political decentralization (Peterson, 1995). The Russian Federation inherited from the former Soviet Union a disproportionally large industrial sector by international standards, which was ridden with severe output distortions, inefficiencies in the use of all resources, and organizational disorders. The Soviet industrial sector absorbed a large share of total employment, investment, and capital stock (Schroeder, 1998). Even though the Russian economy is viewed as undergoing a transition towards a post-industrial model of development characterized by a reduced emphasis on industrial production and a rise of the service sector (Peterson, 1995), the industrial sector remains large in its importance for the Russian economy. For example, in 2002, the GDP of the Russian Federation comprised of 34 per cent by industrial sector, 60 per cent by service sector and six per cent by agricultural sector (World Bank Group, 2003). Industrial transformation in the Russian Federation led to decline in total industrial production in the 1990s. This decline significantly affected the manufacturing industry, while the mining industry has increased its share in overall industrial output (Bond, 1994; Schroeder, 1998). During the period of economic transition in 1991-1999, industrial decline adversely impacted the economic performance of mining industry. For example, production of platinum decreased by 20-35 per cent; and the production of chromium, lead, tin, molybdenum, antimony, niobium, gold and silver decreased by 40-60 per cent. The only exception was the diamond production, which increased by 15 per cent during 1991-1999 (Goskomekologia, 2000).

The economy of Russia is heavily reliant upon exports of mineral resources, especially oil and natural gas. The drop in manufacturing and rise of the mining industry, which commenced in the 1990s, is reflected in the structure of exports from the Russian Federation. Fuels are the leading commodities in Russian exports, providing 50 per cent of total export revenues; whereas ferrous and non-ferrous metals and minerals contribute up to 12 per cent of total Russian exports (see Table 4.1). During 1992-2000, revenues from the export of mineral resources accounted for 65-70 per cent of total foreign currency revenues of the Russian Federation.

Table 4.1 Principal exports of the Russian Federation in 1998-2002

	1998	1999	2000	2001	2002
Exports of goods FOB* (billion US dollars)	74.8	75.7	105.6	101.9	107.2
Exports of goods FOB (%)	100	100	100	100	100
- Fuels and energy (%)	40.2	43.5	52.7	53.7	53.8
- Ferrous metals (%)	9.4	7.8	6.9	6.3	6.7
- Non-ferrous metals (%)	8.0	8.0	6.7	5.6	5.0
- Machinery and transport equipment (%)	11.3	10.8	8.8	10.4	9.4
- Chemicals (%)	3.2	2.9	2.4	2.5	2.3
- Timber and paper (%)	3.5	3.9	3.3	3.4	3.5

* FOB-fright on border

Source: Adapted from Center for Economic Conjuncture under the Government of the Russian Federation Report, 'Russian foreign trade in 2002' (2003).

Gold Mining in the Russian Federation

The Russian Federation is third in the world for its gold reserves and seventh in the world's gold production league (Energia, 2000; Kolmogorov, 2000; Natarov, 2000). The gold mining industry in Russia started to develop intensively in the second half of the 18th century in the Urals and Altai. In the 19th century the 'gold rush' hit Siberia (Toryannikov, 1999). Russia accounts for about one tenth of the total 130,000 tons of the total gold production worldwide in the history of mankind. At the end of 19th century, Russia was the world leader of gold production, and in the 1970s, Russia was producing 200 tons annually and was ranked as the second largest gold producing nation in the world (Benevolsky and Ivanov, 1999) (see Figure 4.1).

More than 314 gold ore and complex gold deposits had been discovered throughout the Russian Federation.[2] Currently, gold mining operations are carried out in 28 regions of the Russian Federation with Magadan *oblast*, Krasnoyarsky *kray*, Republic of Sakha (Yakutia) and Irkutsk *oblast* being amongst the leading gold producing regions (Natarov, 2000) (see Table 4.2). Other large gold producing areas include Amur *oblast*, Chukotsky autonomous *okrug*, and Khabarovsky *kray* (Sagers, 1997; Toryannikov, 1999) (see Figure 4.2).

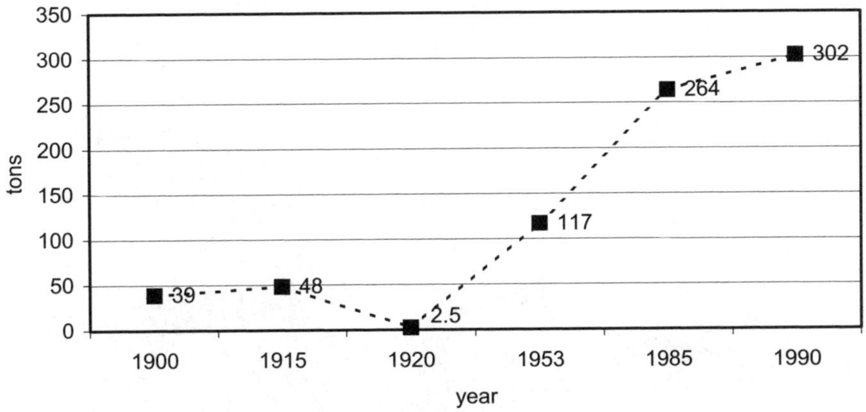

Figure 4.1 Gold production in the Russian Empire and the former Soviet Union, tons

Source: Adapted from Benevolsky and Ivanov, 'Mineral'no-syr'evaya Baza Zolota Rossii na Rubezhe XXI v.' [The Gold Resource Base of Russia at the Turn of XXI Century] (1999).

Table 4.2 Distribution of gold production among Russian regions, %

Region	1995	1996	1997	1998	1999	2000	2002
Magadan *oblast*	22	18	23	29	23	21	19
Krasnoyarsky *kray*	5	5	14	13	14	12	17
Republic of Sakha	23	21	17	11	10	12	10
Irkutsk *oblast*	9	10	9	9	9	11	9
Other regions	41	46	37	38	44	44	45
Total Russian production	100	100	100	100	100	100	100

Source: Adapted from Zoloto: Stranizy Zolotopromyshlennika [Gold: Pages for Gold Miners], URL: http://www.aurum79.ru (accessed on 12/06/02).

In the former Soviet Union, twelve large industrial corporations were involved in gold mining operations. These corporations were structured regionally and reported to one state agency (Glavzoloto of the USSR). During privatization, almost the entire gold mining industry in the Russian Federation was privatized, leading to creation of more than 600 gold mining enterprises in 1999 (Menshikov, 1999; Natarov, 2000). In 2000, 78 large gold mining companies with an annual output of more than 300 kilograms accounted for 74 per cent of total gold production in Russia. Amongst them there are only 16 companies, which have an output of more than 1,000 kilograms of gold per year (see Table 4.3).

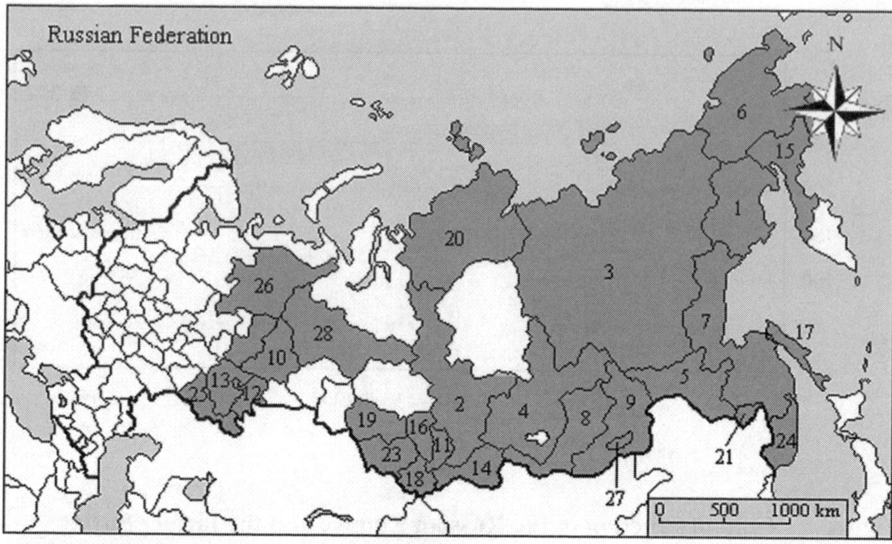

Figure 4.2 Gold-producing areas of the Russian Federation

Source: Based on rating of gold-producing regions in 1998 from Zoloto: Stranizy Zolotopromyshlennika [Gold: Pages for Gold Miners], URL: http://www.aurum79.ru (accessed on 12/06/02).

Although there are many small and medium size gold mining companies, they are experiencing difficulties in obtaining loans from Russian credit institutions as well as from foreign banks. These small and medium size companies lack financial resources to acquire new technology, equipment and machinery and develop new gold deposits. It has become evident that restructuring of the gold mining industry in the Russian Federation and the dissolution of it into numerous small production units was a great strategic failure. During the 1990s, gold production in the Russian Federation was dropping, efficiency of gold mining was decreasing and debts of gold mining companies to the State Budget were growing. The state control and supervision of the gold mining industry has weakened in the 1990s. Since the beginning of 1991, Russian gold production has been declining, but since 1999 the gold production started to grow (Energia, 2000; Kolmogorov, 2000) (see Figure 4.3).

Table 4.3 Distribution of gold mining companies by their production in 2000

Gold production, kg per year	Number of companies	Distribution of companies, %	Share in Russian total gold production, %
Less then 100	389	69	11
100-300	99	17	15
more then 300	78	14	74
including more than 1,000	16	3	46
Total	566	100	100

Source: Adapted from Kolmogorov, 'Zolotodobyvajushaya Promyshlennost Rossii: Promlemy i Perspektivy' [Gold Mining Industry in Russia: Problems and Perspectives] (2000).

Approximately half of Russian gold production comes from placer deposits, although the major gold reserves are found in vein deposits.[3] Due to climatic conditions in Russia mining operations on placer deposits could take place only during summer and some parts of spring and autumn – this seasonal character of mining operations hampers the growth of gold production. This situation is worsened by the fact that 70 per cent of gold is mined in the regions of the Russian Far North, where additional factors such as difficulties in transport routes and where significant proportion of production costs dedicated to fuels and energy prevent the rapid growth of gold production. Consequently, the costs of gold production are rising and the profits of gold mining companies are decreasing (Benevolsky and Ivanov, 1999; Energia, 2000; Kolmogorov, 2000). In order to survive, gold mining companies require loans to conduct mining operations in order to provide for advance delivery of fuels, equipment and resources. In the past, the Government strongly supported the gold mining industry through purchasing gold and distributing credits to gold mining companies. Since 1998, the Government has stopped purchasing precious metals to a State Fund of the Russian Federation, and almost all gold production is purchased by commercial banks, which allocate credits to gold mining companies for the future mining operations (Kolmogorov, 2000).[4]

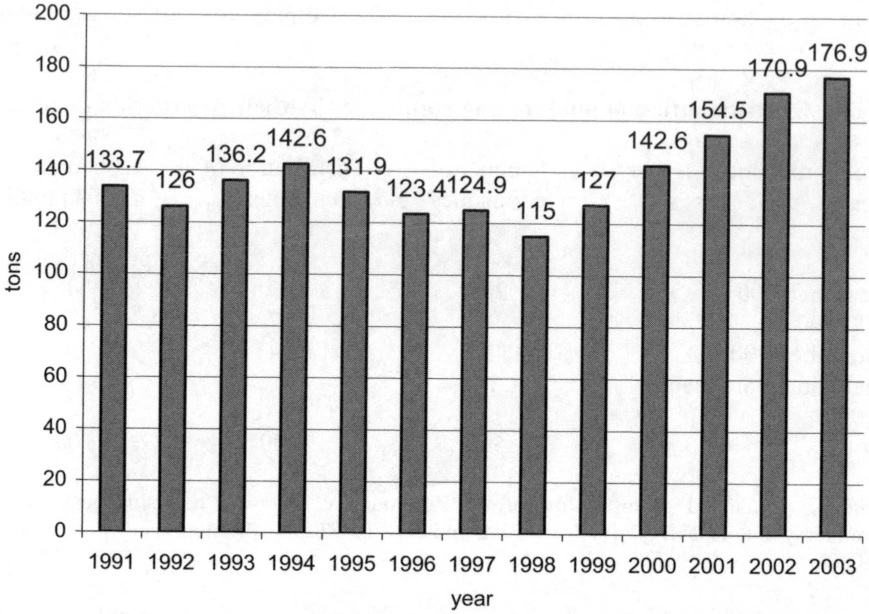

Figure 4.3 Gold production in the Russian Federation in 1991-2000, tons

Source: Adapted from Energia, 'Zolotodobyvajushaya Promyshlennost Rossii' [Gold Mining Industry of Russia] (2000); Zoloto: Stranizy Zolotopromyshlennika [Gold: Pages for Gold Miners], URL: http://www.aurum79.ru (accessed on 12/06/02).

Even though the gold mining industry is in need of foreign investments, a number of foreign gold mining companies in the Russian Federation have reduced from 18 companies in 1996 to nine companies in 2001. Factors that make the Russian Federation unattractive for foreign capital include the overall economic crisis, frequent change of the Government, socio-economic instabilities, inefficiencies of the legal system, and a weak finance and credit system. Although a significant improvement could be brought by legislation on production sharing, due to the legal inefficiencies, this model of joint ventures activity has not yet been properly developed (Kolmogorov, 2000; Rubchenko and Shokhina, 2001).

Diamond Mining in the Russian Federation

The Russian Federation is the world leader in proven reserves of diamonds and annually produces up to 20-25 per cent of world's diamond production (Matveev, 1998; Goskomekologia, 2000). Russian major industrial deposits are situated in three diamond regions: the Republic of Sakha (Yakutia) accounts for 82.4 per cent of diamond reserves of the country, Arkhangelsk *oblast* for 17.5 per cent and Perm *oblast* for 0.1 per cent (Matveev, 1998) (see Figure 4.4). ALROSA Company carries out the major diamond operations in the Republic of Sakha (Yakutia), where 99.8 per cent of Russian diamonds are mined, while Uralalmaz Company in Perm *oblast* provides the remaining 0.2 per cent of the diamond output (Matveev and Cherny, 2000). The most significant area for new diamond developments is considered to be Arkhangelsk *oblast*, where Lomonosov diamond deposit was discovered in 1980.[5]

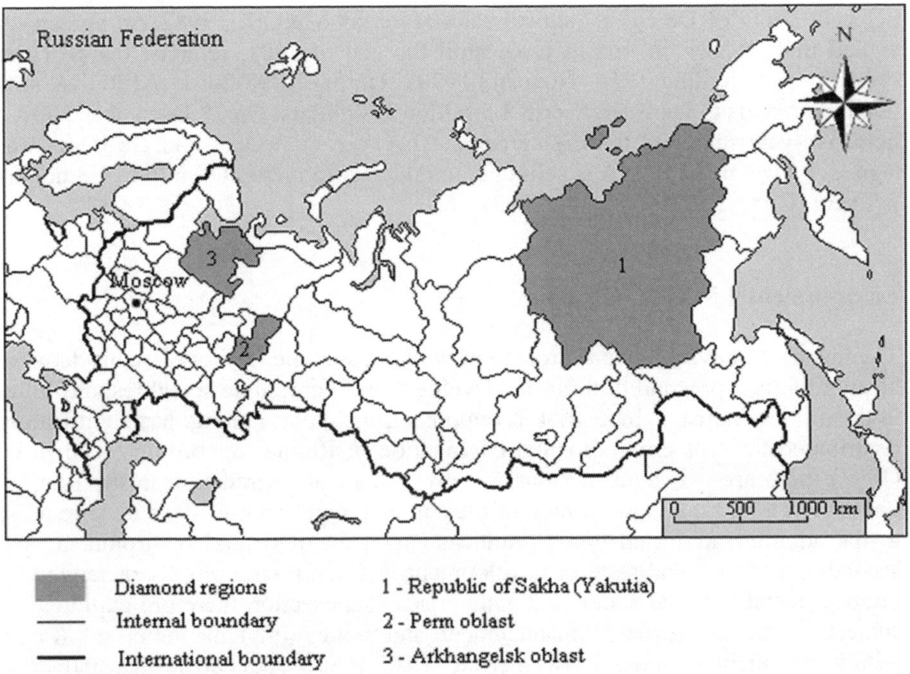

Figure 4.4 Diamond regions in the Russian Federation

During 1998-1999, the Russian Federation produced diamonds worth 1.5 billion US dollars (Matveev and Cherny, 2000). Diamonds are considered to be a strategic resource in the Russian Federation, information on diamond production in carats is a state secret, and therefore data on diamond production is given as an estimate. Diamond mining is a growing sector of the Russian economy; it is the only mining sector that showed an increase in production during the transition period (Goskomekologia, 2000). During 1992-2000, diamond production increased from 18 million carats to 24 million carats (Levine, 1999; Levine and Wallace, 2000). Revenues from diamond mining accounted for 0.3 per cent of the entire Russian GDP in 1993 and for 0.5 per cent in 1997 (Il'ina, 1995; Prokopiev, 1998). The majority of Russian diamonds are exported. The Russian Federation is exporting diamonds in accordance with a trading agreement with De Beers Centenary AG. De Beers had a diamond trading agreement with the former Soviet

Union, which ceased in 1990. In 1990-1997, the trading of diamonds was carried out within a five-year agreement signed in 1990, according to which De Beers had a right to buy 95 per cent of the total Russian uncut diamond production, while Russia retained five per cent of uncut diamond production to sell freely (Levine, 1995). In 1990-1997, a quota of Russia in De Beers' annual sales volume was 26 per cent. According to the next agreement signed with De Beers in October 1997, which was in force until the end of 1998, ALROSA had a right to sell uncut diamonds worth at least 550 million US dollars, but must have stayed within the 26 per cent quota of De Beers' annual sales (Interfax-M&CN, 1998). An agreement signed in 1998, which was in force until the end of 2001, retained the previous conditions of trading (RIA Novosti, 1999). During 1998-2001, ALROSA sold uncut diamonds to De Beers worth 3.5 billion US dollars. On 18 December 2001, a new five-year diamond trading agreement between ALROSA and De Beers was signed, by which ALROSA is selling diamonds worth up to 800 million US dollars per year (Gurov, 2001; Yakutia, 2002).

Environmental Misuse in Russia

The history of environmental protection within a specific territory is considered to be an important precondition for the development of corporate social responsibility in the mining industry. Industrial development, including mining, has significantly contributed towards environmental degradation in Russia. According to Porfiriev (1997) there are two main reasons for environmental degradation in the Russian Federation. Firstly, it is the legacy of the command-and-control systems in relation to the administrative and bureaucratic management of natural environment; and secondly, a transitional state of development in Russia, which is characterized by political instability and socio-economic crises. This section therefore explores the subject of natural resource management and policy implementation in Russia within two main periods: Soviet period (1917-1985), and Perestroika and Post-Soviet period (1985-2001). Furthermore, historical conditions in the Russian environment are explored with specific reference to the environmental implications of industrial development in the Russian North, extending to a study of the current state of environmental policy development and implementation in the Russian Federation.

Soviet Period (1917-1985)

A high level of environmental deterioration in the former Soviet Union has been well documented (Goldman, 1972; Pryde, 1976; Singleton, 1976; Komarov, 1978; Singleton, 1987; Ziegler, 1987; Peterson, 1995). The cause for such degradation of the environment is primarily connected with the model of industrial output during the Soviet period combined with a lack of effective environmental protection measures (Pryde, 1994). The result for such wasteful policy was inefficient use of the natural resources and widespread environmental pollution. During the Soviet period, industrialization in Russia was the most important factor in building the economy. Soviet central planning favored the vast industrial development of the 'eastern regions', which started with the fulfillment of five-year plans, starting from 1928. A distinguishing feature of the economic development of the country was the prescriptive execution of geological, hydrological and other prospecting throughout Russia, especially in sparsely populated areas (Pokshishevsky, 1974). The push to meet production quotas and maintain growth rates placed enormous pressure on reducing operating expenses by ignoring or subverting the existing pollution control legislation (Bond, 1984). The paradox noticed by Lluardo (1995) was that the former Soviet Union, which was the country with very severe environmental problems, was also the first country to cover protection of the environment in its constitutional laws. However, weak implementation of these environmental regulations was one of the main causes for environmental problems (Komarov, 1978; Pryde, 1994). The absence of appropriate economic incentives linked to the environmental regulations meant that Soviet enterprises saw no benefit in adopting environmentally cleaner technologies. Therefore, in order to achieve a cleaner environment, it was necessary to change the socialist methods, which were driving the economy (Komarov, 1978).

Another aspect of natural resource use in the former Soviet Union was the industrial expansion into the Russian North following discoveries of highly valued mineral resources including energy resources, metals, precious metals and other minerals (Pokshishevsky, 1974). The Russian North is a geographical zone that occupies around 65 per cent of the Russian Federation, where the majority of oil, gas, coal, nickel, rare-earth elements, tin, gold and diamond reserves are found (Bradshaw, 1995; Porfiriev, 1997; Alekseev, 1999) (see Figure 4.5). The farthest northern part of this zone is the Arctic, which constitutes approximately a quarter of the Russian North. According to Kotlyakov and Agranat (1999), the Russian North is considered to be an important geopolitical, military and strategic asset of the country due to its vast environmental, territorial and economic resources and cultural heritage. The Russian North is characterized by very fragile environmental conditions (Kriuchkov, 1976; Komarov, 1978) such as:

- Permafrost and severe climatic conditions make natural resource development very problematic.
- Decomposition of wastes is twenty to thirty times slower than many other regions. In general terms, an ordinary river cleans itself of permissible doses of organic matter within 200-300 kilometers. However, in the Russian North even 1,500 kilometers are too few to perform the same function (Komarov, 1978).
- Siberia's lakes and rivers have low regenerative capacities because they ice over for a substantial period of the year (Ziegler, 1987).

On their track to the Russian North, pioneers of the Soviet industrial development had little or no concern for the environment (Singleton, 1976; Lluardo, 1995). The rapid building of the industrial base for communism was an overriding priority of the Communist Party, notwithstanding the environmental consequences of industrialization, urbanization and the technical revolution, which allowed a wasteful exploitation of natural resources and drove the extractive industries far into the Arctic (Koutaisoff, 1987).

Figure 4.5 Map of the Russian North

Source: Adapted from Bradshaw, 'The Russian North in Transition: General Information' (1995).

Another aspect of the wide-scale degradation of the environment lies in the essence of Soviet propaganda. The public was mostly unaware of environmental disruption in the former Soviet Union, due to the lack of official and objective information about the environmental situation (Komarov, 1978). The policy of the Administration was to present the socialist system as the best model for industrial development (Goldman, 1972). Soviet propaganda was based on showing 'achievements' rather than drawbacks of economic development as it could put in danger the authority of the government and the prestige of the country in the international arena (Komarov, 1978). Soviet propaganda in righteousness of the state to ignore environmental issues during industrial development has led to a suppression of public participation in environmental issues and many were too afraid to question or protest government actions (Crotty and Crane, 2000). However, Russian scientists were admitting that environmental problems were evident in the former Soviet Union, which were equally severe as in the US or

Europe (Komarov, 1978). Notwithstanding these facts, socialists in the 1970s had a perception that the Soviet Union as a socialist political-economic system could overcome the problems, while western capitalistic countries, where economic profits play a great role, have no incentive to invest in the protection of the environment (Sarkar, 1999).

Goldman (1972) stated,

> ...it became a tenet of conservationist folklore that environmental disruption will no longer exist in a society where the state owns all the means of production. Most critics have automatically assumed that if all the factories in a society are state-owned, the state will ensure that the broader interests of the general public will be protected. By the contrast pollution was regarded as an inevitable by-product of capitalism... Industrialization and urbanization have been at least as demanding on the environment in the USSR as they have been else where in the developed world (Goldman, 1972).

Reasons for environmental degradation in Russia are summarized in Table 4.4.

Table 4.4 Reasons for environmental degradation in Russia

Sphere	Problems
Economic	• Model for industrial and economic development • Lack of economic incentive for cleaner technologies • Planning system • Disproportional economic structure
Legislation	• Insufficient legislation
Management	• Insufficient management of natural resources • Lack of monitoring over environment • Control over compliance with environmental regulations
Technological	• Technological base of production
Attitudes, values and beliefs	• Underestimating the limits of natural resources • Irresponsible attitudes to the environment • Notion of supremacy of man over nature • Lack of public participation in environmental issues • Vastness of the country can help to conceal the extent of environmental degradation

The natural environment in the Russian Far East has remained the least developed and the most pristine part of the Russian Federation. This has happened due to two main reasons. Firstly, the region hosts a small, highly urbanized population: in 1994, the Russian Far East was populated by eight million people, three quarters of whom were resident in urban areas. Secondly, a harsh climate and inaccessible terrain has discouraged intensive development. Commercial transport throughout the area is limited to river and coastal navigation and to the Trans-Siberian Railway (Peterson, 1995).

The roots for environmental problems in Russia come not only from management inefficiencies, but also from psychological beliefs and perceptions (Singleton, 1987; Lluardo, 1995):

- The former Soviet Union was very rich in natural resources, which were used by a relatively small population that supported the psychological belief in the abundance of natural resources.
- The belief that resources could never be exhausted and waste could be disposed of in unlimited quantities without causing significant damage.
- Irresponsibility in managing high-risk projects such as attempts to divert the flow of Siberian rivers and underground nuclear device testing.
- The doctrine of the supremacy of man over nature provided the philosophical underpinning of the economic development.

Perestroika and Post-Soviet Period (1985-2001)

The breakdown of the socialist system in the former Soviet Union started in 1985 when Gorbachov tried to introduce *perestroika* and *glasnost*, which brought overall political, economic and social changes, and also involved changes to environmental regulations in Russia (Peterson, 1995). When the Russian Federation was formed in 1991, it carried forward regulations adopted during the former Soviet Union by the Supreme Soviet. However, the new state started to form its own environmental legislation and, presently, the State Duma, a legislative body in the Russian Federation, creates legal requirements for environmental protection (Protasov and Molchanov, 1995). Since 1991, the environmental laws and statutes, as well as the administration system for environmental protection, have been undergoing changes. The Russian Federation has developed a new national environmental policy. National environmental policy implementation in the Russian Federation is carried out through the system of State agencies for environmental protection. State agencies for environmental protection is frequently changed, merged and reformed.

Veber (1999) argues that many factors including economic, socio-economic and ecological that are present in the Russian Federation dictate the necessity to move towards a model of sustainable development. It is stressed that sustainable social and economic development in the Russian Federation and its territories, particularly in highly industrialized areas, depends on better environmental protection and the rational use of natural resources (Makarov et al, 1995). At the United Nations Conference on Environment and Development held in Rio de Janeiro in June 1992, the Government of the Russian Federation along with other national governments expressed an interest in implementation of Agenda 21. In August 1992, the Government of the Russian Federation formed an inter-agency committee for the development of decisions adopted in Rio for the Russian Federation. However, the project to develop a national plan on sustainable development has merely resulted in the production of documents, which summarized existing programs (Tysyachnyuk, 1998). In February 1994, the

President of the Russian Federation signed a Decree No.236 'Strategy of Russian Federation on environmental protection and provision of sustainable development'. On 1 April 1996 the President approved the 'Conception for transition to sustainable development', which was based on the main principles adopted on the United Nations Conference in Rio de Janeiro. However, the Conception remained largely inactive, as Veber (1999) agues that recommendations expressed in the Conception did not come together with the 'course of reforms'. On the basis of the Conception, the Government adopted a 'State strategy for sustainable development in the Russian Federation' in December 1997, which has not yet been adopted by higher state authority.

State monitoring of environmental pollution is based on statistical reporting by all economic enterprises, which produce negative environmental impacts or use natural resources. These enterprises are required by law to submit quarterly or/and annual reports on water use and derivation, air emissions, waste disposal and handling of hazardous materials. On the basis of these statistical reports the Government compiles Annual State Report on the Environmental Situation in the Russian Federation, which is a comprehensive statistical report on environmental pollution caused by major industries.

The former Minister for Environmental Protection of the Russian Federation describes the situation in the Russian Federation as an environmental crisis, which is caused by high consumption of non-renewable resources and inefficient industrial processing (Makarov et al, 1995). Although environmental pollution in the Russian Federation has reduced in recent years (see Table 4.5), this reduction is rooted in an overall drop in production and does not reflect investments in pollution abatement (Knaus and Klarer, 1997; Danilov-Danilyan, 1999; Tanner, 1999). However, the rate with which production has dropped has not been accompanied by a corresponding rate in the reduction of environmental pollution (Porfiriev, 1997), i.e. it can be argued that in reality the environmental performance of industry is worsening (Danylov-Danilyan, 1999). Economic crisis, from which the Russian Federation has been suffering since the 1990s, has largely prevented environmental improvements (Danylov-Danilyan, 1999). Indeed, existing systems of pollution control and management have become less effective, while the rapidly growing private sector is poorly monitored and controlled (Makarov et al, 1995).

Air pollution in the Russian Federation shows a decreasing trend of total discharge from stationary sources, dropping from 21.2 million tons in 1995 to 18.8 million tons in 2000. The industrial sector, as a major pollutant of the environment, contributed up to 80 per cent of the total air pollution from stationary sources in 2000. The most air polluting industrial sector is the electric energy sector, accounting for 25 per cent of total air pollution, followed by non-ferrous and ferrous industries (which contribute 23 per cent and 16 per cent accordingly) and the oil extraction industry (which accounts for 11 per cent of total air pollution). Exploitation of mineral resources including non-ferrous, ferrous metallurgy, oil extraction, coal and gas contribute to 42 per cent to the total air pollution in the Russian Federation.

Table 4.5 Air pollution from stationary sources in the Russian Federation in 1996-2000, thousand tons

Industry	1996	1997	1998	1999	2000
Russian Federation, total	20,274	19,332	18,662	18,540	18,819
Industry, total	16,661	15,852	14,950	14,704	15,221
Electric power engineering	4,748	4,428	4,346	3,936	3,857
Non-ferrous metallurgy	3,598	3,622	3,292	3,312	3,477
Ferrous metallurgy	2,536	2,380	2,189	2,330	2,396
Oil extraction	1310	1,325	1,385	1,329	1,619
Oil processing	824	819	770	748	736
Coal industry	596	535	545	560	604
Gas industry	542	451	428	456	501
Machinery & metal processing	602	543	460	454	440
Building materials	528	468	397	417	433
Chemical industry	413	415	388	415	427
Wood and paper industry	434	383	352	367	379
Food industry	250	224	198	198	182
Atomic power engineering	151	139	105	92	89
Light industry	64	56	50	51	45

Source: Adapted from Ministry for Natural Resources of the Russian Federation Report, 'State report on environmental situation in the Russian Federation in 2000' (2001).

Data from State reports on the environmental situation in the Russian Federation shows that the mineral industry is a significant contributor to water pollution in the Russian Federation (see Table 4.6). The non-ferrous industry accounts for 22 per cent of total industrial wastewater discharge in the Russian Federation, followed by ferrous industry, accounting for 16 per cent, oil processing (five per cent) and coal industry (four per cent). The data on the generation of toxic waste indicates again that the extractive industry is the most hazardous, with non-ferrous, ferrous and coal industries being the most wasteful (see Table 4.7). For the last five years, pollution from industrial activities has been fairly stable with variations within one or three per cent.

It is evident from the data on air and water pollution in the Russian Federation that the extractive industries, particularly non-ferrous and ferrous metallurgy, along with electrical engineering, oil and gas are the industries which significantly impact upon the state of the Russian environment. However, the intensity of environmental pollution in the Russian Federation varies from region to region and also depends on the level of industrial development (Peterson, 1995). Therefore, it is important to look further at the particular details of specific regions, which host extractive industry developments. The State report on the environment situation in the Russian Federation provides an analysis of the quantity of air pollution per capita in regions of the Russian Federation as well as the quantity of industrial wastewater discharge and quantity of toxic waste (see Table 4.8).

Table 4.6 Discharge of polluted industrial wastewater into surface watercourses in the Russian Federation in 1996-2000, million cubic meters

Industry	1996	1997	1998	1999	2000
Russian Federation, total	22,414	23,043	21,986	20,657	20,291
Communal and housing	12,072	12,053	12,127	12,082	12,133
Agriculture	2,574	3,264	2,597	1,769	1,407
Industry	7,444	7,335	6,868	6,445	6,514
Wood and paper industry	1,443	1,323	1,221	1,348	1,472
Chemical industry	1,363	1,322	1,240	1,249	1,280
Electric power engineering	1,073	1,326	1,448	995	946
Ferrous metallurgy	705	692	677	699	755
Machinery and metal processing	640	624	552	597	510
Coal industry	658	620	442	396	393
Non-ferrous metallurgy	483	425	378	364	380
Atomic power engineering	241	266	253	297	254
Oil processing	228	193	185	164	152
Light industry	150	139	120	98	129
Food industry	124	116	98	97	88
Building materials	123	114	112	122	86
Oil extraction	25	21	11	4	10
Gas industry	6	3	3	3	7

Source: Adapted from Ministry for Natural Resources of the Russian Federation Report, 'State report on environmental situation in the Russian Federation in 2000' (2001).

Amongst the regions of the Russian Federation with a high index of air pollution are: Yamalo-Nenetsky *orkug* (1,079 kg per person), Krasnoyarsky *kray* (882), and Khanty-Manskiyský autonomous *okrug* (815). Among the regions of the Russian Federation with high sewage discharge index are: Arkhangelsk *oblast* (368 cubic meters per person), Murmansk *oblast* (362), and Irkutsk *oblast* (343). Among the regions of the Russian Federation with high toxic waste index are: Kemerovo *oblast* (4,700 kg per person), Vologda *oblast* (4,747) and Krasnoyarsky *kray* (4,700). The Russian Far East region unites the Republic of Sakha (Yakutia), Primorsky *kray*, Khabarovsky *kray*, Evreyskaya autonomous *oblast*, Amur *oblast*, Kamchatka *oblast*, Koryaksky autonomous *okrug*, Magadan *oblast*, Chukotsky autonomous *okrug* and Sakhalin *oblast*, is a host for non-ferrous metallurgy and provides the country with gold and diamonds, rare earth metals, etc.

Table 4.7 Generation of toxic waste by industrial sectors in the Russian Federation, million tons

Industry	1996	1997	1998	1999	2000
Russian Federation, total	82	89	107	108	132
Industry, total	78	84	102	102	124
Non-ferrous metallurgy	19	21	20	27	35
Ferrous metallurgy	32	33	34	30	32
Chemicals/petrochemical	7	9	8	11	13
Coal industry	0.1	0.1	11	7	12
Building materials	5	7	7	8	9
Electric energy	8	8	9	8	8
Oil extraction	0.7	0.8	5	2	6
Machinery/metal processing	2	2	3	3	3
Food industry	0.9	0.6	0.6	0.9	1.5
Wood and paper industry	1	0.9	1	1	1
Oil processing	0.9	1	1	1	1
Gas industry	0.08	0.06	0.03	0.06	0.1
Light industry	0.06	0.07	0.06	0.1	0.08

Source: Adapted from Ministry for Natural Resources of the Russian Federation Report, 'State report on environmental situation in the Russian Federation in 2000' (2001).

Table 4.8 Regional per capita air pollution, industrial wastewater discharge and toxic waste indices in Far East region of Russia in 1999

Region	Air pollution, kg per person	Industrial wastewater discharge, cubic meters per person	Toxic waste disposal, kg per person
Amur *oblast*	50-100	100-150	less than 100
Chukotsky autonomous *okrug*	400-700	50-100	less than 100
Evreyskaya autonomous *oblast*	50-100	50-100	250-500
Kamchatka *oblast*	50-100	50-100	100-250
Khabarovsky *kray*	100-200	150-200	250-500
Koryaksky autonomous *okrug*	200-400	less than 50	100-250
Magadan *oblast*	100-200	150-200	4,000 and above
Primorsky *kray*	100-200	150-200	1000-2000
Sakhalin *oblast*	100-200	100-150	100-250
Republic of Sakha (Yakutia)	100-200	50-100	less than 100

Source: Adapted from Goskomekologia Report, 'State Report on Environmental Situation in the Russian Federation in 1999' (2000).

Development of CSR in the Russian Federation

During the transition period, the Russian industry has increased its environmental impact for several reasons including outdated technology, reduction in the effectiveness of wastewater and air pollution purification, creation of secondary sources of environmental contamination, an increased number of accidents, a worsening of industrial and technological discipline and a low level of effectiveness in relation to state environmental control (Vukina et al, 1999; Crotty et al, 2001; Makarov, 2001).

Although there is a growing awareness amongst businesses for the need to change management of environmental issues in the Russian industry, businesses often lack the incentive to change their institutional approaches to environmental responsibility and often refer to lack of financial resources to improve their environmental performance. There is a narrow perception of mitigation of environmental impacts and a general approach amongst the majority of Russian companies to environmental operations as 'end-of-pipe' operations (Vinichenko et al, 2001). Russian industry clearly underestimates or sometimes disregards the contemporary changes in the global approach to the environment, which leads to an inevitable loss of economic opportunities not only for companies, but also for the country as a whole (Makarov, 2001). Environmental responsibility, which often manifests through the development of corporate environmental policies, establishment of corporate structures for implementation of environmental stewardship and adoption of environmental management systems, has not yet become a popular practice amongst Russian businesses. Environmental protection authorities are often incapable to sufficiently enforce environmental standards in the industry either through closure, or through administering payments of fines and charges. Fines and charges for breaching environmental regulations do not create a significant economic or management incentive for Russian businesses to improve their environmental performance.

Moreover, the development of mechanisms for improvement and systematization of environmental performance by the industry has not received due attention in the environmental legislation (Chernogrivov and Bareisha, 2001). For example, environmental management is not a requirement in the Russian Federation although according to article 71 of the Law of the Russian Federation 'About Environmental Protection' which is the principal document for environmental regulation in the Russian Federation:

> Industrial environmental control is carried out by environmental services of enterprises, organizations and agencies and has an aim to control the implementation of plans and actions for environmental protection and improvement of the environmental situation, rational use of natural resources, compliance with standards for quality of the natural environment and compliance with requirements of the environmental protection legislation.

Environmental management and audits are viewed as voluntary instruments for the reduction of environmental impacts. Current developments in the area of EMS include the introduction of the certification of EMS according to international standards. In 1998, voluntary state standards such as GOST R ISO 14001, ISO 14004, ISO 140010, ISO 140011 and ISO 140012 have been adopted by National Standards Organization in the Russian Federation, which in essence are the Russian versions of the international standards for EMS (Goskomekologia, 1998b). By 2002, only 14 Russian companies have been certified to GOST R ISO 1400.[6] In relation to environmental auditing, Goskolekologia (1998b), a principal environmental protection agency in the Russian Federation, has issued several Decrees for the establishment of environmental auditing in the Russian Federation and created a system for certification and accreditation of environmental auditors and consultants.[7]

Introduction of proactive methods for improvement of environmental performance (including elements of environmental management such corporate environmental policy, environmental mapping and environmental audits) in the industry is often associated with separate initiatives undertaken in several regions of the Russian Federation, involving several large industrial enterprises, regional governments and various environmental NGOs. Several regions of the Russian Federation including Tomsk, Nichegodskaya, Ryazanskaya, Vladimirskaya, Kaliningradskaya *oblasts* and Republic of Bashkortostan, the regional environmental protection agencies have developed systems for the promotion of environmental operations by business and industry (Goskomekologia, 1998b). Nevertheless, there is no evidence of pervasive development of mechanisms for adoption of environmentally responsible practices in the Russian Federation (Volostnov and Bareisha, 1999; Soldatenko, 2001).

Low level of general management within Russian companies leads to a slow development of environmental responsibility, which should be an integral part of general corporate strategy. The concepts of organizational mission, differentiation of corporate goals and objectives, implementation of systematic planning and development of procedures, are new and unusual for Russian companies. A low level of management expertise in Russian companies is shown in the insufficient use of contemporary approaches to planning and analysis of effectiveness, low attention to personnel motivation and a very formal approach to personnel training. Russian industry, which has been recently privatized, has not yet acquired a positive attitude towards voluntarism in relation to environmental operations. The legacy of state regulation method prevents Russian companies to perceive the nature and benefits of voluntary standards and practices; they are prone to consider only obligatory standards and instruments of state regulations as a direction in their management practices. (Vinichenko et al, 2001).

Decline in industrial production during the 1990s has significantly affected Russian communities through decline in employment, decline of living standards and health deterioration related to environmental pollution in urban areas (Porfiriev, 1997; Ellman, 2000). The closure of large industrial enterprises in Russia that were principal supplier of employment, housing, childcare and various social services for particular towns were immense (Crotty and Crane, 2000). During the Soviet era, industrial activity was under state control. Economic developments of many Russian regions were associated with the establishment and operation of large industrial projects, which provided employment to a great number of people, often causing significant migration from Western parts of Russia to Eastern part of the country. Large industrial projects such as transport projects or mineral exploitation projects along with associated industrial and social development caused significant changes to prior underdeveloped areas. New projects were usually coupled with town planning and construction of residential areas, establishment of transport and service infrastructures and creation of social care and public services.

The Soviet state saw these large industrial enterprises as agents for socio-economic development of the country. In many cases, many social care and public service organizations were an integral part of these large industrial enterprises and funded from their industrial profits. For example, schools, nurseries, hospitals, utility services and food procurement companies in the vicinity of the large industrial project were largely supported by these large industrial enterprises. Soviet socio-economic policy favored creation of 'one factory, one town' model that proved very vulnerable at times of economic crises. Local communities in such towns including local citizens, employees, trade unions, managers and other shareholders of a principal enterprise became economically and socially dependent on its operations.

Economic and political changes in Russia led to the privatization of many industrial enterprises, a drop in industrial production, sometimes closure of large enterprises and formation of new private companies, which could no longer support organizations of social nature such as nurseries and utility services. These organizations, which were dependent on the industrial enterprises, were often transferred to local or regional authorities. Schools, nurseries, hospitals and other organizations stopped receiving support from the principal industrial enterprises and now rely on financing from scarcely budgeted local authorities, which inevitably led to increased prices on their services. This has greatly affected the communities near large industrial enterprises. Many of them not only lost jobs, due to closure of ex-state owned industrial enterprises, but they lost many privileges in housing, childcare and recreation.

The privatization of industrial sectors and the development of corporate business in the Russian Federation, as well as active foreign trade and co-operation with foreign creditors and investors, have led to increased awareness on corporate social responsibility. It has become more popular amongst new Russian industrial enterprises to undertake philanthropic actions. Russian companies are beginning to understand the importance of community relations and social responsibility in relation to their business operations and primarily use it as a method for corporate image improvement. Many large Russian companies are starting to participate in community initiatives, some through charity giving and the establishment of corporate foundations. However, philanthropic activities of Russian companies are unsystematic and uncoordinated. Donations are mainly based on discretion of Russian managers and are not supported by corporate strategies. Philanthropy has not yet become part of a developed corporate agenda. Russian businesses are starting to acknowledge social responsibility, but they lack established and effective mechanisms for engagement with communities and corporate philanthropic activities. Many companies practice direct support of single actions, while large companies in particular support selected organizations and communities on a regular basis (Assoziazia Menedgerov, 2001; Nasledie, online).

Notwithstanding difficult economic conditions, businesses in Russia are engaging in corporate philanthropy and supporting organizations that are oriented to less protected people. Generally, donations are made in money and food. Other forms of donations such as offering professional services, transport, facilities, and information support are less popular, as they are less in demand (Nasledie, online). Although philanthropic practices are increasing, its further development depends upon companies' realization of the benefits of corporate philanthropy and the development of independent non-profit organizations.

Mining in the Republic of Sakha (Yakutia)

The Republic of Sakha (Yakutia) is the largest territorial unit within the Russian Federation, which is situated in the remote northeastern part of the country, and is characterized with severe climatic conditions (see Figure 4.6).[8] The Republic of Sakha covers an area more than three million square kilometers, but has a small population of about one million people. Yakutia first became a part of the Russian Empire in 1632. In 1922, Yakutia received the status of the Yakut Autonomous Soviet Socialist Republic in the structure of the Russian Soviet Federated Socialist Republic. Following the adoption of a Declaration of State Sovereignty in 1990, it was renamed the Republic of Sakha (Yakutia). In this book, the area of the present Republic of Sakha in relation to events before 1990 is referred to as Yakutia. According to the Constitution of the Republic of Sakha of 27 April 1992, the head of the Republic of Sakha is the President, the legislative body is the Legislative Assembly 'Il Tumen', and the Republic of Sakha has its own Citizenship, State Emblem, and carries out internal self-government.

The territory of the Republic of Sakha is abundant in mineral resources. In 1998, as much as 1,249 deposits of 43 kinds of mineral resources were registered in the Republic of Sakha, including gas, oil, coal, gold, diamonds, tin, antimony, and other rare metals (Pravitelsvto RS(Y), 1999). However, only a limited variety of mineral resources is under industrial exploitation, but this is on a large scale (see Table 4.9). In 1998, 158 organizations carried out geological research and mining of mineral resources on 667 sites.

The industrial sector prevails in the regional economy of the Republic of Sakha. Gross Regional Product comprised of 48 per cent by industry, four per cent by agriculture and the remaining share is attributed to service sector. Almost 80 per cent of industrial production refers to extractive industry and only 20 per cent refers to processing (Goskomstat RS(Y), 1994, 1995, 1996). Fuels and non-ferrous metallurgy are the most important extractive sectors (see Figure 4.7). Diamond, gold and antimony mining, as well as diamond cutting, accounted for 60 per cent of the total industrial production in 1995 rising to 73 per cent in 2002 (see Figures 4.8 and Figure 4.9).

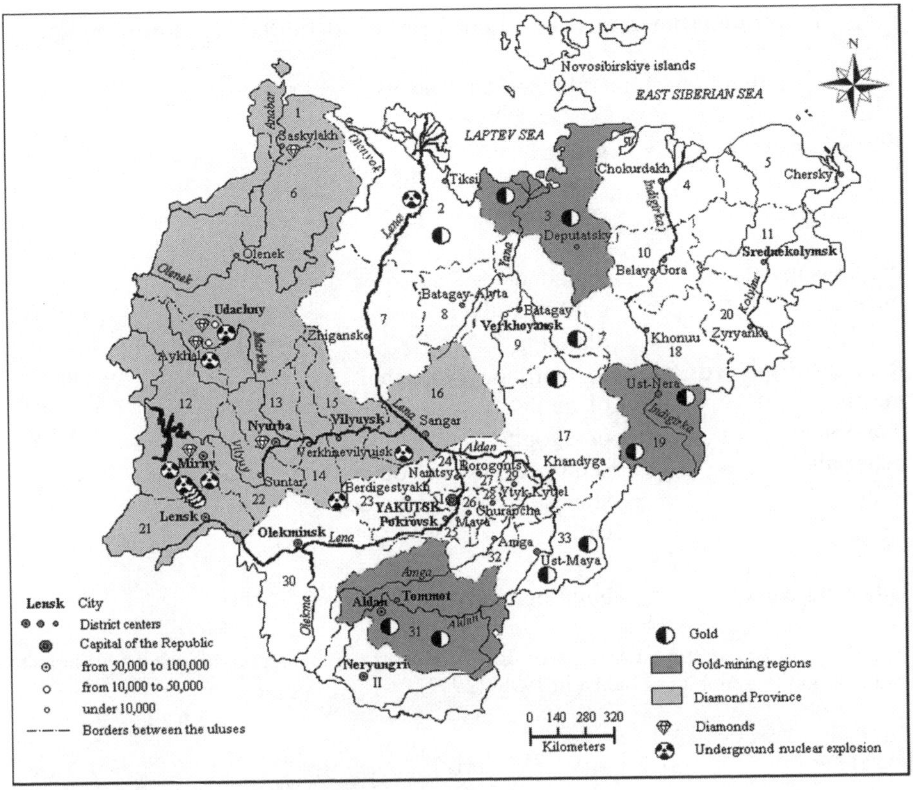

Figure 4.6 Map of the Republic of Sakha

Note: Districts of the Republic of Sakha: 1 – Anabar, 2 – Bulun, 3- Ust-Yana, 4 – Allaikhovsky, 5 – Nizhne-Kolymsky, 6 – Olekminsk, 7 – Zhigansk, 8 – Eveno-Bytantaisky, 9- Verkhonyansky, 10 – Abyi, 11 – Srednekolymsky, 12- Mirny, 13 – Nyurba, 14 – Verkhnevilyuysk, 15- Vilyuysk, 16- Kobyai, 17 – Tompon, 18 – Momsky, 19 – Oymyakon, 20 – Verkhnekolymsky, 21 – Lensk, 22 – Suntar, 23 – Gorny, 24 – Namsky, 25 – Khangalass, 26 – Megino-Khangalass, 27 – Ust-Aldan, 28 – Churapcha, 29 – Tatta, 30 – Olenek, 31 – Aldan, 32 – Amga, 33 – Ust-Maya, I – Territory of the city Yakutsk, II – Territory of the city Neryungri.

The gold and diamond mining industries have long been the leading extractive industries in the Republic of Sakha. However, diamond output is growing while gold output is decreasing (see Table 4.10 and Table 4.11). Furthermore, the extractive industries provide large employment in the Republic of Sakha (see Table 4.12).

Table 4.9 Information on use of mineral resources in the Republic of Sakha

Groups of minerals	Mineral resource deposits	Total	Licensed
Energy resources	• oil	12	3
	• natural gas and condensate	31	5
	• coal	47	20
Precious metals and minerals	• diamonds	39	18
	• gold and silver	835	451
	• platinum	1	1
Rare and dispersed metals	• tin, antimony, tungsten, bismuth	62	26
Non-metallic materials	• phlogopite	22	2
	• semi-precious and other stones (chrome diopside, etc)	7	5
	• salt	1	1
Other minerals	• building materials	186	63

Source: Adapted from Pravitelstvo RS(Y) Report, 'State report on the environmental situation in the Republic of Sakha in 1998' (1999).

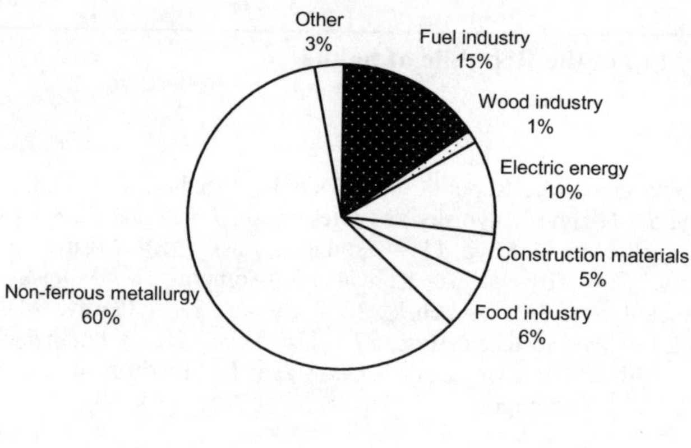

Figure 4.7 Structure of industrial sector in the Republic of Sakha in 1995, %

Source: Adapted from Goskomstat RS(Y) Report 'Socio-economic situation in the Republic of Sakha (Yakutia): January-December 1999' (2000).

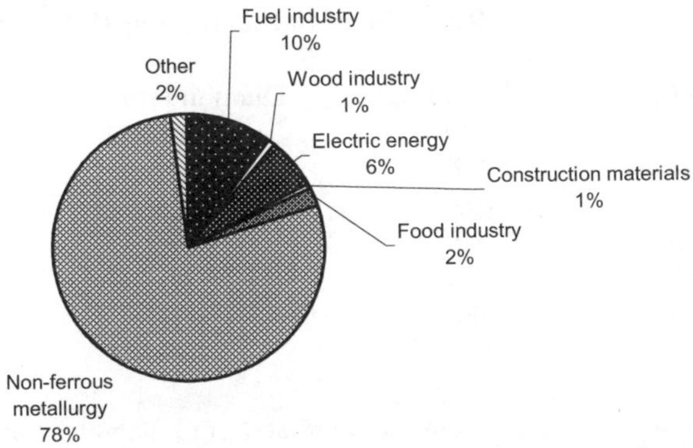

Figure 4.8 Structure of industrial sector in the Republic of Sakha in 1999, %

Source: Adapted from Goskomstat RS(Y) Report, 'Socio-economic Situation in the Republic of Sakha (Yakutia): January-December 1999' (2000).

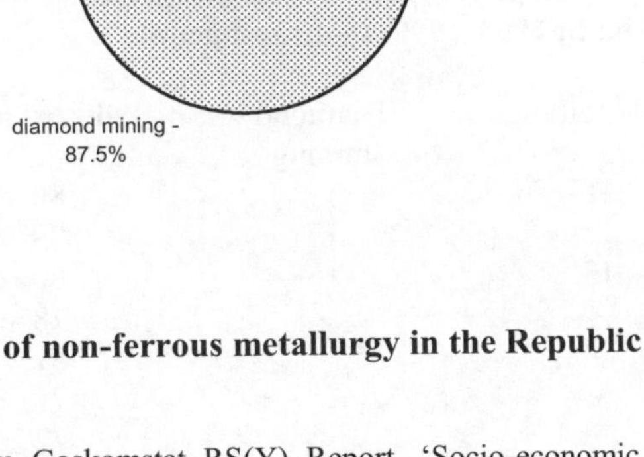

Figure 4.9 Structure of non-ferrous metallurgy in the Republic of Sakha in 1999, %

Source: Adapted from Goskomstat RS(Y) Report, 'Socio-economic Situation in the Republic of Sakha (Yakutia): January-December 1999' (2000).

Table 4.10 Index of physical value added in precious metals minerals mining in the Republic of Sakha during 1993-1999, % to previous year

Year	Gold mining	Diamond mining	Industry total
1993	100	95	95
1994	99	-	101
1995	89	105	101
1996	89	105	102
1997	78	99	95
1998	60	116	103
1999	118	104	106

Source: Adapted from Goskomstat RS(Y) Reports, 'Socio-economic Situation in the Republic of Sakha (Yakutia)' (1994, 1995, 1996, 1997, 1998, 1999, and 2000).

Table 4.11 Share in industrial production of the Republic of Sakha in 1993-1999, %

Year	Gold mining	Diamond mining	Industry total
1993	17	51	100
1994	11	46	100
1995	12	47	100
1996	10	43	100
1997	9	42	100
1998	4	60	100
1999	6	68	100

Source: Adapted from Goskomstat RS(Y) Reports, 'Socio-economic Situation in the Republic of Sakha (Yakutia)' (1994, 1995, 1996, 1997, 1998, 1999, and 2000).

Table 4.12 Employment provided by the industrial sector of the Republic of Sakha in 1993-1999, thousand people

Year	Gold mining	Diamond mining	Industry total	Republic's total
1993	15	17	80	-
1994	15	17	79	553
1995	13	18	80	536
1996	-	-	78	526
1997	-	-	81	498
1998	-	-	71	486
1999	9	36	72	459

Source: Adapted from Goskomstat RS(Y) Reports, 'Socio-economic Situation in the Republic of Sakha (Yakutia)' (1994, 1995, 1996, 1997, 1998, 1999, and 2000).

According to the Constitution of the Republic of Sakha, all natural resources belong to the people of the Republic of Sakha. The Agreement between the Russian Federation and the Republic of Sakha that distinguishes the property rights, states that the diamond mining deposits belong to the Republic of Sakha. In March 1992, the Federal Treaty was signed, which included a protocol stating that the Government of the Republic of Sakha retains ten per cent of precious metals and 20 per cent of the diamonds produced in the Republic of Sakha (Kempton 1996; Lynn, 1999). In June 1995, the ownership and control over natural resources was again formalized in the bilateral agreement between the Russian Federation and the Republic of Sakha, according to which the Government of the Republic of Sakha has a right to retain 25 per cent of diamonds and 30 per cent of gold produced in the Republic of Sakha (Lynn and Fryer, 1998; Solomov, 1998).

Gold Mining in the Republic of Sakha

The first gold mining area to be explored was the Aldan district in south Yakutia, where gold was found as early as 1851. However, the absence of roads and the remoteness of the deposits from settlements made it economically not viable to develop these placer deposits. In 1923, the expedition supported by the newly formed Soviet Yakut Government discovered placer deposits, giving a push to the industrial development of gold mines in Aldan.[9] In 1924, Aldanzoloto (Aldan gold) was created and later in 1926 trust of the Soviet Union, Yakutzoloto Corporation (Yakut gold) was formed to overlook the gold mining operations on the territory of Yakutia. The Yakutzoloto Corporation consisted of four gold mining complexes including Aldanzoloto, Indigirzoloto, Kularzoloto, Djugdjurzoloto and the tin mining complex Deputatsky (Aldanzoloto, 1994). Production processes related to gold mining slowly improved and the gold production expanded.[10] The development of gold mining in Aldan has brought an overall economic development to Aldan district. Large numbers of people migrated to the area, building new settlements for miners. Mining development motivated the building of electric energy stations (Seligdarskaya, Yakokutsaya and Chulmanskaya) and helped to contribute to the local and regional infrastructure via the building of roads, and the Aldan-Yakutskaya railroad, an extension of the Baikalo-Amurskaya railroad (Motyashov and Rastorguyev, 1991).

The re-organization of the gold mining industry in Yakutia began in the beginning of the 1990s. In 1991, the gold mining industrial association 'Yakutzoloto' was transformed into joint stock concern 'Zoloto Yakutii' (Gold of Yakutia). In the same year, 'Aldanzoloto' became independent of 'Yakutzoloto' and transformed into joint-stock concern. In 1992, all mining complexes within 'Zoloto Yakutii' were transformed into closed joint-stock companies. In 1995, a holding and industrial company Sakhazoloto (Sakha gold) was established on the basis of the joint-stock company Zoloto Yakutii (Gold of Yakutia) and all mining enterprises of Yakutia. In 1996, according to a Decree of the President of the Republic of Sakha, Sakhazoloto was assigned to offer credit to the gold mining industry. The privatization of the gold mining industry thus dissolved the unity of gold mining in Yakutia. However, the independent gold mining companies lacked financial support and this led to a difficult financial situation in the whole of the gold mining industry of the Republic of Sakha. The gold mining complex Kularzoloto was declared bankrupt and was liquidated in 1999. Other gold mining companies decreased their industrial output and accumulated huge debts (Desyatkin, 1999).

In 1999, there were 57 gold mining enterprises operating on the territory of Yakutia, 50 in, 1998 and 21 in 1994. There are several gold mining regions within Yakutia: Aldan district, Dzhugdzhur fields, Indigirka area, and the Yana River. The largest gold mining companies are: Aldanzoloto Company, which operates in the Aldan district and accounted for 20 per cent of the total gold production in the Republic of Sakha in 1999 (Goskomstat RS(Y), 2000); and Indigirzoloto Company which operates in the Ust-Maysky district. Gold production in the Republic of Sakha has declined by 20 per cent since 1991 (see Table 4.13 and Table 4.14).

Table 4.13 Gold production in the Republic of Sakha in 1991-1996, tons

Year	Production in the Republic of Sakha, tons	Index, 1991-100%	Total Russian production, tons	Share of RS(Y) in total Russian production, %
1991	32.8	100	161.8	20
1992	30.5	93	150.6	20
1993	33.4	102	149.5	22
1994	30.5	93	142.6	21
1995	28.8	88	131.9	22
1996	23.4	71	123.3	19

Source: Adapted from Sagers, 'Regional Trends in Russian Gold Production' (1997).

Table 4.14 Gold production in the Republic of Sakha in 1991-1998, tons

Year	Gold production, tons	Index, 1991-100%
1991	32.8	100
1992	30.5	93
1993	33.4	102
1994	31.3	95
1995	28.9	88
1996	22.9	70
1997	19.9	61
1998	11.1	34

Source: Adapted from Zoloto: Stranizy Zolotopromyshlennika [Gold: Pages for Gold miners], URL: http://www.aurum79.ru (visited on 12/06/02).

Diamond Mining in the Republic of Sakha

Exploration for diamonds in Yakutia near river Vilyuy started in 1947, and the first diamond was already found in 1949. Alluvial diamond deposits were discovered later in 1950 and in 1954 the first kimberlite pipe called Zarnitsa was discovered near the river Daldyn, a tributary of the river Vilyuy (Belykh, 1993; Vecherin, 1997; Kiseleva, 1999).[11] When geologists discovered a very rich kimberlite pipe in 1955 near the river Irelyakh, they sent a telegraph to the authorities in Moscow stating, 'We smoke pipe of peace, excellent tobacco'. This very rich diamond pipe later called Mir ('peace' in Russian) was the start of the diamond mining industry in Yakutia (Tanin, 1999).

In January 1957, a state owned Yakutalmaz Corporation (Yakut diamond) was established under the supervision of the Ministry of Non-ferrous Metallurgy of the former Soviet Union in order to conduct diamond-mining development in Yakutia (Belykh and Vybornov, 1984; Belykh, 1993). The first carats of gem quality diamonds were extracted on 10 June 1957 (Belykh and Vybornov, 1984), and two enrichment plants, which were supplied with ore from the Mir pipe, were launched in 1958 (Belykh, 1993). The state owned corporation 'Yakutalmaz' was the controlling diamond mining company in Western Yakutia for the period 1957-1992, exploiting both alluvial and ore diamond deposits. Industrial development in the Western Yakutia provided the impetus for new urban settlements for miners, for example Mirny (1959), Chernyshevsky (1961), Aykhal (1962), Almazny (1963), and Udachny (1968) (Belykh and Vybornov, 1984). A labor force for this industrial development mainly came from western parts of the former Soviet Union and caused large migration into what was hitherto a sparely populated area.

Table 4.15 Diamond output of ALROSA in 1993-2000, million US dollars

Year	Diamond production	Index, 1993-100%	Diamond sales	Index, 1993-100%
1993	1,115.0	100	1,121.5	100
1994	1,217.8	109	1,268.7	113
1995	1,285.6	115	1,350.6	120
1996	1,383.6	124	1,441.1	128
1997	1,368.7	124	1,315.5	117
1998	1,507.6	135	1,424.6	127
1999	1,521.0	136	1,416.9	126
2000	1,516.0	136	1,444.0	129

Source: Adapted from Matveev and Cherny, 'ALROSA Company: Strategy for Development' (2000).

Changes in diamond mining began in 1993, when Yakutalmaz Corporation was liquidated and a new joint-stock company Almazy Rossii-Sakha (Diamonds of Russia-Sakha, shortened into ALROSA) was formed. ALROSA is the mother company for the Almazy Anabara Company, which operates in Anabar district and for the ALROSA-Nyurba Company, which operates on the Botuobinskaya and Nyurbinskaya kimberlite pipes in Nyurba district. Currently, diamond-mining operations are carried out in Anabar, Mirny and Nyurba districts, and the industrial divisions of the diamond mining company are located in Lensk district. A vast area in Western Yakutia, including eight districts: Anabar, Lensk, Mirny, Verkhnevilyuysk, Vilyuysk, Nyurba, Olenek and Suntar, is called the diamond province (see Figure 4.9). Despite an overall drop in the industrial production of the Republic of Sakha, the diamond mining industry is a fast growing sector of the economy (see Table 4.15).

Environmental and Social Impacts of Mining in the Republic of Sakha

The Republic of Sakha is an important environmental center of the Arctic. Various and vulnerable water, tundra and taiga systems have only slow rates of self-renewal and self-cleaning. It is essential to consider the slow rate of ecological renewal in the North versus the impacts of anthropological interference, especially during intensive industrial development (Makarov and Shats, 2000). Savvinov (1993, 1994, 1999) and Poiseev (1999) characterize the area of the Republic of Sakha as having:

- slow ability of the soil and natural watercourses for self-cleaning;
- small potential for absorption and dispersion of air pollution;
- slow rehabilitation of vegetation;
- industrial contaminants when they enter the natural environment only decompose slowly which can lead to the accumulation of hazardous elements including heavy metals throughout all sections of the ecological system.

Historically, traditional natural resource use in Yakutia primarily focused upon cattle farming, reindeer farming, hunting and fishing, and was thus dependent upon high levels of environmental quality (Burtseva, 1993; Poiseev, 1999). Unfortunately, the history of the industrial development in Yakutia clearly demonstrates a rapid exploitation of mineral resources started in the Soviet period and had no regard for either environmental or social concerns (Savvinov, 1994; Marples, 1999). Moreover, the economic policy of the former Soviet Union for the northern territories such as Yakutia was oriented towards extraction of mineral resources, which were then exported to other parts of the country. Therefore, the northern territories were merely regarded as provinces for the supply of mineral resources. There was no compensation for Yakutia, no concern for nature conservation, and no concern for more traditional branches of economy or indigenous people (Kempton, 1996; Poiseev, 1999). The extraction of mineral resources was carried out under the supervision of the Soviet Union Departments, and mining enterprises had no obligations or responsibilities to the area. Moreover, the former system of administration and control, when different departments were supervising an enterprise, had no integrated policy for natural resource use (Poiseev, 1999).

Because of unfavorable factors in Yakutia that influence the cost of production, e.g. severe climate, transport underdevelopment, absence of an appropriate workforce, and energy supply difficulties, highly profitable extraction of non-ferrous metals and fuel resources were developed initially in order to underpin less lucrative industrial development (Poiseev, 1999). The industrial development in Yakutia started in the 1920s with the exploitation of gold mines in southern Yakutia. In the 1930s, the development of coal deposits took place and in the 1950s, the development of the diamond mining industry began. Along with these mining developments, high priority was given to the development of overall infrastructure through the creation of appropriate transport systems and electric power generation (Poiseev, 1999; Yakovleva et al, 1999).

Poiseev (1999) describes natural resource use in Yakutia during the Soviet period as irrational, destructive and inconsiderate of environmental impacts. The exploitation of mineral deposits was primarily via relatively cheap extraction methods, which in most cases meant open-pit mining (Poiseev, 1999) and caused a significantly negative impact on the environment (Savvinov, 2000). In some regions of Yakutia, irrational and irresponsible resource exploitation led to negative changes in watercourse structure and function and to a fall in land productivity and biological diversity (Savvinov, 1994). Several zones of environmental contamination, caused by mining developments, have been identified in Yakutia (Makarov and Shats, 2000; Savvinov, 2000), including southern Yakutia, where coal and gold mining are developed; western Yakutia, where diamond mining is developed; and the rivers Yana and Indigirka where tin and gold mining are developed.

The extractive industry significantly contributes to the total air pollution; however its share in total air pollution of the Republic of Sakha has diminished (see Table 4.16 and Figure 4.10). The decrease in air pollution from the extractive industry may reflect an improvement in environmental activity of enterprises; however it may also reflect a decrease in production due to the closure of many gold mines and other mineral mines. The gold and diamond mining industries are the largest contributors to environmental pollution in the region. For example, gold mining accounted for 71 per cent and diamond mining for 29 per cent of total air pollution from stationary sources from the non-ferrous industry of the Republic of Sakha in 1994 (Pravitelstvo RS(Y), 1995).

Discharge of industrial wastewater from mining enterprises is also major environmental problem in the Republic of Sakha. Even though the number of enterprises in the gold mining sector has risen from 21 in 1994 to 50 in 1998, the collection of water from surface watercourses has decreased from 42 million cubic meters in 1994 to 13 million cubic meters 1998. These changes are related to the reduction in gold production, especially to the fall in production by companies Aldanzoloto, Indigirzoloto and Zoloto Djugdjura. The creation of new enterprises largely relates to the reworking activities on deposits previously explored by bigger companies. Although the total discharge of industrial wastewater to surface watercourses decreased from 25.2 million cubic meters in 1994 to 9.5 million cubic meters in 1998, the percentage of contaminated industrial wastewater increased from 45 per cent in 1994 to 93 per cent in 1998 (see Table 4.17). This situation almost certainly reflects a decrease in environmental standards by the new, smaller enterprises.

Table 4.16 Air pollution from stationary sources in the Republic of Sakha in 1996-1999, thousand tons

Industry	1996	1998	1999
Total air pollution from stationary sources	175.9	148.3	136.1
Electric engineering	38.8	38.4	83.9
Non-ferrous metallurgy	31.3	21.0	12.8
Transport	-	16.1	14.1

Source: Adapted from Pravitelstvo RS(Y) Reports, 'State Report on Environmental Situation in the Republic of Sakha (Yakutia)' (1999, 2001).

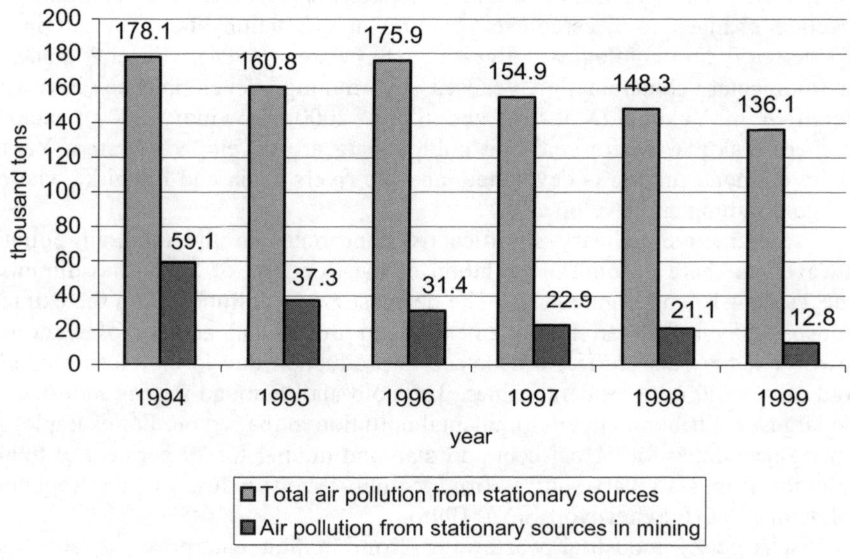

Figure 4.10 Air pollution from stationary sources in mining in 1994-1999, thousand tons

Source: Adapted from Pravitelstvo RS(Y) Reports, 'State Report on Environmental Situation in the Republic of Sakha (Yakutia)' (1995, 1996, 1997, 1998, 1999, 2001).

Democratic developments in the Russian Federation have given the regions more independence and opportunities to solve their own economic, social and environmental problems. According to the Constitution of the Russian Federation, natural resource use, nature conservation, environmental security, and specially protected natural areas are under the mutual authority of the Russian Federation and the region. In the 1990s, following the adoption of the Declaration of Sovereignty, the Republic of Sakha has acquired a right for independent policy development in nature conservation and natural resource exploitation. The Republic of Sakha has managed to adopt some protective environmental measures following the destructive consequences of the previous central regulatory dictate and thus create a positive momentum on the road to environmental protection. The reforms cover legislation and governmental and regulatory structure, and attention has been given to public awareness on environmental issues and environmental education (Zakharova, 1996).

Table 4.17 Water use in gold mining in the Republic of Sakha

	1994	1995	1996	1997	1998
Number of reporting enterprises	21	42	48	51	50
Collection of water, million cubic meters	42.0	30.2	26.6	22.3	13.0
Derivation of industrial wastewater, million cubic meters	25.2	20.5	17.4	16.0	9.5
Contaminated industrial wastewater, million cubic meters	11.4	12.3	12.5	10.4	8.8
Contaminated industrial wastewater, %	45	60	72	65	93
Normatively cleaned industrial wastewater, million cubic meters	13.8	8.2	4.9	5.6	0.7
Normatively cleaned industrial wastewater, million cubic meters	55	40	28	35	7

Source: Adapted Pravitelstvo RS (Y) Reports, 'State Report on Environmental Situation in the Republic of Sakha (Yakutia)' (1995, 1996, 1997, 1998).

According to Article 71 of the Constitution of the Russian Federation, federal state property including mineral resources is under the jurisdiction of the Russian Federation. Article 72 of the Constitution stipulates that legislation concerning mineral resources and environmental protection is under the mutual jurisdiction of the Russian Federation and the regions (Republics, *krays*, *oblasts*, *and okrugs*). Article 12 of the Law of the Russian Federation on mineral resources No. 2395-1 adopted on 21 February 1992, repeats the statement of the Constitution that mineral resources are state property. Issues related to possession, *usufruct* and use of mineral resources are under mutual jurisdiction of the Russian Federation and the regions.

The Republic of Sakha was one of the first regions of Russia to unite separate environmental departments under the Ministry for Nature Conservation. The 'Program for the Implementation of the State Environmental Policy of the Republic of Sakha for the Improvement of Biological Diversity for 1996-2001' is one of the central documents of environmental regulation in the Republic of Sakha. The Republic of Sakha accepts the need for more efficient use of natural resources, the importance of environmental protection and the principles of sustainable development. The National Revival Foundation and the Target Fund for Future Generations have been established to finance priority areas in the restructuring program. They are financed by a tax on the use of non-renewable natural resources and represent unique foundations within the Russian Federation. The funds are used for investment in the future, i.e. emerging industries, infrastructure, training and environment.

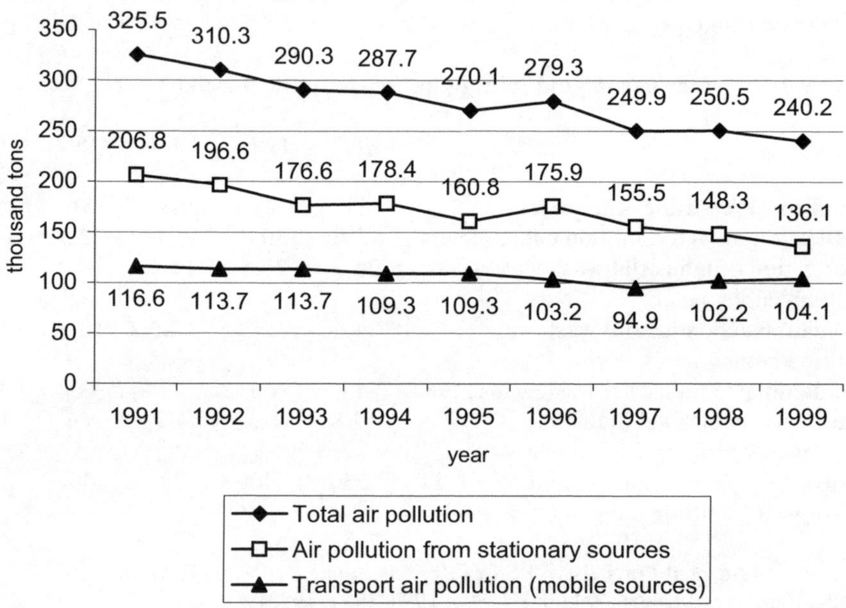

Figure 4.11 Dynamics of air pollution in the Republic of Sakha in 1991-1999, thousand tons

Source: Adapted from Pravitelstvo RS(Y) Report, 'State Report on Environmental Situation in the Republic of Sakha (Yakutia)' (2001).

Moreover, the Republic of Sakha is actively involved in international co-operation, which includes participation and mutual projects with international and intergovernmental organizations. The partners of the Republic of Sakha are the World Wide Fund for Nature (WWF), Northern Forum, United Nations Educational, Scientific and Cultural Organization (UNESCO), World Conservation Union (IUCN), Department for Nature Conservation of Alaska (US), the French organization 'Nature Without Borders' and others. Current international co-operation is oriented towards environmental education, establishment of environmentally protected areas, environmental monitoring and research (Pravitelstvo RS(Y), 1998).

The development of nature conservation in the Republic of Sakha was given new impetus in 1996, when the President of the Republic of Sakha initiated a creation of the system of specially protected natural areas, which would cover 25 per cent of the territory of the Republic of Sakha. The planned system of specially protected natural areas in the Republic of Sakha would become the world's largest nature reserve. Since 1994, the Government of the Republic of Sakha then started to issue regular official reports on the environmental situation open to public access and complied by the Ministry for Nature Protection. These reports incorporate information on air and water pollution, radiation, data on flora and fauna in the region, health status of the population and industrial impact on the environment. Data from these State reports show that air pollution in the Republic of Sakha has decreased during 1991 to 1999 (see Figure 4.11).

Social Impacts of Mining in Yakutia

Mining operations and accompanying industrial development have greatly affected the lives of peoples in Yakutia. There are several major social impacts of mining and accompanying industrial development on indigenous population in Yakutia:

- relocation from historic area of habitation, disruption to traditional economic activities and loss of skills associated with traditional lifestyles;
- loss of cultural heritage and language of indigenous people, associated with the disruption caused to traditional lifestyle;
- negative environmental impacts from mining and other industrial operations that affected the traditional economic activities of indigenous peoples (reindeer breeding, hunting and fishing);
- difference between earnings of industrial workers and agricultural workers, leading to differences in quality of life.

Industrial development in Yakutia has led to a significant migration from other parts of Russia, changing the proportion in the number of indigenous population of Yakutia (Yakuts, Evenks, Evens, Chukchis, Yukagirs and Dolgans) in proportion with the migrant population. In 1926, before the beginning of wide-scale industrial development in Yakutia, the share of migrant population in the region's total population (285 thousand people) was 12 per cent, it has significantly increased with the beginning of gold mining operations in Aldan and tin mining in near Yana River. In 19391 the proportion of migrant population has increased up to 40 per cent of the total population (413 thousand people). The next rise in population relates to the beginning of diamond mining operations in Vilyuy region, in 1959 the proportion of migrant population has increased up to 50 per cent of the total population (487 thousand people). In the 1970s, the share of migrant population increased up to 60 per cent and in 1989 it amounted to 64 per cent. Currently, small nations of the North (Evenks, Evens, Chukchis, Yukagirs and Dolgans) constitute only two per cent of the entire population of the Republic, while dominant indigenous ethnic group (Yakuts) constitutes 30 per cent of the total population (Poiseev, 1999).

According to Poiseev (1999), the industrial development such geological exploration and mining operations, in Yakutia has significantly affected the life of small nations, as the development entered the area of traditional livelihood of these peoples, many indigenous peoples were forced to live their historic areas of habitations and leave their traditional economic activities. According to data in 1989, about 16 million hectares of rain-deer pastures were removed (Gorokhov, 1989). This practice has led to displacement of indigenous population from their traditional economic activity – rain-deer breeding. Small nations were not politically protected in the Soviet Union. Until recently rights of small nations were not legalized. Indigenous peoples of the North did not have right for their territory, natural resources and areas of habitation.

Indigenous peoples were continued to be displaced from hunting and fishing, as migration has caused wide spread poaching, hunting grounds were disturbed by uncontrolled transport links. Significant impact on fishing, the traditional economic activity of the indigenous peoples, was caused by industrial wastewater pollution of the river system in Yakutia. The very existence of indigenous population and their culture was threatened by the advances of industrial development. The number of indigenous peoples has reduced in the areas of mining developments. For example, according to Poiseev (1999), in 1931 the number of Evenks living in Sadynsky district was 1,600, while currently, in Mirny district, formed on the territory of the former district, is 177. Indigenous population was not initially recruited in the industry, due to significant differences into traditional lifestyles, connected with traditional natural resource use, and new industrial activities, environmentally destructive practices. Educated part of the indigenous population were qualified in social and public sector (teachers, agronomists, leisure, medics, etc) and the beginning of massive industrial development, indigenous population did not have qualified technical professionals to step into the industry.

Industrial development has caused the significant gap in the earning levels between people engaged in agriculture (where most indigenous population was engaged in) and mining industry. In 1985, a wage of an agricultural worker constituted 82 per cent of a wage of an industrial worker. The gap has reduced during the 1990s, and in 1994 agricultural worker's wage constituted 43 per cent of an industrial worker's wage (Poiseev, 1999).

Environmental Impacts of Gold and Diamond Mining in the Republic of Sakha

Historically, the extractive industry in Yakutia operated under a slogan 'Give more gold and diamonds', while issues of environmental protection were neglected. The gold and diamond mining industries have caused significant negative impacts to the environment in the areas of these mining operations including land disturbance, water contamination, and a decrease in biological diversity. According to statistics on disturbed lands, out of 36,000 hectares of total disturbed land in the Republic of Sakha in 1998, diamond and gold mining accounted for 24,700 hectares, which comprise 68 per cent of the total. Not surprisingly, Aldan district, the center of gold mining, has 7,600 hectares of disturbed land and Mirny district, the center of diamond mining, has 9,800 hectares of disturbed land (Goskomstat RS (Y), 1999). Industrial developments in Aldan district via expansion of disturbed lands, air and water pollution and infrastructure developments have all had a negative impact upon traditional economic activities (hunting and reindeer farming) of the indigenous population (Evenks) due to the decrease of quality and quantity of reindeer pastures and hunting grounds (Volotovsky and Kuznetsova, 1993).

The state of the environment in the diamond-mining region of Yakutia attracted the attention of scientists at the end of the 1980s and several expeditions were conducted. There are various scientific works available on the topic of environmental impacts of diamond mining development in Western Yakutia and environmental contamination in Vilyuy region, including monographs: 'Ekologia Basseina Reki Vilyuy: Promyshlennoe Zagryaznenie' (1992); 'Ekologia Reki Vilyuy' (1993); and 'Ekologia Vilyuya: Materialy k Ozenke Ekologicheskogo Sotoanyia' (1996). Conclusions drawn from these works were devastating and clearly demonstrated that industrial pollution from these diamond-mining developments crucially contributed to the contamination of the area (Savvinov et al, 1992; Savvinov, 1993).

During the early years of operation, the diamond mines operated without any environmental protection constraints and consequently contaminated the river systems. Until the beginning of the 1980s, industrial wastewaters, drainage from spoil heaps as well as groundwater pumped from diamond mines were disposed without any purification and contaminated local water streams (Labutina and Smolenko, 1990; Kopylov et al, 1993; Petrova and Volozhin, 1996). This was very hazardous because the diamond mining industry used toxic liquids in ore processing, including thallium (Kopylov et al, 1993). Technical waste and drainage from rock dumps contaminated water sources not only near to the mine, but also far away from the immediate area of mining activity (Savvinov et al, 1992). In 1981, the Party made an attempt to mitigate environmental impacts of industrial enterprises in the North of Russia via issuing a Decree (Decree of the Central Committee of Communist Party of the Soviet Union and the Council of Minister of the USSR No. 65 'About actions for enforcement of protection of seas, rivers and other watercourses of the Arctic basin from contamination' issued on 15 January 1981), which requested an improvement in the effluent treatment facilities in order to eliminate discharge of untreated wastewater into watercourses and to reduce the discharge of contaminants into the air by 1995. In 1982, according to the Government decision, Yakutalmaz Corporation developed the first plan for elimination of discharge of untreated wastewater into watercourses, which was later revised in 1988 and was in effect until 1993.

Environmental research of the Vilyuy region documented that the environmental degradation was linked not only to diamond mining operations, but also to the operations of the Vilyuy hydroelectric power station, which was built to supply the diamond mining developments with electrical power (Savvinov et al, 1992). Diamond mining developments and hydroelectric engineering occupy large tracks of land for operations, dumps, roads, construction, tailings, reservoirs and social infrastructure. By 1 January 1997, the total area of disturbed land on the territory of the diamond province was 8,332 hectares, while the largest part of disturbed land (8,159 hectares) lies within the territory of Mirny district (Shipkov, 1999).

Vilyuy Hydroelectric Power Station

The embryonic diamond mining industry needed electricity and in 1958 the decision was made to build a hydroelectric power station upstream on the river Vilyuy, 100 kilometers away from Mirny (Belykh, 1993). The building of a hydroelectric power station created a reservoir with the initial flooding starting in 1969 and finishing in 1973. The reservoir covered an area of 2,170 square kilometers, with an average width of 4.6 kilometers, and the total water volume amounting to 36 cubic kilometers. The area of flooding consisted of large masses of wood and swamps, which has subsequently affected the quality of the water in the reservoir (Kirrilov, 1990).

As it was stated in the 'Note' of the commission of the Siberian Branch of the Academy of Sciences of the USSR of 19-25 August 1983, water quality in the reservoir during the initial two to three years after its creation was unsatisfactory (Petrova and Volozhin, 1996). The concentration of phenols increased up to 30-40 times above the maximum permissible concentrations (MPCs). According to the Russian environmental standards, the MPC for phenol is 0.001 milligrams per liter (Kurlyandsky and Sidorov, 1998), which means that the concentration of phenol in the Vilyuy water reservoir reached 0.003-0.004 milligrams per liter.

More than ten years were needed for the stabilization of the oxygen regime in the Vilyuy water reservoir. This artificial regulation of the river Vilyuy flow significantly changed the hydrological, hydro-chemical and hydro-biological characteristics of the flooded area, as well as the ice regime, and the reservoir itself became the source of hydrogen sulfide and phenols that contaminated the environment (Kirrilov, 1990). The negative impact of this development has been identified as affecting fish resources of the river Vilyuy. A study carried out during 1975-1990 showed that in separate sections of the river there has been a complete loss in natural productivity of fish (Savvinov et al, 1992; Savvinov, 1993; Shipkov, 1999). The catch of fish in the river Vilyuy has been dramatically reduced, and there also had been a change in fish diversity (Egorov, 1990).

Furthermore, the creation of the water reservoir and the diamond mining developments affected the life style of the indigenous people. An indigenous community of Evenks, about 650 people, lived on Upper Vilyuy. They were forced to leave their long lived-in lands as a result of the diamond mining developments and construction of the water reservoir (Egorov, 1990). Some indigenous people settlements from Syulyukarskyy, Chuoninskyy and Botuobiynskyy Councils were forced to leave their land, which was then submerged (Shipkov, 1999). Reindeer farming has ceased to exist, hunting has decreased, and agriculture was significantly affected by environmental degradation, which has resulted in over 200 Evenks almost loosing their traditional skills, language and culture (Egorov, 1990).

Nuclear Explosions in Yakutia

During the Soviet period, most information about nuclear explosion tests were classified as top secret and no evidence were revealed to the public. However, as many as 715 nuclear tests were carried out in the former Soviet Union during 1949-1990 including nuclear weapon tests and peaceful nuclear explosions. Nuclear explosions were carried out on the surface, in the air, under water and underground. Major nuclear weapon tests were carried out on established sites in Semipalatinsk in Kazakhstan and at the Arctic Archipelago Novaya Zemlya in the Barents Sea. During 1965-1988, under the program of underground nuclear tests for peaceful activities in the Soviet Union, a total of 128 peaceful nuclear explosions and 32 tests were carried out (Egorov et al, 2000).

Yakutia was considered as one of the purest regions in Russia from a radio-ecological point of view, although recent studies have shown that this is now untrue. At the end of the 1980s, information about nuclear explosion tests was open to the public. This important data about the radioactive situation in the Republic of Sakha was collected and presented at the Republican Scientific and Practical Conference in Yakutsk in 1993 (Chomchoev et al, 1993). According to the report by Chomchoev et al (1993), during 1967-1988, a total of 12 underground peaceful nuclear explosions were carried out in Yakutia (see Table 4.18).

Two underground explosions back-fired resulting in substantial nuclear fallout that compares with levels of contamination in areas surrounding Chernobyl. In addition, there are deposits of radioactive metals extracted in the 1950s and 1960s, which have additionally contaminated the environment with uranium and thorium elements and other related decay products. Yakutia was also influenced by radioactive fallout from the Novaya Zemlya nuclear shooting-range. Seven out of the 12 underground explosions were made in the territory of the diamond province and two of them had fallout of radioactive elements on the surface. Two underground nuclear explosions 'Kraton-3' and 'Kristall' backfired, creating fallout of radioactive elements (Burtsev and Kolodeznikova, 1997; Egorov et al, 2000). The failed underground nuclear explosion 'Kristall' was directly connected to the diamond mining development. It was carried out for the creation of a river dam for tailings from an enrichment plant on stream Ulakhan-Basyttakh, left tributary of the river Daldyn. The underground nuclear explosion 'Kristall' was carried out on the 2 October 1974 near the diamond mining pipe 'Udachnaya', 2.5 kilometers to the north of the settlement Udachny-2. The charge had a power of 1.7 kilotons and was located at 98 meters under the surface. The customer for the nuclear explosion was the Ministry of Non-ferrous Metallurgy of the USSR and the executor was the Ministry for Average and Heavy Machinery of the USSR. A preliminary project included eight underground nuclear explosions, but owing to the failure of the 'Kristall' explosion, the execution of further works was rejected (Burtsev and Kolodeznikova, 1997).

Table 4.18 Nuclear explosions in the Republic of Sakha in chronological order

Name of explosion	Distance from the settlement	Date	Power, Kiloton	Depth, Meters	Purpose
Kristall	town Udachny (Mirny district), 2.5 km	02.10.74	1.7	98	building of the dam on the river
Gorizont-4	settlement Kusur (Bulun district), 20 km	12.08.75	7.8	483	seismic exploration
Oka (slit No.42)	settlement Taas Yuryakh (Mirny district), 38 km	05.11.76	11(+/-)1	1520	search for gas and oil
Kraton-4	lake Nijili (Kobyai district), 12 km	09.08.78	22(+/-)3	567	seismic exploration
Kraton-3	settlement Aikhal (Mirny district), 38 km	24.08.78	19(+/-)1	577	seismic exploration
Vyatka (slit 43)	settlement Taas Yuryakh (Mirny district), 26 km	07.10.78	16(+/-)2	1550	search for gas and oil
Kimberlite-4	settlement Toybuya (Verkhnevilyuysk district), 27km	12.08.79	6	1986	seismic exploration
Sheksna (slit No.47)	settlement Taas Yuryakh (Mirny district), 7.2 km	10.10.82	15(+/-)3	1540	search for gas and oil
Neva-1 (slit No.65)	settlement Taas Yuryakh (Mirny district), 31.5 km	10.10.82	16	1510	search for gas and oil
Neva-2 (slit No.68)	settlement Taas Yuryakh (Mirny district), 40.5 km	06.07.87	13	1526	search for gas and oil
Neva-3 (slit No.61)	settlement Taas Yuryakh (Mirny district), 42.3 km	24.07.87	13	1515	seismic exploration
Slit No.101	settlement Taas Yuryakh (Mirny district), 41.4 km	12.08.87	3	815	construction of oil reservoir

Source: Adapted from Chomchoev et al 'Geologicheskaya Obstanobka i Vosmozhnie Ekologicheskie Posledstviya Podzemnykh Yadernykh Vzryvov: na Primere 'Kraton-4' [Geological Situation and Possible Environmental Consequences of Underground Nuclear Explosions: a Case of 'Kraton-4'], (1993).

A study of environmental consequences of the explosion was conducted by a research expedition in 1990. On the site of the explosion the expedition discovered a crater 60 meters in diameter and six meters in depth, filled with water (Burtsev and Kolodeznikova, 1997). The expedition undertook several radiation measurements of water and soil samples from the site, in neighboring areas and downstream at Ulakhan Basyttakh. Results of the radiation studies showed that the level of beta radiation at the immediate site of the underground nuclear explosion was five to six times above the natural background level. The level of gamma and beta radiation in the settlements of Polyarny, Novy Gorod, Nadezhny, Udachny-2 and in the quarry 'Udachny' did not exceed the natural background level (Burtsev and Kolodeznikova, 1997). The underground nuclear explosion 'Kristall' destroyed the forest over an area of five hectares near the site (Shipkov, 1999).

Rehabilitation works on the site of the underground nuclear explosion 'Kristall' were conducted during the winter of 1992, using barren rock from the quarry 'Udachny'. The sarcophagus in the form of a capped core was built on the site, seven meters high and 150 meters in radius. A neighborhood area of five hectares was also covered with barren rock (Burtsev and Kolodeznikova, 1997). On 22-31 January 1996 in Moscow, the State commission of the Republic of Sakha (Yakutia) examined the data on the underground nuclear explosions 'Kristall' and 'Kraton-3'. The commission attracted a range of specialists from the Ministry for Atomic Energy of the Russian Federation, the Ministry for Emergency Situation of the Russian Federation, the Russian Academy of Sciences and the Academy of Sciences of the Republic of Sakha. The commission officially recognized these two explosions as disasters, and that the environment was contaminated by the radionuclides which are dangerous for people's health and lives (Shipkov, 1999).

Health Problems in the Vilyuy Region

The Vilyuy region occupies the river Vilyuy valley and the territory to the north of the river, uniting five districts of Yakutia: Mirny, Nyurba, Suntar, Verkhnevilyuysk and Vilyuysk with a total population of 192,000 people, which accounts for 18 per cent of the total population in the Republic of Sakha. Indigenous people account for 64 per cent of the region's population and represent 35 per cent of the total number of indigenous people in the Republic of Sakha. The lifestyle of indigenous people in the Vilyuy region is based on cattle breeding, fur farming and agriculture. Migrants constituting 36 per cent of the population of the Vilyuy region mostly live in the industrial settlements related to the diamond mining industry (Yakovleva et al, 2000). Because of the low levels of precipitation, less than 200 millimeters per annum, and the pervasive permafrost, the Republic of Sakha is classified as an area with a relatively low content of actual drinking water, hence the reason why settlements are set close to waterways. The river Vilyuy serves as the drinking water supply, as well as being used for fishing, hunting and transport (Petrova and Volozhin, 1996).

As part of an ongoing study into the health status of the population in the Republic of Sakha, a method of medical geographical division, considering bio-climatic and geographical factors, was introduced, dividing the territory of Yakutia into five medical-geographical regions: Central, Polar, Southern, Transpolar and Vilyuy (Petrova and Volozhin, 1996). Statistical analysis of health indices within these five districts has attracted the attention of medics who are concerned with the extraordinary health conditions in the Vilyuy region and the findings of their research has clearly sharpened the focus towards trying to forecast the health status of the population (Savvinov et al, 1992).

Studies undertaken by Krivoshapkin et al (1990), and Petrova and Volozhin (1996) made a significant contribution to the medical research in the Vilyuy region. Studies have shown higher morbidity rates in diseases of the cardiovascular system, gastrointestinal system, respiratory organs, urinary system, locomotion and nervous systems and malignant tumors in comparison to the morbidity rates in the Republic of Sakha as a whole. This feature has been attributed to environmental degradation in the region. High rates of viral hepatitis, gastrointestinal infections and dysentery in the Vilyuy region were linked to the unsatisfactory quality of drinking water, which included high concentrations of aluminum, manganese, nickel, chromium and titanium (Petrova and Volozhin, 1996). Excessive concentration of these elements in a human organism causes pathologies of the locomotion and nervous system, gastrointestinal system, malignant tumors and anomalies amongst children. High concentrations of cadmium, copper and nickel were found in the blood of citizens in settlements in the Vilyuy region as well as high concentrations of nickel in hairs (Petrova and Volozhin, 1996).

A relationship between morbidity level and the distance from the diamond mining operations and the Vilyuy water reservoir was also established. Morbidity rates are significantly higher close to these industrial sites (Petrova and Volozhin, 1996). The medical-ecological research strongly indicates that there is a link between the industrial pollution from diamond mining operation and the state of health of the indigenous population in the Vilyuy region (Krivoshapkin et al, 1990; Petrova and Volozhin, 1996). Initial results of medical and ecological studies indicate that health degradation may be more closely correlated with chemical pollution than with the radiation fall-out. Nevertheless, research must be expanded and developed in order to identify more clearly the precise nature of these linkages (Petrova and Volozhin, 1996). Furthermore, concern is growing that the pollution from these industrial developments is so significant that there may be negative consequences for the gene bank of the Vilyuy region itself (Petrova and Volozhin, 1996).

Summary

- Territorial specifics for the development of corporate social responsibility depend upon environmental regulations, local environment, level of environmental protection, affected communities, property rights and economic development on a territory where mining operations are conducted. Due to the existence of several levels of authorities in some countries, such as the Russian Federation, which impact on enforcement of environmental and economic regulation, there are three contextual levels (national, regional and local) within which the corporate social responsibility in the mining industry develops.

- Gold and diamond mining industry in the Russian Federation plays an important role in economic development of the Russian Federation. The development of gold and diamond mining in the Russian Federation reflects the changes that are taking place during the period of transition from planned economy to market economy, including privatization of the industry, liberalization of foreign investment regimes, decentralization of economic policy and decline in industrial production. Environmental degradation in the Russian Federation is caused by ruthless exploitation of mineral resources during the period of Soviet Era industrial expansion into the North and decline of environmental protection during the period of transition of economic, legal and political system.
- Regionally, the gold and diamond mining industries in the Republic of Sakha significantly contribute towards economic development of the territory. The changes in center-periphery relations in the 1990s led to retaining of substantial benefits from exploitation of mineral resources by the Republic of Sakha. Gold and diamond mining considerably add to regional state budget, employment and exports revenues. Gold and diamond mining accounts for the largest share of environmental pollution, including land degradation, water and air emissions in the Republic of Sakha.
- The launch of gold and diamond mining projects in the Republic of Sakha during the era of Soviet industrial development has changed the livelihoods of previously rural communities through creation of industrial and transport infrastructure and town-building, and causing large migration to prior scarcely populated areas. Mining of gold and diamond has negatively affected local environmental systems in Aldan and Mirny mining districts, contributing to relocation of indigenous communities, health problems associated with water pollution and degradation of natural environments. Recently privatized mining companies are now facing challenges of legacy of Soviet industrial developments such as economic dependency on mining, environmental degradation, social and health problems associated with mining operations.

Notes

[1] In 1991, following the disintegration of the former Soviet Union, the Russian Federation was formed retaining its boundaries within the Russian Soviet Federated Socialist Republic. The Russian Federation, which later is referred to as Russia, with its present boundaries covers an area of over 17 million square kilometers, which account for 11 per cent of the Earth's total dry land. According to the census in 1994, the population of the Russian Federation is about 148 million people. The Russian Federation comprises 89 entities: 21 Republics, one autonomous *oblast*, ten autonomous *okrugs*, six *krays*, 49 *oblasts* and two federal cities: Moscow and St. Petersburg. The Russian Federation possesses 12 per cent of world's oil reserves, 32 per cent of gas, 26 per cent of iron, 15 per cent of zinc, 11 per cent of coal and ten per cent of lead. The potential value of confirmed mineral reserves in Russia exceeds 28,000 billion US dollars (Interfax-M&C, 1998). Out of twenty thousand mineral resource deposits documented in the Russian Federation 37 per cent are under industrial exploitation (Goskomekologia, 2000).

[2] These gold deposits are scattered from the western part of Russia near the Baltic region to the eastern part in Chukotka. The main gold reserves are located in Eastern Siberia, the Far East of Russia and the Urals (Natarov, 2000). Due to nature of the Russian political system, there are predictable difficulties in accessing data on gold production. Only since 1993 the Russian State Committee on Statistics has started to report regularly on the Russian gold production in official monthly statistical reports on economic performance (Sagers, 1997). There are still some constraints in accessing any information on gold mining; for example, the information on reserves of gold deposits is still under a state secret (Rubchenko and Shokhina, 2001).

[3] Gold is usually found in placer and ore deposits. Placer or alluvial gold consists of the weathering of auriferous rocks, which have been washed into riverbeds or similar situations (Puddephatt, 1978). Ore or vein deposits are found in the rock.

[4] In 1996, the Government of the Russian Federation adopted a Federal program 'Production of gold and silver in Russia until 2000', which was developed for launching new gold mining companies and the reconstruction of existing ones. The program aimed to reduce costs via the introduction of new technologies, and the improvement of the socio-economic and environmental situation in the gold mining regions (Kolmogorov, 2000).

[5] Lomonosov deposit consists of six diamond pipes: Arkhangelskaya, In honor of Karpinskiy-1, In honor of Karpinksiy-2, In honor of Lomonosov, Pionerskaya and Pomorskaya. In 1992, a technical-economic assessment proved the deposit to be economically viable for development and development plans passed the State EIA (ecological expertise). SeverAlmaz Joint-Stock Company was established which received a license for deposit development (Levine, 1996; Levine, 1999). A mining complex is currently under construction at the Lomonosov diamond deposit, the finished mining complex should operate at full project capacity, equivalent to 14 per cent of the current Russian output (Khan, 1999).

[6] Obshetvenny Registr Certificazii System Ecologicheskogo Menedgmena [Public Register for Certification of Environmental Management Systems], URL: http://www.14000.ur/register (accessed on 5/07/02).

[7] Decree No. 181 of 20 March 1998 'About environmental auditing in the system of State Environmental Committee of Russia' and Decree No. 436 of 16 July 1998 'About conducting of works for introduction of environmental auditing in the Russian Federation'.

[8] The Republic of Sakha (Yakutia) is an area where Sakha people (designated by Russians as Yakuts) historically lived. Yakutia is a historic name for the current territory of the Republic, called after Yakuts. The Republic of Sakha (Yakutia) lies between 74°00'- 55°30' North latitude and 105°00'-165°00' East longitude, stretching 2,000 km north to south and 2,500 km east to west (Petrova and Volozhin, 1996). The Republic covers an area of more than three million square kilometers, which is about 18 per cent of the Russian territory. It is divided to 33 administrative districts (*ulus* – administrative district), where the main cities are Yakutsk, Mirny and Neryungri. Yakutsk, the capital of the Republic of Sakha, is 8,468 km east of Moscow. Over 40 per cent of the Republic of Sakha lies within the Arctic Circle and all of the territory lies in the zone of pervasive permafrost, stretching to three geographical zones: middle taiga, northern taiga and tundra (Poiseev, 1999). The population of the Republic of Sakha is 949,280 according to the census data of 2002. Ethnic Sakha people (Yakuts) comprise 45.5 per cent of the Republic's population, whereas Russians comprise approximately 41 per cent of the total population. The remaining portion comprises Ukrainians, Tatars, Belorussians and the indigenous people of the North (Chukchis, Evens, Evenks and Yukagirs). Two-thirds of the total Republic's population lives in the cities. The climate of the Republic of Sakha is sharp continental: seasonal temperature variations range from +38° C in the summer to -62° C in the winter. The

Republic of Sakha is the coldest part of the Russian Federation, and contains the 'Pole of Frost' for the Northern Hemisphere (Verkhoyansk – Oymyakon), where the minimum temperature has been registered at -72° C (Mostakhov et al, 1980).

[9] In 1923, an expedition supported by the Yakut Government under the leadership of V. Bertin, explored the rivers Tommot and Orto-Sala. The expedition attracted local hunter M. Tarabukin and succeeded in discovering the placer deposit Nezametny, which gave birth to the industrial development of gold mines in Aldan, described by Sagers (1997) as the only true 'gold rush' in the former Soviet Union. Thorough exploration for gold deposits in Aldan started in 1925 and as a result more gold deposits were found: Lebedinoe (1925), Orochinskoe (1928), Leninskoe (1929), Tyrkadinskoe (1930) and Savvinskaya (1945) (Desyatkin, 1998).

[10] At the beginning, gold was simply washed by hand and mechanical mining started with the launch of a steam driven drag in 1926. Shaft mining was launched in 1928 at Orochinskoe, and new electric drags started to operate in 1931 along with the hydraulic methods. Ore mining started in 1933 with the development of 1st Lebedinskaya, 2nd Savvinskaya and Krutaya veins (Aldanzoloto, 1994).

[11] Assumptions about the possibility of diamond discoveries in Northern Russia were made as early as in the 19th century (Vecherin, 1997). However, exploration for diamonds was only considered seriously following the Second World War. In 1946, the Tungusska Geological Diamond Survey Expedition, which included four geological parties, was established for the exploration of diamonds on the Siberian Platform, the territory between the rivers Yenisey and Lena and started exploring diamonds in 1947 (Sobolev, 1981; Kiseleva, 1999). In 1950, within the Ministry of geology of the Soviet Union, the Amakinskaya expedition was formed which in 1950-1951 discovered several rich alluvial diamond deposits in the basin of the river Vilyuy (Vecherin, 1997). Only in August 1954 was the first kimberlite pipe called Zarnitsa discovered near the river Daldyn, a tributary of the river Vilyuy (Vecherin, 1997). Following these initial discoveries, additional commercial diamond deposits were found in Western Yakutia, including the famous diamond pipes called: Uduchnaya (1955), Sytykanskaya (1955), Aykhal (1960), Internatsionalnaya (1969), and Jubileinaya (1975) (Vecherin, 1997). Up to present, c. 600 kimberlite deposits are known in Western Yakutia covering an area of over 600,000 square kilometers. Among the 600 deposits, 150 are believed to be diamond-bearing although only a few are feasible for economic exploitation (Interfax-M&CN, 1998). Currently, 39 diamond deposits are documented in the State register of mineral resources of the Republic of Sakha, while 18 of them are licensed for economic development (Pravitelstvo RS(Y), 1999).

PART II

CORPORATE SOCIAL RESPONSIBILITY STRATEGIES IN MINING

Chapter 5

CSR Strategies in Growing and Declining Mining Sectors

Introduction

Public concerns about environmental, social and economic impacts of companies within the global metals and minerals mining industry is constantly increasing. Mining companies are beginning to respond to these accelerating social and environmental concerns via incorporating the notion of corporate social responsibility in their business practices (see Chapter 2). The analysis of the global top gold and diamond mining companies shows that these companies started to develop environmental and social strategies involving concepts of sustainable development and corporate social responsibility, conduct environmental operations in accordance with international management standards and respond to concerns of local communities via a range of targeted initiatives (see Chapter 3).

The Russian Federation plays a significant role in the provision of metals and minerals to the world market including gold and diamonds (see Chapter 4). However, it is widely recognized that transition economies such as the Russian Federation and developing countries generally delay incorporation of the corporate social responsibility concept into their business strategies; often reflecting their prevailing political and economic conditions (see Chapter 2). Many metals and minerals developments in developing countries are under the control of multinational companies whose origins are in developed countries. These multinational companies face various pressures from local communities, local government, NGOs and international organizations with regard to their environmental and social impacts. Although multinational mining companies have expanded their operations into countries with transition economies, the mining industry in the Russian Federation mostly operates under the control of domestic companies, which is particularly true for the precious metals and minerals sector.

Chapter 5 now aims to critically assess case study companies in the Republic of Sakha in the context of wider environmental and social responsibility. This Chapter consists of three main parts. Section one explores the case study of Aldanzoloto, although the largest gold mining company in the region, a declining mining sector. Section two presents a case study of ALROSA, the largest diamond mining company in the Republic of Sakha and a growing mining sector. Each case study explores analyzes companies' structure and economic performance and the nature of corporate strategies concerning environmental protection, health and safety and employee relations. The final section of this Chapter presents a comparison of environmental and social strategies between Aldanzoloto and ALROSA with a reference to strategies of global mining companies.

Development of Environmental and Social Strategies in Aldanzoloto

Structure and Economic Performance of Aldanzoloto

Aldanzoloto Company, later referred to as Aldanzoloto, is the largest gold mining company in the Republic of Sakha and is situated in Nizhny Kuranakh of Aldan district, 450 km south from the capital of the Republic of Sakha, the city of Yakutsk (see Figure 5.1).

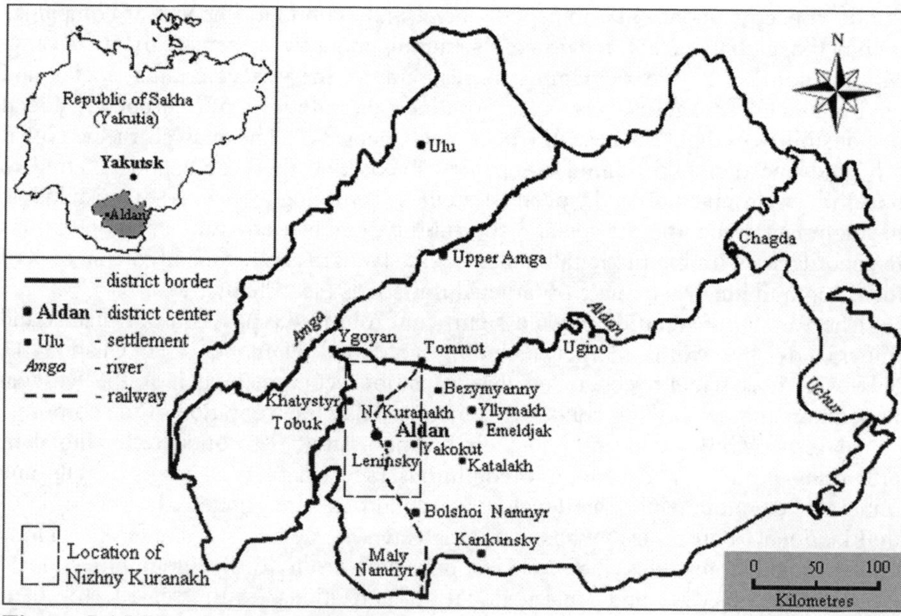

Figure 5.1 Map of Aldan district

Aldanzoloto is the oldest gold mining enterprise in the Republic of Sakha, which traces its origin in 1924. During 1924-1991, Aldanzoloto was a part of Yakutzoloto Corporation, a large state-owned gold mining enterprise which controlled the majority of gold mining production in the Republic of Sakha. In 1991, Aldanzoloto left Yakutzoloto and was privatized; the main shareholders of Aldanzoloto are the State Committee for Precious Metals and Minerals of the Republic of Sakha (Yakutia), Erel Nongovernmental Pension Fund, local authority administration (the Administration of the Aldan district), employees of Aldanzoloto and Neftel Investment Fund. Aldanzoloto is primarily involved in gold and silver mining, geological exploration and jewelry production, but also holds an interest in construction and timber business, transport and services and other gold mining companies such as Kuranakhskaya Gold Mining Company and Gold Placers of Aldan. Aldanzoloto is a large size company employing about 2,300 people (see Figure 5.2), which conducts its mining operations on several gold deposits (Novoe, Yakokutskoe, Kanavnoe and Delbe) and Kuranakh ore field, all situated within the limits of the Aldan district.

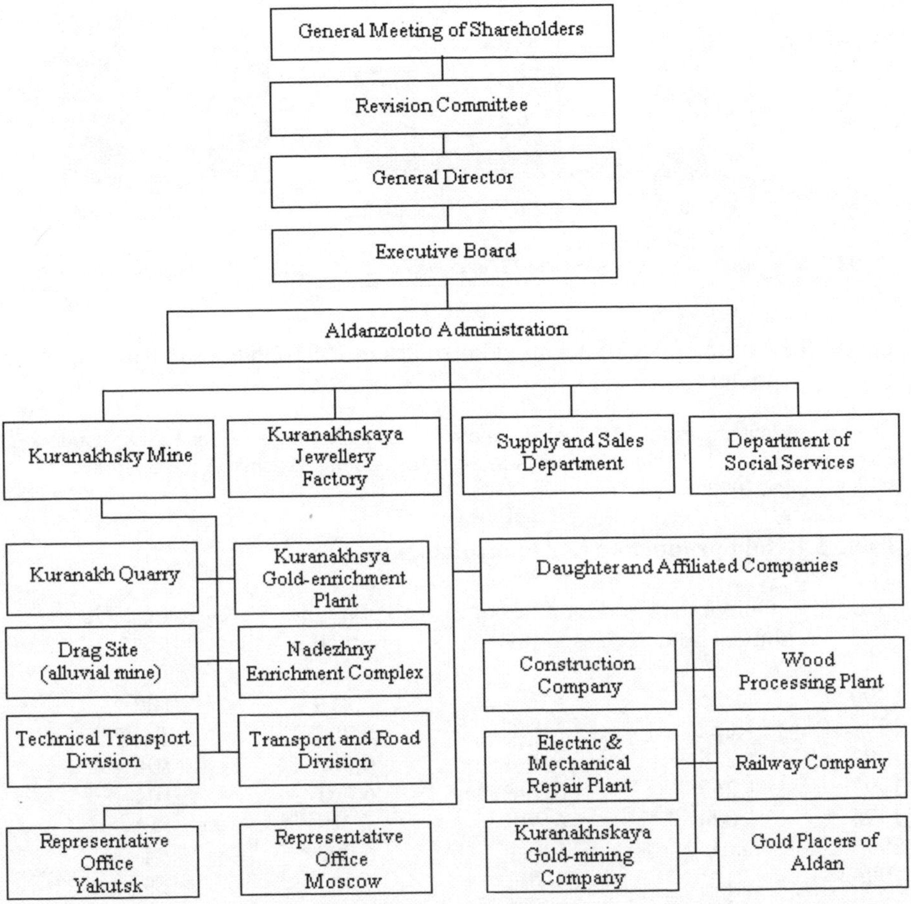

Figure 5.2 Structure of Aldanzoloto

During the 1990s, Aldanzoloto has substantially decreased its sand-washing operations and increased ore extraction (see Figure 5.3). In 1990, Aldanzoloto operated twelve drags, whilst in 1996, the number of drags has been reduced to six and finally in 2000, there was only one operating drag. A dramatic reduction in the number of drags relates to the low profitability of mining operations on placer deposits. During the 1990s, Aldanzoloto experienced an overall drop in gold output; in 2000, gold production decreased by 40 per cent in comparison to 1990 (see Table 5.2), leading to substantial drop in production sales (see Figure 5.4).

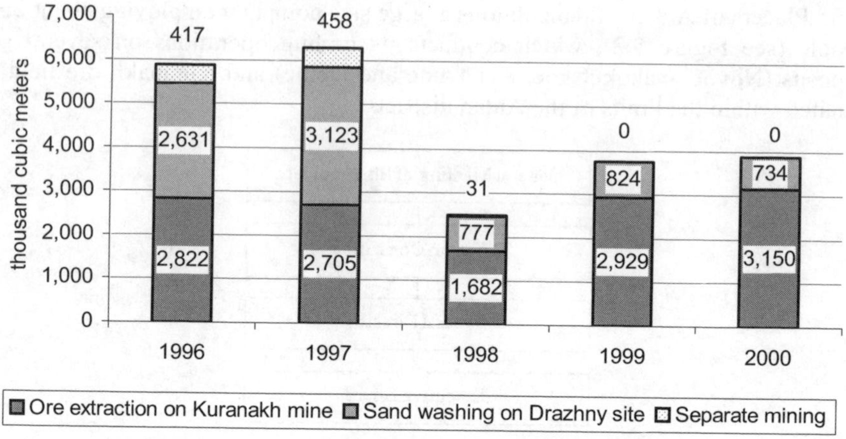

Figure 5.3 Mining operations in Aldanzoloto in 1996-2000, thousand cubic meters

Source: Adapted from AK Aldanzoloto, 'Poyasnitel'naya Zapiska k Godovomu Bukhgalterskomu Balansu' [Explanation Note to Annual Balance Sheet] (1996a, 1997a, 1998a, 1999a, 2000a).

Table 5.1 Gold production in Aldanzoloto, kg

Year	Output of alluvial gold	Output of ore gold	Total gold production	Growth, 1990-100%
1990	1,205	5,214	6,419	100
1991	1,062	5,140	6,202	96
1992	1,274	5,388	6,662	104
1993	1,263	5,398	6,661	104
1994	1,024	6,248	7,272	113
1995	861	5,141	6,002	94
1996	712	3,709	4,421	69
1997	618	3,265	3,883	60
1998	99	1,422	1,521	24
1999	84	2,740	2,824	44
2000	50	3,735	3,785	59

Source: Adapted from AK Aldanzoloto, 'Poyasnitel'naya Zapiska k Godovomu Bukhgalterskomu Balansu' [Explanation Note to Annual Balance Sheet] (1996b, 1997a, 1998a, 1999a and 2000a).

Net revenues decreased from 82 million US dollars in 1997 to 32 million US dollars in 2000, leading the company into a deep financial crisis (AK Aldanzoloto, 1996a, 1997a, 1998a, 1999a, 2000a). Although reserves of the Kuranakh ore field are very large, Aldanzoloto is unable to sustain profitable extraction and processing of gold due to outdated production facilities and lack of new investments. In 1999, Aldanzoloto contributed to 20 per cent of total gold production in the Republic of Sakha (Goskomstat RS(Y), 1999, 2000), it was the single largest gold mining company, contributing to 1.3 per cent of total industrial production and 0.6 per cent to Gross Regional Product of the Republic of Sakha.

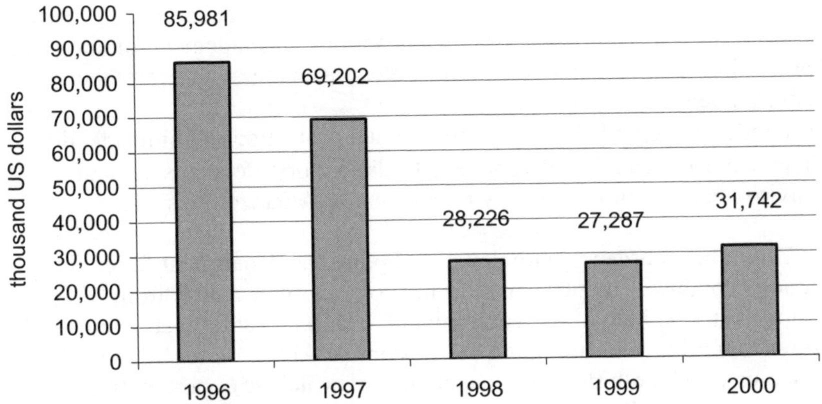

Figure 5.4 Production sales in Aldanzoloto in 1996-2000, thousand US dollars

Source: Adapted from AK Aldanzoloto, 'Poyasnitel'naya Zapiska k Godovomu Bukhgalterskomu Balansu' [Explanation Note to Annual Balance Sheet] (1996a, 1997a, 1998a, 1999a, 2000a).

Environmental Protection in Aldanzoloto

Aldanzoloto has established a department for environmental protection at its major production division, Kuranakhsky mine, which coordinates overall conduct of operations relating to environmental protection, natural resource use and reclamation in Aldanzoloto. The department incorporates an environment, health and safety laboratory. Each division of Aldanzoloto has an assigned environmental protection officer, who reports to the department for environmental protection at Kuranakhsly mine.

Although Aldanzoloto has not developed a formal corporate environmental policy or long-term environmental action program, the company establishes its environmental protection operations on a series of plans that are developed annually by different departments of the company and approved by a local environmental protection agency, Upper Aldan Territorial Committee for Nature Protection. These plans are implemented at each division of Aldanzoloto and co-coordinated by a respective department which developed them:

- The major document on environmental protection in Aldanzoloto is an annual action plan which is developed by the department for environmental protection at Kuranakhsky mine to comply with environmental protection legislation in the area of Aldanzoloto operations. This document outlines company's plans in the areas of compliance, environmental payments, reporting to state agencies for environmental protection, environmental monitoring, tailings maintenance, health and safety, water protection, state statistical reporting, waste, land rehabilitation and environmental education. In addition to the annual action plan a series of detailed plans are produced particularly concerning environmental protection actions, management of air pollution, protection and efficient use of water resources, and protection of the environment from oil spills.
- Engineers develop plans for environmental protection of tailings facilities.
- The environment, health and safety laboratory develops annual plans for environment, health and safety monitoring of Aldanzoloto sites.

Air pollution In accordance with developed plans the department for environmental protection conducts regular monitoring of air emissions from Aldanzoloto operations with the help of the environment, health and safety laboratory and takes measures to minimize air pollution at divisions of Aldanzoloto. In 2000, there were 279 sources of air pollution in Aldanzoloto, of which 156 were monitored, where major sources of pollution were boiler houses and transport divisions. Data for 1996-2000 shows that a drop in industrial production of Aldanzoloto has led to a corresponding decrease in air emissions (see Table 5.2).

Table 5.2 Air emissions from Aldanzoloto operations in 1996-2000, tons

	1996	1997	1998	1999	2000
Sulfur anhydride	482	373	256	302	215
Nitrogen dioxide (NO2)	548	382	440	515	422
Carbon dioxide (CO2)	5,684	4,889	1,871	2,164	1,584
Coal ashes	2,144	1,022	933	873	604
Wood dust	457	67	68	133	0.1
Non-organic dust	338	279	870	828	797
Cement dust	0	0	0	0.1	0.2
Fluoric hydrogen	0.05	0.02	0.01	0.01	0.01
Manganese compound	0.09	0.03	0.02	0.01	0.02
Chromium compound	0.06	0.02	0.05	0.05	0.004
Total air emissions	9,771	7,030	4,543	4,931	3,683
Permitted limit for air emissions	11,525	no permit	no data	5,254	5,497

Source: Adapted from AK Aldanzoloto, 'Vozdeistvie Deyatel'nosti AK Aldanzoloto na Prirodnuyu Sredu' [Impact of Aldanzoloto Operations on the Natural Environment] (1996b, 1997b, 1998b, 1999b, 2000b).

Water management Management of water resources in Aldanzoloto is conducted in accordance with an annual permit for special water use and water collection for consumer and industrial purposes from natural water sources issued by the Ministry for Nature Protection of the Republic of Sakha. In the 1990s, Aldanzoloto has decreased water collection four times due to the drop in industrial production (see Table 5.3).

Table 5.3 Water collection and discharge in Aldanzoloto in 1996-2000, thousand cubic meters

	1996	1997	1998	1999	2000
Permitted limit for water collection	12,846	12,147	9,210	3,341	3,904
Actual water collection, including for:	12,237	9,791	4,211	2,437	2,862
- industrial purposes	8,208	8,626	2,843	1,438	1,849
Index of water collection per ore extracted, cubic meter/cubic meter	1.4	1.4	1.1	0.4	0.5
Discharge to watercourses, including:	8,420	8,575	3,285	2,021	227
- not normatively cleaned	4,124	6,377	3,181	2,021	227

Source: Adapted from AK Aldanzoloto, 'Vozdeistvie Deyatel'nosti AK Aldanzoloto na Prirodniju Sredu' [Impact of Aldanzoloto Operations on the Natural Environment] (1996b, 1997b, 1998b, 1999b, 2000b).

Consequently, discharge of industrial wastewater to watercourses has been reduced to a quarter of its original amount in the period from 1996 to 1999, but the total volume of contaminant content in the discharged wastewater has been reduced only by half of its original amount during the same period (see Table 5.4). The environment, health and safety laboratory monitors water discharge and pollution on placer deposits, where drag lines operate, and near enrichment complex according to prescribed schedules. Analyses of water samples are carried out for content of different chemicals including iron, copper, zinc and oil products. The main contaminants in the wastewater from the gold mining facilities are cyanide and petroleum products. Major water protection facilities in Aldanzoloto are tailings of Kuranakhskaya gold enrichment plant which are monitored all year round. The capital construction department of the company conducts monitoring of tailings for technical purposes. Aldanzoloto carries out environmental risk assessment of tailings. Plans for emergency situations are developed for each site of Aldanzoloto, including tailings and boiler-houses.

Table 5.4 Discharge of contaminants to watercourses by Aldanzoloto in 1996-1999

Contaminant	Unit	1996	1997	1998	1999
Suspended substances	Tons	41.4	52.5	4.9	2.9
Sulfate	Tons	79.9	58.2	47.1	46.0
Oil products	Tons	0.1	0.1	0.01	0.002
Iron	Kg	653.9	715.3	313.1	233.1
Copper	Kg	4.6	8.7	0.7	0.01
Zinc	Kg	15.3	34.2	5.8	0.0
Mercury	Kg	0.0	0.0	0.0	0.0
Cyanide	Kg	9.1	5.3	10.2	5.7
Total	Tons	122.1	111.5	52.3	49.1
Index of total contaminants per gold extracted	Kg/Kg	27.6	28.7	34.4	17.4

Source: Adapted from AK Aldanzoloto, 'Vozdeistvie Deyatel'nosti AK Aldanzoloto na Prirodniju Sredu' [Impact of Aldanzoloto Operations on Natural Environment] (1996b, 1997b, 1998b, 1999b).

Energy resources Although use of energy resources is monitored in Aldanzoloto, there is poor integration of this activity with the department for environmental protection. The Chief power-engineering specialist is in charge of energy management and monitors and produces regular reports on energy consumption (see Table 5.5).

Table 5.5 Energy use in Aldanzoloto in 1995-2000

	1995	1996	1997	1998	1999	2000
Electric energy use, million Kilowatts	238	222	139	103	118	124
Index of electric energy use per gold extracted, thousand Kilowatts/kg	39.7	50.2	35.8	67.7	41.8	32.7
Coal consumption, thousand tons	202	133	104	67	63	55
Change in coal consumption (1995-100%)	100	65	51	33	31	27

Source: Adapted from AK Aldanzoloto, 'Vozdeistvie Deyatel'nosti AK Aldanzoloto na Prirodnuyu Sredu' [Impact of Aldanzoloto Operations on the Natural Environment] (1996b, 1997b, 1998b, 1999b, 2000b).

During 1995-2000, consumption of coal, used for generating heat in boiler-houses, dropped by 70 per cent and electric energy use reduced by 50 per cent, which is also attributed to the fall in industrial production. Aldanzoloto does not show signs of improvement in efficient electric energy use as indices for electric energy use per unit of production show.

Waste Waste generation and disposal is monitored at each division, workshop and production site of Aldanzoloto. Waste generation in Aldanzoloto has almost halved in the period from 1996 to 2000 (see Table 5.6). At the end of the year each division supplies information on waste generated to the department for environmental protection at Kuranakhsky mine, which later submits a statistical report to the local environmental protection agency. Aldanzoloto has two authorized sites for waste storage, including a site for scrap metal, approved by the local environmental protection agency. Some secondary use of resources takes place, for example, some petroleum products and wood are recycled on site. Classes 3 and 4 of wastes are transported to authorized storage sites. Scrap metal is transported to special areas for temporary storage and later transferred to other organizations via railroad.

Table 5.6 Utilization of consumer and toxic waste in Aldanzoloto in 1996-2000, tons

	1996	1997	1998	1999	2000
Total waste, including:	54,337	19746	19,871	21,493	21,947
Class 1 waste[1]	1	16	19	11	10
Class 2 waste	355	180	84	136	181
Class 3 waste	17	8	3	3	3
Class 4 waste	52,165	18,476	18,782	20,667	20,886
Class 5 waste	1,799	1,067	894	676	867
Second use on site	592	1278	43	42	17
Neutralized on site	360	2	13	8	2
Handed over to other organizations	348	-	39	148	1,515
Transported to organized dumping and storage grounds	34,602	18,100	37,506	58,272	77,477
Transported to authorized dumping grounds of solid and consumer waste	18,547	235	281	529	651

Source: Adapted from AK Aldanzoloto, 'Vozdeistvie Deyatel'nosti AK Aldanzoloto na Prirodnuyu Sredu' [Impact of Aldanzoloto Operations on the Natural Environment] (1996b, 1997b, 1998b, 1999b, 2000b).

Reclamation Land management and reclamation is under the supervision of the department for environmental protection and mine-surveyors. In 2000, about 2,020 hectares of disturbed land referred to placer deposits and 2,752 hectares to ore field. Each year a small fraction of the disturbed land is restored, mostly on placer deposits, where drags stopped operating (in 1996 – 12 hectares, in 1997 – 168 hectares, in 1998 – 50 hectares, in 1999 – 57 hectares and in 2000 – 63 hectares of disturbed land were restored) (AK Aldanzoloto, 1996b, 1997b, 1998b, 1999b, 2000b).

Plans for reclamation of disturbed land are developed by mine-surveyors which are later approved with the local environmental protection agency. Several placer and ore deposits have been mothballed and therefore their reclamation is delayed. These mothballed deposits are awaiting new technologies for ore extraction as at present it is economically not viable to extract gold from these deposits due to their low grade. However, these deposits are not entirely closed and reclamation will take place only after the full exploitation of their resources which is indefinitely delayed due to the lack of financial investments to upgrade technological facilities. Land reclamation is hampered by lack of machinery and resources, as environmental protection employee of Aldanzoloto stated:

> On drag field, where drags had finished their operations, the reclamation of spoil heaps is conducted. But at the moment we have a lack of machinery. We even lack machinery for the production process.

According to Upper Aldan Territorial Committee for Nature Protection, the delay in reclamation of disturbed land in Aldanzoloto could result in long-term impacts on local watercourses and habitat. In 1992-1993, Aldanzoloto commissioned the Institute of Biology of the Republic of Sakha to carry out an experimental re-vegetation of five hectares of disturbed land on two experimental sites. The research showed that by applying a layer of soil, rehabilitation took place much quicker; however, the common practice within the company for land reclamation is leveling of the ground and leaving it for self re-vegetation. The research of the Institute of Applied Ecology of the North demonstrated that spoil heaps that were treated similarly some 70 years ago have just started to have a thin layer of soil and vegetation, thus indicating an extremely slow pace for natural rehabilitation of the natural environment.

Environmental reporting The review of environmental performance in Aldanzoloto is carried out during the development of monthly, quarterly and annual internal reports on environmental protection operations that are presented to the General Director of Aldanzoloto. Once a year, Aldanzoloto reports on environmental protection activities to the Ministry for Nature Protection of the Republic of Sakha. As with other industrial organizations Aldanzoloto produces state statistical reports according to the following statistical forms: 2-ТП (air), 2-ТП (water use), 2-ТП (toxic waste), 3-OC (construction of environmental protection facilities) and 4-OC (environmental payments) and submits them to local statistical authorities.[2] Although Aldanzoloto produces Annual Reports, which are distributed only to company's shareholders, since its establishment in 1991, the company started to include information concerning environmental protection in Annual Reports only in 1996. Aldanzoloto lacks corporate initiative to disseminate information on environmental protection to the wider public apart from the state authorities. The company does not produce external Environmental Reports, nor does it include environmental protection information on its corporate website.

Environmental expenditure Due to reduction in production base and the lack of resources, Aldanzoloto has decreased its expenditure on environmental protection operations and facilities (see Table 5.7).

Table 5.7 Aldanzoloto environmental expenditure in 1998-2000, thousand US dollars

	1998	1999	2000
Current expenses for environmental protection	3,872	100	105
- protection and rational use of water resources	3,745	59	41
- air protection	29	35	58
- protection from waste	26	4	2
- land rehabilitation	72	3	4
Current expenses for capital repair of environmental facilities	47	139	25

Source: Data provided by Aldanzoloto.

In 1997, Aldanzoloto spent over one million US dollars for construction of environmental protection facilities, but during 1998-2000 the company on average spent 80 thousand US dollars per year (AK Aldanzoloto Annual Reports, 1996b, 1997b, 1998b, 1999b, 2000b).

Environmental performance The overall environmental impact of Aldanzoloto has decreased due to the drop in industrial production, improved environmental protection and closer control of the company's operations from the local environmental protection agency, as evidenced by officer from local environmental protection agency, Upper Aldan Territorial Committee for Nature Protection:

> At the moment, the impact is minimal due to several reasons. Firstly, there is a reduction in industrial base. Secondly, the Company has a serious attitude towards the compliance of environmental protection operations. Thirdly, Aldan Inspectorate of Upper Aldan Territorial Committee for Nature Protection closely controls environmental protection activity in Aldanzoloto.

Indeed, cooperation between the company and environmental protection agencies has strengthened. Starting from 1992, the local environmental protection agency and Aldanzoloto sign annual agreements with regards to environmental payments, which have substantially decreased in the 1990s (see Table 5.8). The Upper Aldan Territorial Committee for Nature Protection supervises the environmental protection activity in Aldanzoloto: approves the environmental protection actions developed by Aldanzoloto on an annual basis and also approves the projects of norms for maximum permissible discharge and maximum permissible emissions, which are also approved by the Ministry for Nature Protection. The local environmental protection agency also controls the process of obtaining the necessary licenses and permits for all projects of Aldanzoloto relating to the environment.

Table 5.8 Pollution payments of Aldanzoloto in 1996-2000, thousand US dollars

	1996	1997	1998	1999	2000
Payment for air pollution	36	13	14	7	4
Payment for water pollution	8	6	2	0.2	2
Payment for pollution from mobile sources	7	6	1	0.8	1
Payment for waste utilization	1	0.05	0.1	0.5	1
Payment for pollution to underground	2	0.3	-	-	-
Total pollution payments	54	26	17	8.5	8

Source: Adapted from AK Aldanzoloto, 'Vozdeistvie Deyatel'nosti AK Aldanzoloto na Prirodnuyu Sredu' [Impact of Aldanzoloto Operations on the Natural Environment] (1996b, 1997b, 1998b, 1999b, 2000b).

Nevertheless, the main reason for the significant reduction in environmental impacts is linked to the drop in production rates in Aldanzoloto. Although the drop in industrial production has reduced physical volumes of environmental impacts, it has created a lack of financial resources to invest in improvement of environmental facilities. The main environmental impact of gold mining operations of Aldanzoloto is to water resources in the Aldan district, since many old mine developments are not closed, the impact on water resources is continuing.

> Gold mining industry is the oldest industry in the Republic of Sakha. Nevertheless, environmental protection activity of the gold mining industry remains unsatisfactory. Maybe this is partially explained by the specifics of the industry's operations. As a rule, gold is still extracted from channels of rivers and streams. Generally, the work has a seasonal character. At the same time, a certain site is washed for an extended period of time, during four to five years (from interview with a civil servant from the Ministry for Nature Protection of the Republic of Sakha).

> They all mine gold in river basins. They wash banks of rivers and disturb the hydraulic balance. In addition, they often discharge industrial wastewater, which contains a variety of chemical dissolving agents. Their dams are draining and as a result everything is spilled in the rivers. There are a lot of examples of misuse. Practically, no reclamation of disturbed land is conducted. In the recent years, Aldanzoloto is not dealing with environmental problems. They do not have resources. They are almost bankrupted... Many dams were constructed. As a result, they have reduced the impact upon the river

system. However, they have caused the damage. Gold-mining companies simply destroy the rivers (Member of Legislative Assembly 'Il Tumen').

In the period 1993-2000, environmental performance of Aldanzoloto worsened, highlighting the fact that even though the total quantity of pollution decreased following the drop in production output, the intensification of pollution in relation to production levels had increased. This is a problem experienced by many industrial enterprises in the Russian Federation and is attributed to obsolete technology in the transition period.

Health and Safety in Aldanzoloto

Health and safety issues in Aldanzoloto are coordinated by the health and safety department established at Kuranakhsky mine, the main production division of the company. The Head of the health and safety department reports to the Director of Kuranakhsky mine, who also has a position of Deputy General Director for Production at Aldanzoloto. Every division and workshop of Aldanzoloto has an employee assigned with health and safety responsibilities, who report to workshop managers and also to the Head of the health and safety department at Kuranakhsky mine. Health and safety operations are carried out according to health and safety standards to which the company is accredited. These standards are developed for mining and ore enrichment enterprises that refer to SSBT – system of standards for occupational safety, which set requirements for the organization of operations for labor safety and organizational and methodical basis for standardization in the field of work safety; requirements for safety of industrial equipment; requirement for safety of industrial processes; requirement for safety appliances of the employees; requirement for safety of buildings and facilities.[3] Aldanzoloto uses a number of state sanitary standards for industrial premises that prescribe air quality norms at workplace and electric safety.

Health and safety issues are also regulated by Collective Agreement, which Aldanzoloto Company concludes with the Trade Union of Aldanzoloto employees every year. The Collective Agreement contains several clauses according to which the company agrees to take responsibility for health and safety of its employees via providing:

- Safe and healthy working conditions for employees of Aldanzoloto in accordance with standards for labor protection.
- That the company will not allow an employee to undertake a shift, if an employee refuses to undergo the medical examination.
- Professional training and education for employees as well as training on health and safety issues for all employees in accordance with current legislation.

The health and safety department develops annual plans for health and safety actions which are oriented towards the improvement of working conditions and approved with the Trade Union of Aldanzoloto employees. These plans are developed for each workshop and include, for example, measures for improvement of ventilation or purchase of new uniforms. Moreover, Aldanzoloto conducts accreditation of workplaces according to health and safety standards (see Table 5.9). In 2001, only half of workplaces in Aldanzoloto complied with these standards.

Table 5.9 Employee working conditions in Aldanzoloto in 1997-2001

	1997	1998	1999	2000	2001
Total number of employees	3,835	3,308	2,077	2,083	2,296
Number of employees working in conditions, non compliant with health and safety standards	2,375	1,621	1,399	1,073	1,413
Including number of women	239	250	195	226	407

Source: Adapted from AK Aldanzoloto, 'Poyasnitel'naya Zapiska k Godovomu Bukhgalterskomu Balansu' [Explanation Note to Annual Balance Sheet] (1996a, 1997a, 1998a, 1999a, 2000a).

The company has a system in place for regular training on health and safety issues such as workplace hygiene, working conditions and safety, prophylactics of occupational diseases and first aid. Health and safety training is conducted once every three to six months in all workshops of Aldanzoloto. All employees of Aldanzoloto undergo regular medical examination. A medical examination before and after shifts is conducted for certain groups of employees, particularly those who work in hazardous conditions. For this purpose four medical points are organized in Aldanzoloto: Kuranakhskaya gold-enrichment plant, technical transport division, transport and road division and Kuranakh quarry.

In addition, the environment, health and safety laboratory carries out monitoring of health and safety standards in workplaces (air quality, ventilation, fire safety, etc). The health and safety department of Aldanzoloto produces annual statistical reports and submits them to local statistical authorities. The health and safety department also produces half-year reports on health and safety operations in Aldanzoloto for the purpose of internal analysis. At the end of the year a report on health and safety issues is presented to the General Director of Aldanzoloto; this information is included in Aldanzoloto Annual Reports. Moreover, the health and safety department reports at the annual meetings of Trade Union of Aldanzoloto employees.

Mines, enrichment plant and the transport division have the most frequent occurrences of health and safety accidents (AK Aldanzoloto, 2000a). After each industrial accident, a commission on labor protection and industrial control which is established at Kuranakhsky mine holds a meeting to examine the course of the accident and develops measures for improvement of working conditions to prevent further accidents. Common reasons for such accidents are low levels of employee discipline, non-compliance with instruction for labor protection and low control of company's managers over working conditions (AK Aldanzoloto, 2000a). Although the number of employees in the period from 1997 to 2000 reduced by 45 per cent, the expenses (in real terms) on health and safety operations in the same time period decreased by 80 per cent (see Table 5.10).

Table 5.10 Accidents in Aldanzoloto in 1998-2000

Years	Number of accidents	Number of fatalities	Frequency rate[4]	Severity rate[5]	Health and safety expenditure, thousand US dollars
1996	29	2	5.3	46.8	369
1997	14	0	4.3	36.5	443
1998	17	0	8.1	28.8	365
1999	16	0	7.8	31.8	117
2000	12	2	5.4	49.3	68

Source: Adapted from AK Aldanzoloto, 'Poyasnitel'naya Zapiska k Godovomu Bukhgalterskomu Balansu' [Explanation Note to Annual Balance Sheet] (1996a, 1997a, 1998a, 1999a, 2000a).

Employee Relations in Aldanzoloto

Over 2,300 employees were employed in Aldanzoloto in 2000, of which 60 per cent were industrial personnel, 13 per cent were engineers and managers, and 27 per cent of employees worked for supporting divisions. The majority of employees work at Kuranakhsky mine and Kuranakhsky jewelry factory. Aldanzoloto has assigned a Deputy General Director for Personnel Issues at Aldanzoloto, who coordinates overall corporate strategies for personnel and social issues through wage and personnel services.

The wage and personnel services develop annual plans for personnel development, approved by the General Director of Aldanzoloto, which cover issues on personnel strategy, disciplinary practices, relations with the Trade Union, personnel training, working hours flexibility, health and safety issues, shareholders consultation, social guarantees, advantages for company employees, holidays, medical treatment, compensation, etc. Some personnel related issues are outlined in the Collective Agreement, signed with the Trade Union of Aldanzoloto employees. Collective Agreement usually covers issues related to wage and payment conditions, social guarantees and compensations, medical treatment, sport and social activities and annual leave arrangements. Average salaries in Aldanzoloto have substantially decreased in the period of 1996 to 2000. The average (nominal) salary of Aldanzoloto employee in 1996 amounted to 493 dollars per month, in 1997 – 465, in 1998 – 250, in 1999 – 114, and in 2000 – 137 (AK Aldanzoloto, 1996a, 1997a, 1998a, 1999a, 2000a).

The company pays considerable attention to employee training via implementation of an annual training and development program. Training is offered both in a training and education centre, which was established within the structure of Aldanzoloto in the 1970s, and other education institutions. Moreover, Aldanzoloto offers internship and job training for students from local colleges and scholarships for students to get higher education degrees. Since the 1980s, Aldanzoloto, when it used to be a part of Yakutzoloto Corporation, started to undertake attempts to recruit members of the indigenous population to its mining operations. Recent attempts to recruit members of the indigenous population, i.e. Sakha and Evenks from the neighboring districts are not always successful due to low rate of retention of the recruits after the completion of training courses.

> I do not know why they do not stay. However, we provide with accommodation and everything. It is a specific work in the Company. This is a labor intensive work, a lot of labor. Not everyone can sustain. Not only indigenous people leave, others also leave us. There is a large flow of personnel. We recruit 900 people a year, and 900 people leave the Company in the course of the year. This is out of 2,100 people, total number of the Company employees. The flow is almost 50 per cent. However, we provide with accommodation. 70 per cent of the accommodation in Nizhny Kuranakh is with conveniences. I do not know why they leave. Maybe it's because of a distant location from the large cities (interview with senior manager of Aldanzoloto).

The reasons for failure in attracting large numbers of members from the indigenous population relates to generic problems with personnel retention. Indeed, during 1998-2000, every year up to 50-60 per cent of personnel in Aldanzoloto are renewed (AK Aldanzoloto, 1998a, 1999a, 2000a). Nevertheless, Aldanzoloto plans to continue recruitment of indigenous people from neighboring districts for wood processing, jewelry and souvenir production within the company.

Several incentive schemes for employees are working in Aldanzoloto. Almost all employees hold shares of Aldanzoloto, and in principle participate in corporate profits. The company provides extra holidays for employees and developed a bonus system. In the past, Aldanzoloto used to assign a corporate pension to employees retiring from the company, however, this practice was stopped in 1998 due to financial problems in the company. In 1998, about 1,300 employees got made redundant; all redundancies were conducted in accordance with the legislation and financial payments were carried out. Aldanzoloto has developed a system for providing distinguished retired personnel with accommodation in western parts of Russia, rendering relocation assistance.

Welfare of employees is a high priority in Aldanzoloto and it is evident in the company's policies. The company maintains several sports and recreation facilities including a club, sports and health centers, two swimming pools, a sanatorium, a hall of residence and a hotel. In this respect, Nizhny Kuranakh, which is the host settlement of Aldanzoloto, is distinctly better equipped than other settlements in the Aldan district. Employees of Aldanzoloto have more advantages than other citizens and probably have the best leaving conditions in the Aldan district; this includes comfortable housing, food provision, recreation and regular income.

In the past, Aldanzoloto had a system for assisting in house purchases, according to which the company covered 80 per cent of a house price. Again the practice was stopped due to financial problems in Aldanzoloto, but in exceptional cases Aldanzoloto still acquires houses for certain employees upon the discretion of the corporate managers. During 1991-1995, Aldanzoloto had a system for providing education to members of employee families, 88 students (members of employees' families) were fully sponsored to study in various higher education institutions within and outside the Republic of Sakha. However, in 2000, the company changed its strategy and since supports more than two hundred students, providing them with industrial internships and training, and covering their travel expenses to and from the place of study, all of which was agreed in the Collective Agreement.

Development of Environmental and Social Strategies in ALROSA

Structure and Economic Performance of ALROSA

ALROSA was created in 1992 and became a successor of the Yakutalmaz Corporation. ALROSA started its operations from 1 January 1993. ALROSA is a joint-stock company with shares distributed amongst its founders: the State Committee of State Property Administration of the Russian Federation holds 37 per cent of shares; the Ministry of State Property Administration of the Republic of Sakha (Yakutia) holds 32 per cent; the Administrations of eight districts of the diamond province in the Republic of Sakha (Anabar, Lensk, Mirny, Nyurba, Olenek, Suntar, Verkhnevilyuysk and Vilyuysk) each hold one per cent of shares; and employees of ALROSA hold 23 per cent of shares, which are managed by the company's Board of Directors on behalf the employees.

Company management comprises of the Supervisory Council, which identifies the general directions of the company's operations, consists of representatives from ALROSA, the Government of the Russian Federation, the Government of the Republic of Sakha, local authority and other organizations; while the President and the Board of Directors run current management of the company (AK ALROSA, 1996a). ALROSA has 30 structural divisions, 32 daughter and ten dependent companies (see Figure 5.5), as well as affiliated companies and divisions in different areas of the Russian Federation and several representative offices abroad (see Figure 5.6).

ALROSA runs five diamond mining complexes: Anabar, Aykhal, Mirny, Nyurba and Udachny, and also has a daughter company, Almazy Anabara, which output is included in the total diamond production of ALROSA (see Table 5.11). During the 1990s, the most productive diamond mining complexes were Aykhal, Udachny and Mirny mines. In 1998, Aykhal accounted for 47 per cent of extracted ore, Udachny accounted for 39 per cent, Mirny accounted for 13 per cent and Anabar accounted for one per cent of extracted ore. In 2000, the newest diamond mining complex, Nyurba, started to mine diamonds from Nyurbinskaya and Botuobinskaya diamond pipes, with hope to become a leading mining complex, while Mirny mining complex is expected to reduce its operations significantly due to planned closure of Mirny mine, the oldest diamond pipe under industrial exploitation in the company (AK ALROSA, 2000a).

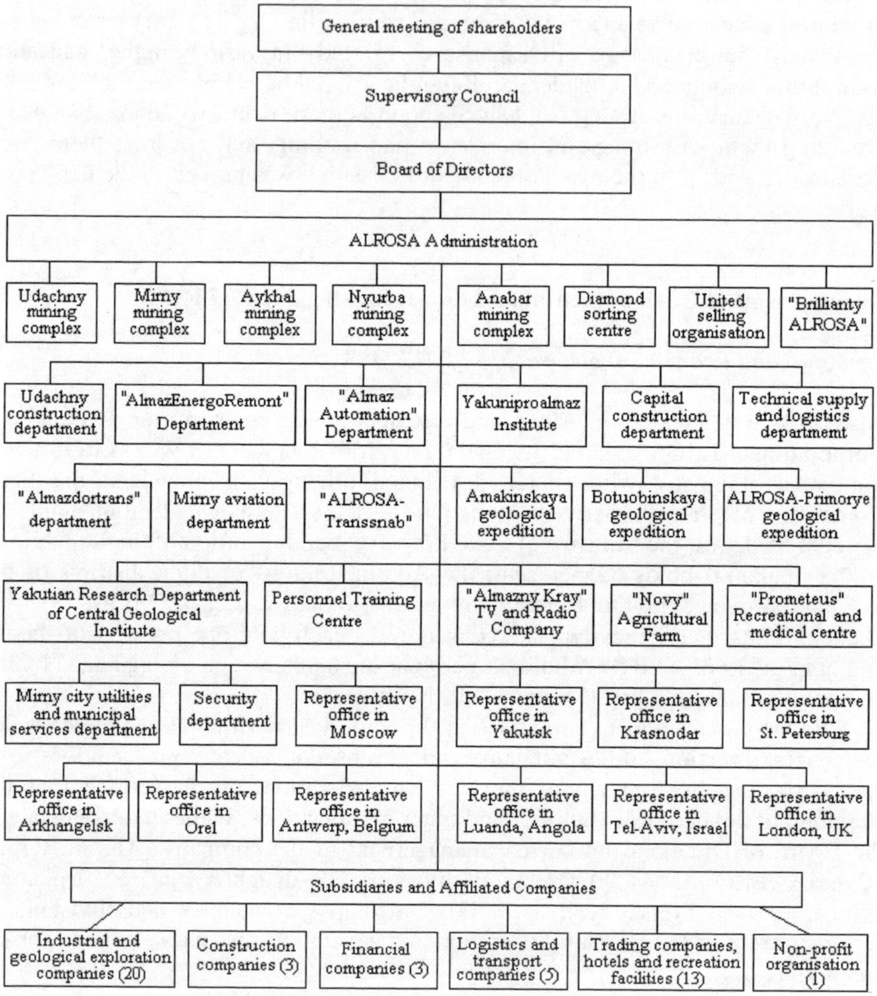

Figure 5.5 Structure of ALROSA

Source: Adapted from AK ALROSA, 'Godovoi Otchet 2000' [Annual Report 2000] (2000a).

ALROSA is an exceptional example of production growth during the period of economic transition in the Russian Federation, which is characterized by a typical drop in industrial production. During the period of 1996-2000, ore extraction in ALROSA increased by 60 per cent and ore processing increased by 56 per cent (see Figure 5.7), leading to growth of diamond production by 17 per cent (see Figure 5.8). During 1993-2000, the company's diamond sales increased by 24 per cent (Figure 5.9).

Figure 5.6 Location of ALROSA operations in the Russian Federation

Source: Adapted from AK ALROSA, 'Godovoi Otchet 2000' [Annual Report 2000] (2000a).

Table 5.11 Mining complexes of ALROSA

Mining Complex	Name of diamond deposit	Location
Almazy Anabara	Alluvial deposits on the river Mayat	Anabar district
Anabar	Alluvial deposit Ebelyakh	Anabar district
Aykhal	Jubileynaya and Sytykanskaya diamond pipes	Town Aykhal, Mirny district
Mirny	Mir and Internatsionalnaya diamond pipes	Town Mirny, Mirny district
Nyurba	Nyurbinskaya and Botuobinskaya diamond pipes	Nyurba district
Udachny	Udachnaya and Zartinsta diamond pipes	Town Udachny, Mirny district

Source: Adapted from AK ALROSA, 'Godovoi Otchet 2000' [Annual Report 2000] (2000a).

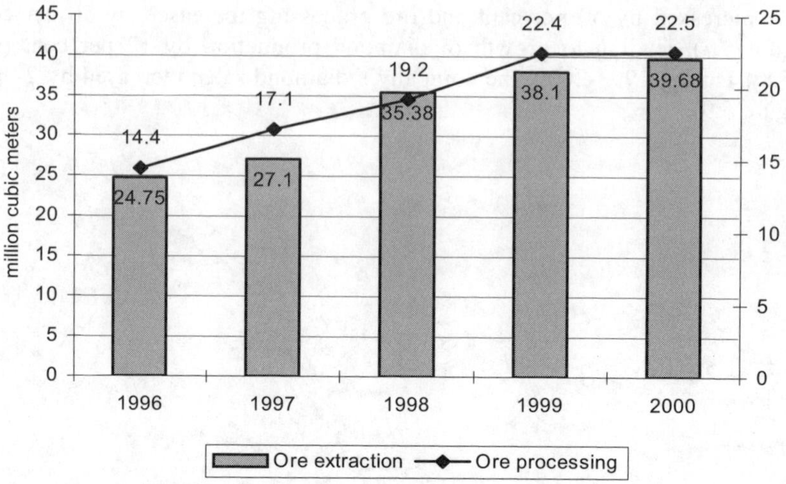

Figure 5.7 Ore extraction and ore processing in ALROSA in 1996-2000

Source: Adapted from AK ALROSA, 'Godovoi Otchet 2000' [Annual Report 2000] (2000a).

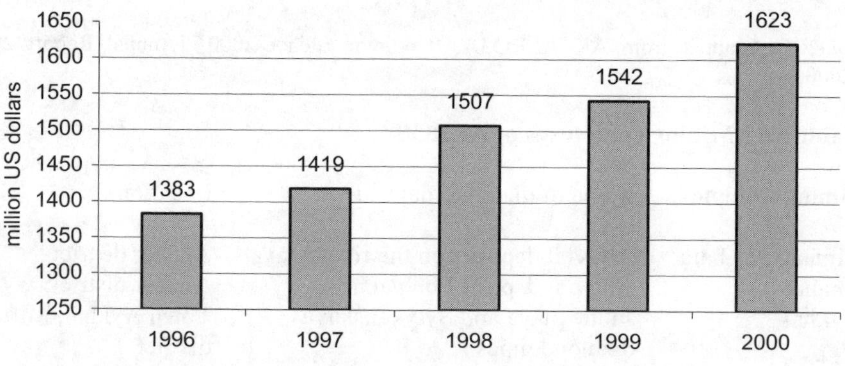

Figure 5.8 Diamond production in ALROSA, million US dollars

Source: Adapted from AK ALROSA, 'Godovoi Otchet 2000' [Annual Report 2000] (2000a).

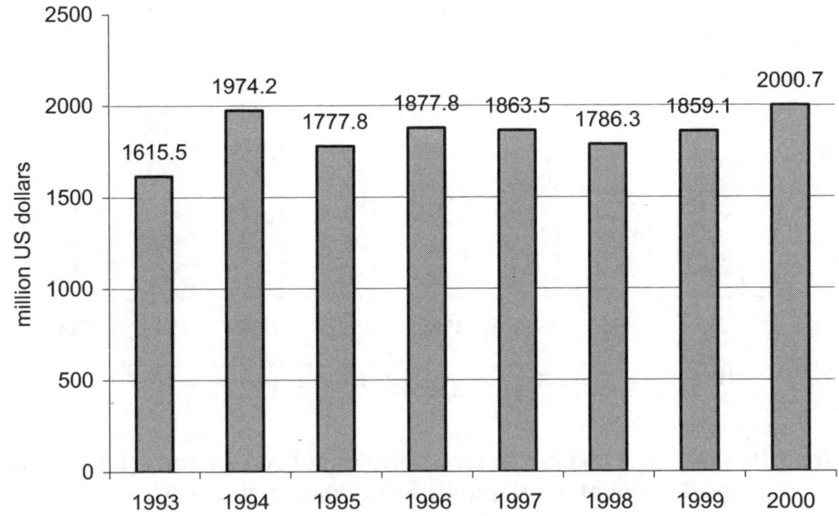

Figure 5.9 Diamond sales of ALROSA, million US dollars

Source: Adapted from AK ALROSA, 'Godovoi Otchet' [Annual Report] (1998a, 2000a).

ALROSA substantially contributes to the total industrial production and Gross Regional Product of the Republic of Sakha as a whole (see Figure 5.10). During the 1990s, the dependency of the regional economy of the Republic of Sakha on diamond mining has increased. ALROSA has leased the diamond deposits from the Government of the Republic of Sakha for period of 50 years. On 19 January 1993, ALROSA and the Government of the Republic of Sakha signed the lease agreement 'Lease of land sites, diamond deposits, other natural resources and fixed and current assets of Yakutalmaz Corporation'. On the basis of the lease, ALROSA is subject to several types of leasing payments including: recovery of mineral resources; use of mineral resources; lease of property assets of Yakutalmaz Corporation; environmental fund. Rates for these leasing payments were initially set by the Lease agreement between ALROSA and the Government of the Republic of Sakha in 1993 and were subsequently revised in 1998 and 2000. Leasing payments from ALROSA in 1993-2000 are shown in the Figure 5.11, and Figure 5.12 shows the share of leasing and tax payments from ALROSA in the State budget revenues of the Government of the Republic of Sakha.

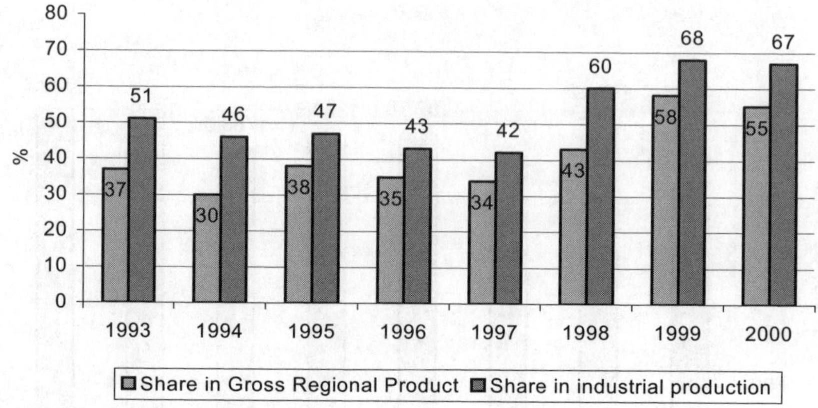

Figure 5.10 Share of ALROSA in Gross Regional Product and total industrial production of the Republic of Sakha in 1993-2000

Source: pers. comm.

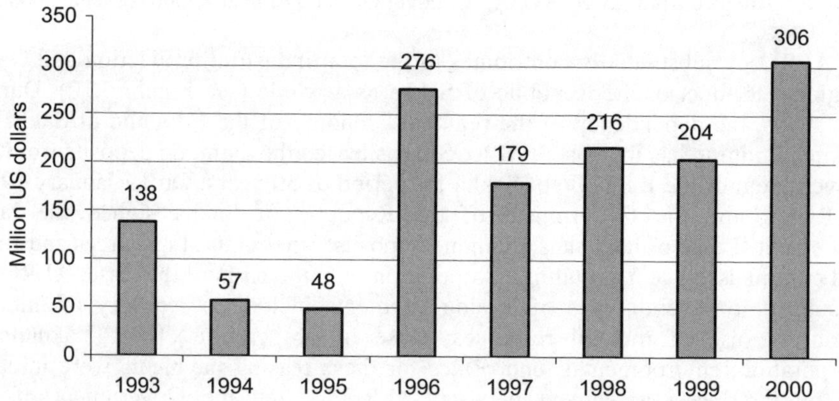

Figure 5.11 Leasing payments from ALROSA in 1993-2000, million US dollars

Source: pers. comm.

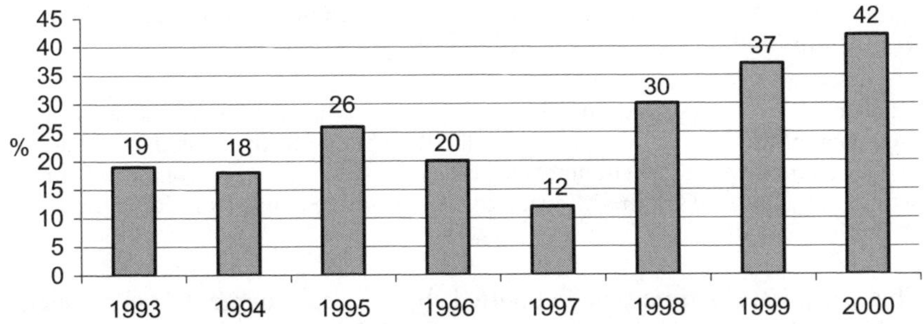

Figure 5.12 Share of leasing payments and other taxes from ALROSA in the State Budget revenues in 1993-2000, %

Source: Data provided by the Ministry of Finance of the Republic of Sakha (Yakutia).

Environmental Protection in ALROSA

Environmental protection operations in ALROSA are carried out through an established structure of environmental services. A corporate environmental manager with responsibilities for overall coordination of environmental protection operations of the company is appointed within ALROSA's Head Office in Mirny, who has a position of a Deputy chief engineer for environmental protection of ALROSA (while Chief engineer ALROSA is directly lined under the General Director of ALROSA). Several divisions of ALROSA carry out operations on environmental protection: five mining complexes (Anabar, Aykhal, Mirny, Nyurba, Udachny), two geological exploration parties (Amakinskaya and Botuobinskaya), the capital construction department, the transport department (Almazdortrans), the utilities and municipal services department, Mirny airlines and the agricultural farm. Each mining complex of ALROSA (Anabar, Aykhal, Mirny, Nyurba and Udachny) has an established environmental department, whilst other industrial divisions of ALROSA have only environmental officers. The Head of environmental department at a mining complex as well as environmental officer at an industrial division have positions of Deputy Chief Engineers for environmental protection of a mining complex or a division. The corporate environmental manager of ALROSA coordinates the activities of all environmental departments and environmental officers within the structural divisions of ALROSA. In total, there are 12 assigned environmental officers in ALROSA. In addition, each mining complex has an industrial and sanitary laboratory to carry out environment, health and safety monitoring.

Environmental protection in ALROSA is guided by a number of internal documents, as the company has not adopted a unified formal corporate environmental policy. For example, an internal document 'Environmental protection activity of ALROSA' issued in 2001 contains the major long-term environmental commitments of ALROSA in relation to the environment. The company's environmental operations are mostly oriented towards compliance with environmental protection legislation, as environmental protection officer from ALROSA admitted:

> The Company is working in strict conformity with legislation for environmental protection of the Russian Federation and the Republic of Sakha (Yakutia) and according to the requirements of environmental protection standards…Environmental protection actions are approved by the Ministry for Nature Protection of the Republic of Sakha (Yakutia).

This corporate attitude is supported by all environmental personnel, for example, employee of environmental laboratory of ALROSA stated:

> The Company is aiming to meet with regulatory requirements and standards. All of it is stated in projects of norms for maximum permissible emission and maximum permissible discharge, ecological passports of the enterprises of ALROSA…The Company obeys to all recent Decrees of the Russian Federation and the Republic of Sakha (Yakutia).

Long-term corporate objectives of ALROSA in the field of environmental protection are contained in a number of documents such as:

1. Norms for maximum permissible emission and maximum permissible discharge of contaminants developed for each industrial division.[6] The laboratory for environmental analysis at Yakutniproalmaz Institute develops 'Project of norms for maximum permissible emission and maximum permissible discharge' for each division, for a period of five years. These projects go through approval with the Mirny Territorial Committee for Nature Protection, Sanitary and Epidemic Inspectorate and finally with the Ministry for Nature Protection of the Republic of Sakha. These projects document all sources of pollution, define a sanitary safety zone around industrial facilities, and define whether industrial facilities discharge and emit contaminants within maximum permissible limits.
2. Ecological passports developed for each industrial division.[7] The laboratory for environmental analysis develops an ecological passport for each industrial division of ALROSA for the period of one year on the basis of the above-mentioned project. Ecological passports are approved by division managers and then approved with the Mirny Territorial Committee for Nature Protection. An ecological passport of a division includes information about input and use of raw materials, impacts on the environment, and sources of pollution. An ecological passport examines issues such as discharge of industrial wastewater into watercourses and onto the surface; waste utilization; tailings; disturbed land; transport; and environmental payments. It also includes a permit for the release of contaminants. If a release of contaminants exceeds the permissible limits then a plan for environmental protection actions is developed in order to reduce emissions down to permissible limits.

Environmental commitments Analysis of corporate documents reveals that ALROSA's commitments to the environment are found in four main documents, namely: Annual Internal Environmental Reports, Additional Internal Environmental Reports, Annual Reports and Tariff Agreements. It should be noted that neither Annual Internal Environmental Reports nor Additional Internal Environmental Reports are open to public and therefore considered as internal corporate documents. A common definition 'direction for environmental protection activity' is used in these corporate documents when referring to environmental protection operations. This definition combines notions of a longer-term goal and a particular objective for environmental protection activity of ALROSA and represents strategies of ALROSA in the field of environmental protection.

The content analysis of directions for environmental protections of ALROSA stated in these four main documents in the period of 1994 to 2002 shows that Annual Internal Environmental Reports contain the most detailed outlines of corporate environmental strategies (see Table 5.12). The only formal commitments to the environment made by ALROSA is stated in the Tariff Agreement, an agreement between ALROSA and the Trade Union of ALROSA employees (Profalmaz) on labor conditions, payments, social partnership and other issues, which is usually signed for one or two concurrent years.

Management of environmental protection operations ALROSA does not develop an environmental action plan for the company as a whole, instead, each industrial division develops an individual plan of operations for environmental protection every year, which is oriented towards reduction of discharges and emissions of contaminants and typically covers the following issues:

- water collection;
- water discharge;
- air pollution;
- waste disposal;
- land rehabilitation;
- environmental protection operations;
- expenditure for environmental protection;
- payments for discharge of contaminants;
- environmental research and development;
- construction of environmental protection facilities.

The environmental departments and/or environmental officers in divisions are responsible for the development of such plans that are later approved by division managers and the corporate environmental manager. Prepared plans are also approved with the local environmental protection agency. Not all environmental issues are under integrated control of the company's environmental services. For example, environmental services of ALROSA are not responsible for monitoring the use of materials and resources. Electric energy use is within the authority of the Chief engineer of ALROSA.

Table 5.12 Directions of environmental protection operations of ALROSA

Area of environmental protection operations/Statements	1	2	3	4
Minimization of environmental impacts				
• Reduction of contaminant discharge into atmospheric air	✓	✓	✓	
• Reduction of rock extraction and extraction works	✓		✓	
• Elimination of discharge of industrial wastewater in watercourses	✓	✓	✓	
• Reduction of discharge of contaminated consumer sewage in watercourses	✓			
• Restoration of disturbed land and technological deposits after closure	✓	✓	✓	
Specific statements on environmental protection activity				
• Reverse discharge of drainage water from open pits into underground aquifers	✓	✓	✓	
• Reduction of fresh water consumption for industrial purposes	✓	✓	✓	
• Increase internal cycle water supply	✓	✓	✓	
• Provision of industrial wastewater storage	✓	✓		
• Extension of environmental protection facilities	✓	✓		
• Modernization of sewerage networks and wastewater treatment facilities	✓	✓	✓	
• Modernization of neutralization and purification of air pollution	✓	✓		
• Clean-up of industrial and consumer sewage up to standards	✓	✓	✓	
• Reduction of industrial and consumer waste generation and its disposal	✓	✓		
• Re-use of waste within enterprises	✓			
• Protection of soil during waste disposal			✓	
Compliance				
• Timely execution of compulsory payments in the State Budget		✓		
• Provision with environmental protection licenses	✓	✓		
• Compliance with environmental legislation requirements			✓	✓
• Reduction of penalties for breach of environmental protection legislation through reduction of accidents	✓	✓		
• Licensing of industrial and sanitary laboratories			✓	
• Development of project documentation for environmental protection	✓			
• Environmental monitoring of industrial activities as required by legislation	✓	✓	✓	
• Conduct environmental impact assessment	✓		✓	
Environmental education, Research and development				
• Improvement of skills of employees' engaged in environmental protection; promotion of environmental education among other employees and population	✓	✓		
• Timely execution of scientific and technical works and conformation of project decisions in agencies of state supervision and control	✓	✓		
Environmental technologies				
• Use of environmentally sound technologies	✓		✓	
• Improvement of resource-preserving technologies to reduce negative environmental impacts and increase value of production via use of high extraction rates and preservation of minerals	✓	✓	✓	
Environmental expenditure				
• Increase of current expenditure on environmental protection	✓			
Environmental reporting				
• Regular dissemination of information among population about environmental situation and its impact on health in area of corporate operations through mass media				✓

Note: 1- Annual Internal Environmental Reports for 1994, 1995, 1996, 1997, 1998, 1999, 2000; 2- Additional Internal Environmental Reports for 2000, 2001; 3- Annual Reports for 1998, 1999, 2000; and 4- Tariff Agreements for 1998, 1999-2000, 2001-2002.

Source: Adapted from AK ALROSA (1994a, 1995a, 1996b, 1997a, 1998a, b, c, 1999a, b, c, 2000a, b, c, 2001a, b).

Operations relating to control of overburden are under the control of mine-surveyors of ALROSA, who monitor extraction and storage of overburden, no environmental monitoring of drainage from overburden storage sites is conducted. Industrial and sanitary laboratories, incorporated in each mining complex, conduct monitoring of environmental, and health and safety operations. Information collected by the industrial and sanitary laboratories is collected and analyzed by the laboratory for environmental analysis at Yakutniproalmaz Institute.

Air pollution Air pollution is monitored by industrial and sanitary laboratories at 11 industrial divisions of ALROSA (see Table 5.13). ALROSA uses purification systems in order to reduce air pollution on its industrial sites. Trucks are installed with clean-up filters. Since 1999, the company started to purchase trucks and other transport that meets environmental standards for air pollution.

Table 5.13 Air pollution in ALROSA during 1993-1999, tons

	1993	1994	1995	1996	1997	1998	1999
Total air pollution	16,500	13,100	12,500	16,800	10,900	13,400	9,000
Air pollution from mining complexes	7,411	7,538	7,930	8,236	7,853	11,244	4,671
Air pollution from other divisions	9,065	5,579	4,621	8,521	3,091	2,128	4,333

Source: Data provided by ALROSA.

Water management The company manages water collection and discharge in accordance with a permit issued by the Ministry for Nature Protection of the Republic of Sakha. Industrial and sanitary laboratories conduct regular monitoring of water consumption, use and discharge of wastewater (see Table 5.14 and Table 5.15). The main source of water collection is surface water, while only 25 per cent of water is collected from underground sources. More than 60 per cent of water collected by the company is used for industrial purposes, the rest is for consumer purposes (AK ALROSA, 2000b).

The majority of used water is discharged to surface watercourses. Industrial wastewater and consumer wastewater are treated for effluents. Consumer wastewater is treated with systems of biological and mechanical clean up at the stations for biological cleaning in cities of Aykhal, Mirny and Udachny. Over the period of 1993-2000, water collection and discharge has increased due to an increase in industrial production, but the treatment of wastewater going to surface watercourses still remains low. In fact, during 1999-2000, there was no treatment of wastewater at all.

Table 5.14 Water collection and discharge in ALROSA during 1993-1999, million cubic meters

	1993	1994	1995	1996	1997	1998	1999
Permitted limit for water collection	42.7	42.5	41.4	44.5	49.8	52.7	52.2
Actual water collection, including:	40.8	40.7	41.3	44.2	47.5	52.9	48.7
- for industrial and consumer purposes	31.9	32.2	33.7	35.5	37.8	43.9	36.3
Water collection per ore extracted, cubic meter per cubic meter	-	-	-	1.8	1.8	1.5	1.3
Water discharge to surface watercourses	25.1	24.4	26.1	25.7	26.6	26.2	25.7
- % of contaminated water	99.6	59	61	99	99	61	100

Source: Adapted from AK ALROSA (2000b, 2000c).

Table 5.15 Discharge of contaminants to surface watercourses by ALROSA in 1993-2000, tons

	1993	1994	1995	1996	1997	1998	1999	2000
Petroleum products	14	10	11	13	13	14	10	11
Suspended substances	631	640	539	809	732	692	646	903
Solid sediments	n/a	n/a	n/a	n/a	n/a	5,689	7,024	6,235
Other contaminants	n/a	n/a	n/a	n/a	n/a	3,665	3,938	2,598
Total	n/a	n/a	n/a	n/a	n/a	10,496	12,038	10,241

n/a – data not available

Source: Adapted from AK ALROSA (1999b, 2000b, 2000c, 2001a).

There are several hydro-technical facilities for environmental use, including: six active tailings facilities, three closed tailings facilities and a collector for water pumped from Mir diamond quarry. All tailings facilities in ALROSA operate in accordance with *TEO* (Feasibility Study), which include a section of *OVOS* (environmental impact assessment).[8] Tailings serve as a system where industrial wastewater from enrichment plants is stored and ensures protection of watercourses from contamination. After sedimentation of industrial wastewater in tailings, it is reused in the enrichment process. Special technology for tailings construction helps to ensure elimination of tailings drainage into watercourses. The tailings are stored for 20 years; after the end of tailings dumping, they represent a 'technological mineral deposit' and remain for secondary processing until the technology to exploit them is developed.

ALROSA has a system for emergency situations in place; each site has a plan for the elimination of accidents. There are separate plans for the elimination of accidents on tailings facilities and there is a service for traffic safety, which efficiently responds to traffic accidents and is engaged in the prevention of oil spills. Plans are developed for the release of spring floods in the area of tailings facilities that are accordingly approved with the Administration of the Mirny district.

Waste ALROSA regularly monitors waste generation and utilization at each structural division. The company handles waste in several ways: recycles waste on site, neutralizes waste within the company, hands over to other organizations for recycling, stores waste on corporate dumping sites, and hands over waste to other organizations for storage (see Table 5.16).

Table 5.16 Waste disposal in ALROSA in 1999-2000, tons

	1999	2000
Waste on sites at the beginning of the year	4023.2	459.5
Generation of waste within a year	3950.9	1618.7
Class 1 hazardous waste[9]	0.0002	0.00046
Class 2 hazardous waste	1007.8	1607.6
Class 3 hazardous waste	3943.1	8.7
Utilization of waste on site	1116.7	818.7
Neutralized on site	1.3	443.6
Transferred to other companies for recycling	191.7	194.4
Transferred to other companies for storage	-	1.3
Disposed on own dumping grounds	212.75	277.7
Waste on sites at the end of the year	6664.5	620.1

Source: Adapted from AK ALROSA, 'Poyasnitel'naya Zapiska o Vozdeistvii Khozyaistvennoi Deyatel'nosti AK ALROSA na Prirodnuyu Sredu' [Explanation Note about Impact of Industrial Operations of ALROSA on the Natural Environment] (1999b, 2000b).

Reclamation Environmental services monitor land disturbance and control land rehabilitation works. ALROSA has disturbed lands in seven districts: Anabar, Bulun, Lensk, Mirny, Nyurba, Olenek, and Verkhnevilyuysk. On 1 January 2001, ALROSA had 9,478 hectares of disturbed land within the territory of company's operations, of which 36 per cent referred to tailings, 19 per cent to dead rock dumps, 17 to hydro-technical and other facilities, 15 per cent to alluvial deposits, nine per cent to diamond quarries, three per cent to quarries of construction minerals, half per cent to sites of geological exploration and half per cent to special dumps. In 1999, in the structure of total disturbed land 39 per cent referred to Udachny Mining Complex, 37 per cent to Mirny Mining Complex, 14 per cent to Aykhal Mining Complex and ten per cent to other sites (AK ALROSA, 1999b, 2000b).

Usual practice for reclamation of disturbed land is leveling of the ground (see Table 5.17), re-vegetation practices are not widely used, and only six per cent of total disturbed land is re-vegetated. On new developments, Nyurbinskaya and Botuobinskaya diamond pipes the soil layer was collected and was stored separately for later use in land rehabilitation.

Table 5.17 Restoration of disturbed land in ALROSA during 1993-2000, ha

	1993	1994	1995	1996	1997	1998	1999	2000
Totally disturbed land	-	-	-	-	-	-	8,684	9,478
New disturbed land	-	-	-	-	-	-	158.2	325.4
Restored land	130.8	164.6	125.1	122.7	n/a	108.7	160.9	107.9

Source: Adapted from AK ALROSA (1994a, 1995a, 1996b, 1997a, 1998b, 1999b, 2000b).

Bio-diversity protection and environmental research ALROSA undertakes proactive actions for bio-diversity protection in the area of its industrial operations. Since 1998 the company has been cooperating with the Ministry for Nature Protection on the subject. In 2000, ALROSA signed an agreement with the Ministry for Nature Protection of the Republic of Sakha in the field of environmental protection, environmental rehabilitation and development of specially protected natural areas. In 2001, ALROSA adopted 'Conception for conservation and rehabilitation of resources of wild habitat in the area of ALROSA operations'. This Conception was developed in cooperation with the Mirny Territorial Committee for Nature Protection and the Department for Biological Resources of the Ministry for Nature Protection. In 2001, ALROSA in cooperation with the Ministry for Nature Protection of the Republic of Sakha also developed a program for the creation of an elk farm in the Mirny district. The program aims to restore elk population through simulated preservation of young elks, breed elks for settling in the territory of the Mirny district and other areas. In 2001, ALROSA also organized a farm for musk oxen in the settlement of Tass-Yuryakh in the Mirny district.

Before the start of any project *TEO* (feasibility study) is developed, which includes *OVOS* (environmental impact assessment). Technical and economic assessment goes through approval with different governmental agencies including the Ministry for Nature Protection of the Republic of Sakha. The Yakutniproalmaz Institute conducts environmental research including:

- environmental assessment of sites for allocation of facilities;
- environmental impact assessment;
- development of documents for obtaining licenses for use of water resources;
- projects of norms for maximum permissible emission of contaminants into atmospheric air;
- projects of norms for maximum permissible discharge of contaminants into watercourses;
- project of norms for generation and limits for disposal of industrial waste;
- ecological passports for enterprises.

Environmental reporting Although ALROSA does not produce a separate publicly available Environmental Report, the company established a system of internal and external environmental reporting. Regular disclosure of environmental operations of ALROSA is found in several media:

- *Annual Internal Environmental Reports* At the end of the year, all industrial divisions of ALROSA submit reports on environmental protection activity to the corporate environmental manager, these reports are summarized with an active involvement of the Yakutniproalmaz Institute and presented as an Annual Internal Environmental Report to the Board of Directors, occasionally this report is presented to the Supervisory Council. A typical Annual Internal Environmental Report includes information on water management, air protection, waste disposal, construction of environmental protection facilities, environmental payments and environmental expenditure. Annual Internal Environmental Reports of ALROSA include information on the division of ALROSA located in the territory of the Republic of Sakha and do not cover environmental protection operations carried out elsewhere (AK ALROSA, 1994a, 1995a, 1996b, 1997a, 1998b, 1999b, 2000b).
- *Statistical reports on environmental operations* Industrial divisions of ALROSA produce statistical reports according to the developed statistical forms such as 2-ТП (air), 2-ТП (water use), 2-ТП (toxic waste), 3-ОС (construction of environmental protection facilities), and 4-ОС (environmental payments). These statistical reports are submitted to local statistical authorities and the local environmental protection agency.
- *Annual Reports* ALROSA started to include information on its environmental protection operations and outline its environmental strategies in its Annual Reports since 1998, although the company has produced Annual Reports since 1993, however information is less detailed than in the Annual Internal Environmental Report. The company's Annual Reports are also available on the company's website and since 2000 have been also produced in English.

Generally, information contained in all three reports follows the structure set out in the forms for statistical reporting, thus indicating a compliance nature of internal and external environmental reporting.

Environmental expenditure ALROSA keeps a record of its environmental expenditure and regularly reports to authorities on its environmental expenditure via submitting statistical reports 4-OC (environmental payments). According to the system of environmental accounting, expenditure on environmental protection of ALROSA includes:

- expenditure for atmospheric air protection;
- expenditure for water protection;
- expenditure for land protection, including land rehabilitation;
- expenditure for capital repair of environmental protection facilities;
- pollution payments within limits of obtained permits;
- pollution payments beyond limits of obtained permits;
- fines;
- payment for natural resources (land, water and mineral resources).

Expenditure on environmental protection is increasing from year to year due to increase in mining operations, update of old environmental protection facilities, such as tailings (see Table 5.18).

Table 5.18 Expenditure on environmental protection in ALROSA in 1993-2000, million US dollars

	1993	1995	1995	1996	1997	1998	1999	2000
1. Expenses for environmental protection	21.7	12.8	29.4	34.2	37.6	47.7	29.9	46.7
1.1 water protection	-	-	25.2	31.9	31.3	40.2	-	40.9
1.2 air protection	-	-	1.1	1.6	2.4	4.3	-	2.4
1.3 land protection	-	-	3.1	0.7	3.9	3.2	-	3.4
1.3.1 land rehabilitation	0.2	0.1	0.02	0.7	0.05	0.05	-	0.8
2. Capital repairs of environmental protection facilities	0.1	4.5	6.9	6.1	6.3	1.0	1.1	1.0
3. Construction of new environmental protection facilities	4.2	8	10.5	17.5	15.1	8.9	4.3	7.1

Source: Adapted from AK ALROSA (1996b, 1997a, 1998b, 1999b, 2000b, 2000c).

Environmental performance Following environmental protection operations undertaken by Yakutalmaz Corporation, the predecessor of ALROSA, in the 1980s, the impact of diamond mining operations on the environment have been considerably reduced and today ALROSA continues to invest a considerable amount of resources in environmental protection.

> Historically important Decree of the Central Committee of the Communist Party of the Soviet Union and the Council of Minister of the USSR about stoppage of untreated wastewater discharge into Arctic watercourses was adopted in 1981. I think that the analogous Decree has been adopted for Russia in 1982. The analogous Decree of the Council of Ministers of Yakutia was adopted after the Russian one. The Decree was accomplished in time only by the Yakutalmaz Corporation, which later become the ALROSA Company. Yakutalmaz carried out huge work for engineering and construction of non-sewage tailings, solving the problem of excluding the direct discharge of mineralized waters from quarries into the river system, construction of sewage treatment facilities in Mirny, Udachny and Chernyshevsky. Due to these efforts, the impact of industrial facilities of ALROSA on the environment has been minimized. Today, ALROSA continues to input considerable amounts into environmental protection activity (from interview with civil servant from the Ministry for Nature Protection of the Republic of Sakha).

ALROSA shows stable levels of environmental performance, while individual divisions show increases or decreases in industrial pollution during 1993-2000, which is related to expansion or reduction in production at those sites. The company always aims to comply with environmental regulation and work within prescribed limits for resource use and industrial emissions. However, it sometimes fails to comply with environmental regulations; common compliance failures include: failures to obtain a permit for collection and discharge of water from watercourses in time and failures to submit statistical reports on environmental operations.

Table 5.19 Pollution payments from ALROSA in 1993-2000, thousand US dollars

	1993	1994	1995	1996	1997	1998	1999	2000
Total pollution payments	350	300	390	297	311	113	53	185
Pollution payment included in production cost	-	-	-	98	69	53	39	91
Pollution payment from income	-	-	-	199	121	60	14	92
- fines	-	-	-	-	-	-	10	10

Source: Adapted from AK ALROSA, 'Godovoi Otchet' [Annual Report] (1996a, 1997b, 1998a, 1999a and 2000a).

According to data on pollution payments, ALROSA regularly exceed its prescribed limits for industrial emission. In 2000, in the structure of pollution payments from ALROSA 45 per cent referred to payments for permissible pollution, 40 per cent to payments for extra pollution and five per cent for fines. Pollution payments from ALROSA increased by 43 per cent during 1993-2000 (see Table 5.19). Data shows that pollution payment that are included in the production costs are within the limits of permits issued by respective authorities, while pollution payments from income indicate payment for pollution outside of the authorized pollution limits. It is evident, that ALROSA regularly emits industrial pollutions that are outside the limits of appropriate permits.

Health and Safety in ALROSA

Since its establishment in 1993, ALROSA has set up corporate services for health and safety. The Department for labor protection at the corporate level of ALROSA coordinates labor protection officers at all divisions of ALROSA. A Head of the department for labor protection in ALROSA has a position of Deputy chief engineer for labor protection of ALROSA and directly reports to the Chief engineer of ALROSA. The Head of the department for labor protection at an industrial division of ALROSA reports to the Chief engineer of the industrial division, and has a position of deputy chief engineer for labor protection. In 2001, there were 138 employees with responsibilities for protecting the health and safety of employees in ALROSA.

The company has developed a corporate policy for workplace safety and improvement of workplace conditions which aims to protect the health and well-being of the employees at work, protect interests of individuals and society from accidents related to potentially hazardous sites, and from the consequences of accidents. Basic principles of the policy include statements on:

- recognition of safety operations as an integral and essential component of successful business;
- provision of a high safety level in accordance with legal requirements and technical standards;
- provision of sufficient means and resources for policy implementation;
- establishment and publication of safety management systems, aims and objectives; familiarization of employees with policy aims and employee involvement with system implementation;
- education of employees of all levels to ensure carrying out their duties and meeting their responsibilities;
- recognition of labor conditions improvement and labor protection as a primary task of all management levels; provision of its understanding, support and implementation in all corporate divisions;
- compulsory regular reporting on policy implementation, effectiveness of labor protection system, evaluation of implementation with set tasks, and

notification of employees with the results of corporate activity in labor protection.

Health and safety policy is revised during the preparation of the Annual Internal Health and Safety reports for the Board of Directors. In addition, ALROSA has developed a program for improvement of labor conditions and protection. The program forms a basis for planning and budgeting of labor protection actions. The latest program was adopted in 1999 and is entitled 'Industrial corporate program of primary actions for improvement of labor conditions and labor protection in ALROSA Company for 1999-2001'. Annual budget of the program was about eight million US dollars. A similar program has been developed for the period of 2002-2005. In addition, health and safety is included in other corporate programs, such as the quality program which is developed for 2001-2005 and incorporates a section for labor protection. Each year ALROSA signs Tariff Agreement with the Profalmaz Trade Union of ALROSA employees, where 21 articles are devoted to the improvement of working conditions, labor protection and health of employees (see Table 5.20). Current objectives of labor protection in ALROSA are:

- Reduction of industrial injury rate and occupational disease rate.
- Implementation of the corporate program 'Program for primary actions for improvement of labor conditions and protection in ALROSA for 2001-2005'.
- Development and implementation of the 'Manual for labor protection management in ALROSA'; Standards for labor protection in ALROSA in conformity with the Russian Federation and international standards of management systems such as ISO 9000, ISO 14000 and OHSAS 18000.
- Introduction of stimulation system for performance and compliance with labor protection requirement within employees.

In 1995, ALROSA adopted a health and safety management system which is the only formal management system that relates to issues of corporate social responsibility in the company. The health and safety management system sets out principles of safety policy; functions of the company's employees with responsibilities in labor protection; control over compliance with labor protection requirements; training and accreditation of staff with health and safety responsibilities; responsibilities of the ALROSA employees for non-compliance with labor protection requirements; and evaluation and promotion of labor protection compliance.

Since 1997, ALROSA conducts accreditation of workplace conditions according to existing workplace standards. In January 2001, out of total of 9,441 workplaces in ALROSA some 8,752 workplaces passed the accreditation. In 2001, ALROSA introduced the industrial control regulations on compliance with industrial safety requirements in hazardous industrial sites which stipulate the structure and responsibilities of ALROSA employees in order to achieve efficient industrial control system.

Table 5.20 ALROSA commitments for health and safety of employees included in Tariff Agreements for 1998-2002

- Comply with legal requirement of the Russian Federation and the Republic of Sakha on labor protection and environmental protection
- Develop a corporate program of actions for improvement of working conditions and labor protection
- Compensate for industrial injury or occupational disease to employees (expenses for medical treatment and travel)
- Conduct accreditation of workplaces on labor conditions
- Supply employees with uniforms, individual safety means and appliances according to standards
- Conduct initial (before employment) and regular (during employment) medical examination of employees which are engaged in hazardous labor conditions
- Provide resort and retreat treatment for special categories of employees (disabled, employees engaged in hard or hazardous labor conditions)
- Supply selected persons with portable express gas-analyzers and supply quarries with automated gas-analyzers for control over atmospheric air
- Supply with communication means (radiophones) according to requests from selected divisions
- Provide professional education and training and instructing on safety for all employees
- Conduct compulsory insurance of employees against industrial accidents and professional diseases
- Conduct a study on additional insurance against industrial accidents
- Conduct independent audit of labor protection, industrial equipment and technological processes on workplaces on demand of state and public bodies for control and supervision
- Provide timely supply of special food for employees engaged in hazardous labor conditions
- Disseminate information on situation with occupational diseases (publish once in three months information about industrial injury rate, morbidity and fire and road safety in corporate newspaper)
- Continuous research of drinking water purification methods
- Acknowledge priority financing of labor protection actions
- In exceptional cases, send employees and their children in medical institutions for examination and/or treatment
- Profalmaz Trade Union is assigned to conduct control over corporate commitments for health and safety
- Conduct mutual meetings of Company and 'Profalmaz' on each case of industrial accident with hard consequences

Source: Adapted from AK ALROSA, 'Tarifnoe Soglashenie mezhdu AK ALROSA i Profsoyuzom Rabotnikov AK ALROSA' [Tariff Agreement between ALROSA and Trade Union of ALROSA Employees] (1998c, 1999c, 2001b).

In 1997, the Joint Committee for Labor Protection was established in ALROSA that holds meetings at least once in three months. Trained officers from the

Proflamaz Trade Union conduct public control over labor protection practice in ALROSA. Assigned managers and professionals of all management levels of ALROSA conduct corporate control over labor protection practice in accordance with developed plans and schedules. Moreover, divisions and departments perform mutual examination on labor protection practices between themselves.

Health and safety education ALROSA conducts four types of employee education on health and safety: introductory instruction; instruction on the workplace; unplanned instruction following accidents; and instruction on shift cover. According to legal requirements, instruction of employees on health and safety is conducted once a year, and the instructing of certain groups of employees including miners and electricians is conducted once every three months.

Apart from the instruction of employees, the employees with labor protection responsibilities go through training in the Academy for Labor and Social Relations in Moscow. Managers of industrial divisions attend training courses on labor protection. ALROSA organizes seminars on labor protection for employees attracting specialists from various higher education institutions. All members of accreditation committees in ALROSA divisions undergo training in the Centre for Personnel Training of ALROSA.

Health and safety reporting ALROSA has a developed system for internal health and safety reporting. The corporate department for labor protection of ALROSA produces a quarterly report on labor protection and industrial safety and presents it to the Chief engineer of ALROSA. ALROSA does not include information on health and safety issues in its Annual Reports, however, information on health and safety performance is presented to ALROSA employees in accordance with the Tariff agreement through publications in a corporate newspaper. Some information on health and safety performance is also included in the Tariff agreement itself. In addition, divisions of ALROSA submit statistical reports: 7-T (accident rate) and 1-T (labor conditions) to appropriate agencies for state statistics.[10] According to these reports, most frequent courses for industrial accidents are tools, automobiles, mining equipment, wood processing equipment and electric equipment.

ALROSA does audit its health and safety performance, but a number of state agencies, including the State Mining Inspectorate, State Labor Inspectorate, State Sanitary and Epidemic Inspectorate, Transport Inspectorate under the Ministry of Transport and State Fire Safety Inspectorate carry out regular state control over health and safety performance in ALROSA. The company is working toward improvement of working conditions and reduction of injuries at workplaces, in 1996-2000, the number of industrial accidents reduced by 30 per cent (see Table 5.21). Health and safety standards in ALROSA are amongst the highest in the industry; in 2000, ALROSA became a winner in the National competition 'Russian organization of high social effectiveness' in a nomination 'Labor conditions and reduction of industrial accidents'.

Table 5.21 Health and safety in ALROSA in 1996-2000

	1996	1997	1998	1999	2000
Number of employees	41,543	35,881	35,924	36,866	39,177
Number of accidents	95	85	77	67	64
Number of fatalities	5	7	5	5	3
Expenditure on health and safety, thousand US dollars	-	7,826	5,160	5,333	6,447

Source: Adapted from AK ALROSA, 'Tarifnoe Soglashenie Mezhdu AK ALROSA i Profsoyuzom Rabotnikov AK ALROSA' [Tariff Agreement Between ALROSA and Trade Union of ALROSA Employees] (1998c, 1999c, 2001b).

Employee Relations in ALROSA

Vice-President for Personnel and Social Issues of ALROSA, a member of the Board of Directors, supervises overall personnel and social policy of ALROSA via personnel department and social department, which at located in ALROSA's Head Office in Mirny. In 2000, some 39,000 employees were working for ALROSA, of which 27 per cent were managers, 73 per cent were permanent workers and two per cent were seasonal workers (see Table 5.22). The largest divisions are: Capital construction department (6,469 employees), Aykhal mining complex (5,373 employees), Udachny mining complex (5,120 employees), Almazdortrans department (4,985 employees), Mirny mining complex (3,556), Utilities and municipal services of Mirny city (1,955 employees).

Table 5.22 Number of employees and average monthly salary in ALROSA in 1993-2000

Year	Total number of employees	Average monthly salary, US dollars
1993	48,510	136
1994	48,284	196
1995	47,479	364
1996	41,547	580
1997	35,813	608
1998	35,738	470
1999	36,937	299
2000	39,175	375

Source: Adapted from AK ALROSA and Yakutniiproalmaz (2001).

Management of personnel related matters is carried out in accordance with one-year plans for personnel development which are approved by the General Director of ALROSA that cover issues of disciplinary practice, labor associations, professional training for employees, and health and safety issues. Personnel issues are also regulated by the Tariff Agreement which highlights issues of time off, accommodation, construction of social facilities, work payments, advantages, bonuses and compensations, employee training, labor protection, recreation, medical treatment, cultural and sports events, social partnership. ALROSA provides training for its employees for upgrading their technical skills and knowledge, many courses are delivered at the company's own Center for Personnel Training.

ALROSA has developed a special program for recruitment of local community members which is implemented in accordance with the Decree of the Government of the Republic of Sakha (Yakutia) No. 552 of 23 November 1998 'About recruitment of employees in ALROSA Company'. According to this Decree, ALROSA is prescribed to recruit citizens from nine districts of the Republic of Sakha (Anabar, Kobyai, Lensk, Mirny, Nyurba, Olenek, Suntar, Verkhnevilyuysk and Vilyuysk) within three types of recruitment: seasonal recruitment; permanent recruitment; and graduate recruitment. Each year about 150 citizens of the diamond province are recruited by ALROSA as seasonal workers on construction sites. The personnel department of ALROSA has assigned recruitment officers to work in the districts of the diamond province on a permanent basis. So far, 520 members of the local community, i.e. citizen of the diamond province, have been through professional training in the Center for Personnel Training. In Nyurba mining complex members of indigenous population comprise 30 per cent of the total number of employees. In ALROSA, Yakuts comprise three per cent and Evenks 0.1 per cent of total number of employees, while ethnic Russian comprise sixty nine per cent of the total number of ALROSA employees (AK ALROSA and Yakutniproalmaz, 2001).

ALROSA has developed different incentive schemes for its employees. ALROSA employees own 23 per cent of ALROSA shares, on which dividends are paid annually, a manager from ALROSA stated:

> Practically all employees, when the Company was privatized in 1993, bought shares and became shareholders of the Company. More than 50 per cent of employees are shareholders.

ALROSA has a system of bonuses for its employees; the size of a bonus depends on employee performance and the overall company's performance and is paid monthly, quarterly and annually. In addition, Diamond Autumn Non-governmental Pension Fund was created for employees of ALROSA. ALROSA has a special program for assistance in relocation of staff to the western parts of the Russian Federation after retirement. Retired staff of ALROSA hands over their accommodation in the Republic of Sakha and get a flat in the houses built by ALROSA outside the Republic. ALROSA built 600 flats in the city of Orel, which is located in the western part of the Russian Federation.

ALROSA provides temporary (hall of residence) and permanent accommodation for its employees. ALROSA provides credits to an employee for

70 per cent of a house price for five years. In 2000, 176 flats were purchased according to this scheme (AK ALROSA and Yakutniproalmaz, 2001). ALROSA maintains several facilities for accommodation, health, recreation and child care (see Table 5.23). ALROSA fully maintains the Department of utilities and municipal services of Mirny city, Center for Personnel Training, TV and *Almazny Kray* Radio Company, Mirny Regional College, and the Health and leisure centre. The Department for utilities and municipal services of Mirny city provides heating, sewage and water supply for communal areas in city of Mirny (AK ALROSA and Yakutniproalmaz, 2001). In 2000, the costs for maintenance of the social infrastructure facilities of ALROSA amounted to 48 million US dollars (see Table 5.24).

Table 5.23 Communal, sports and recreation facilities of ALROSA in 2000

Indices	Total	Capacity, people
Nurseries	27	4,099
Sports halls	12	1,556
Swimming pools	3	162
Cultural facilities	6	1,890
Accommodations served by communal services of ALROSA	1,424 thousand square meters	
Accommodation in property of the Company	228 thousand square meters	
Halls of residence in property of the Company	10.8 thousand square meters	

Source: Adapted from AK ALROSA and Yakutniiproalmaz (2001).

Each year, ALROSA develops plan for the organization of sport and recreation activities for the employees which are implemented at facilities of ALROSA. In addition, ALROSA carries out a program for improvement of the health of its employees, as a manager from ALROSA stated:

> We have a whole health program, this is how the program is called. It embraces virtually all employees of the Company. We have signed agreements for health services to employees of the Company and members of their families, i.e. for those kinds of services, which are not provided in our district hospital. We work with medical centers in Yakutsk, health center, with medical centers in Novosibirsk and Tomsk. That is if an employee needs a specialized medical treatment, if not here, than outside the Republic. The Company undertakes all of it, it's all paid at the Company's expense.

The social department as well as personnel department produce Annual Internal reports on their activities and submit it to the Vice-President for Personnel and Social Issues of ALROSA. Personnel and social issues are covered in ALROSA Annual Reports in the section on Personnel management.

Table 5.24 Expenditure on social development in ALROSA in 1997-1999, thousand US dollars

	1997	1998	1999
Maintenance of social services	98.0	132.2	144.3
- communal services	70.5	100.4	100.7
- nurseries	17.3	17.5	23.7
- recreation facilities	11.0	14.2	19.9
Expenditure on social activities	13.6	10.5	25.3
- education	6.4	4.5	10.9
- health	6.6	5.6	9.5
- other (culture and charity)	0.8	0.4	4.9
Total	112.3	142.7	169.7
Other social expenses	2.47	15.31	31.31
- charity giving	0.00	14.29	28.01
- health program	2.47	1.02	3.30
- road maintenance	28.51	24.33	42.49

Source: Adapted from AK ALROSA, 'Godovoi Otchet' [Annual Report] (1997b, 1998a, 1999a, 2000a).

Comparison of Aldanzoloto and ALROSA

Diamond mining and gold mining are the leading industries in the Republic of Sakha. The importance of the gold mining industry has diminished over the recent years and diamond mining now constitutes the base of the regional economy, providing budget revenues, employment and promoting industrial and socio-economic development in the Republic of Sakha (see Chapter 4). There are several features that unite and as well as differentiate Aldanzoloto and ALROSA.

Similarities between Aldanzoloto and ALROSA:

- Both companies were state owned enterprises and were under the supervision of Soviet Union Departments. The companies were re-organized and privatized during the course of economic and legal reforms in the beginning of the 1990s.
- Both companies have caused substantial negative environmental impacts in the past and have significantly affected local communities, restricted the traditional economic activities and affected the lifestyle of indigenous people of the North.
- Both companies have attracted an external workforce, which has resulted in an increase of migrants. The companies provided the impetus for infrastructure developments such as roads, railroad, electric power engineering, air transport and growth of town building.

- Gold and diamonds are the strategic mineral resources and subject to special regulations.
- Both companies have diversified and participate in other businesses related to mining (mineral processing) and other industries such as manufacturing, farming and transport.

Differences between Aldanzoloto and ALROSA:

- The companies operate in the different geographical zones within the Republic of Sakha: Aldanzoloto operates in the mountain-taiga zone in southern Yakutia whereas ALROSA operations are located in taiga-tundra zone in western Yakutia. This difference has implications for environmental impacts of their mining operations.
- Chronologically, gold mining has been developed earlier (1920s) than diamond mining (1950s).
- The scale of industrial developments is different. Diamond mining operations are larger in terms of production volume than gold mining.
- After privatization, Aldanzoloto is an open type joint-stock company, while ALROSA is a closed-type joint-stock company. After the collapse of the former Soviet Union, gold mining went through decentralization, splitting into a number of smaller companies, while diamond mining retained centralization of the diamond production in the region. Since the 1990s, gold production has dropped with subsequent drop of profits and emergence of financial difficulties, while diamond production has increased and it is a profitable and stable industry.
- Nature of gold and diamond marketing is different. ALROSA ensures its sales via the agreement with De Beers, while Aldanzoloto sells its production on a free market.

Organizational structure Both Aldanzoloto and ALROSA have established extensive internal structures for management of environmental issues, health and safety, and employee relations. ALROSA has a more centralized organizational structure than Aldanzoloto; ALROSA has managers and departments which are responsible for environmental issues, health and safety and employee issues at a corporate level, who coordinate operations of officers and departments at division levels. Aldanzoloto on the other hand has a manager and a department at corporate level for employee and community issues, whilst management of environmental and health and safety issues are coordinated by a manager and a departments at divisional level only. Moreover, managers at the main production division coordinate the operations of managers at parallel divisions, which are not in vertical managerial hierarchy. ALROSA has organized various committees within the company which deal with production issues and some of the environmental, health and safety issues. Nevertheless, the general structure of environmental, health and safety and personnel services within Aldanzoloto and ALROSA shows that environmental and social issues are mainly dictated by production, since the production managers coordinate the environmental, health and safety managers.

Taking an example of the structure of environmental services in both companies, environmental managers at a division level primarily report to division managers for production and then to the corporate environmental manager (ALROSA) or environmental manager in another division (Aldanzoloto), the latter environmental manager has less tools to directly coordinate the environmental issues at division levels (see Figure 5.13). This reflects the low level of consideration of environmental issues within Aldanzoloto and ALROSA in decision-making and a high degree of dependency of environmental decisions based upon production and economic issues. Another pattern is observed with global proactive mining companies, which have a centralized structure for management of corporate social responsibility, often there are managers at corporate level overseeing issues relating to corporate social responsibility and they have committees within the Board of Directors on issues relating to sustainability and corporate social responsibility.

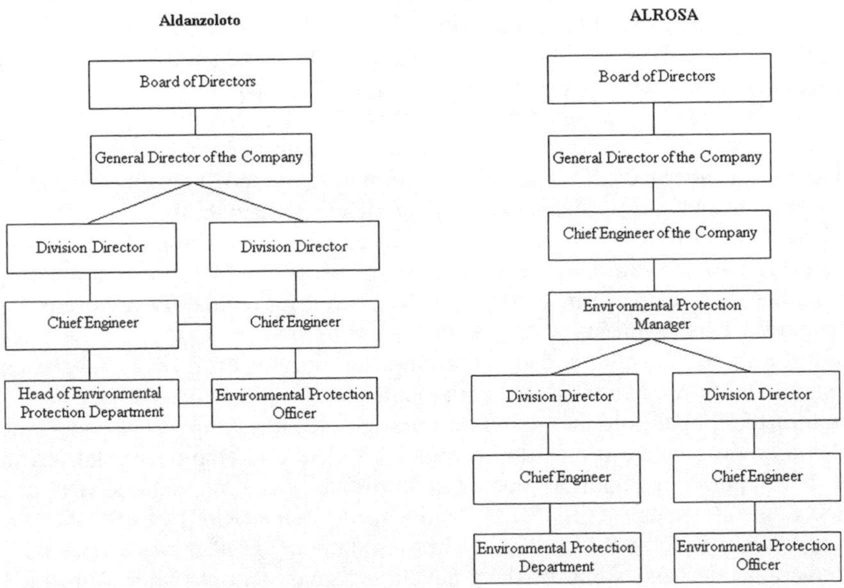

Figure 5.13 Organizational structure for management of corporate social Responsibility issues

Corporate policies The majority of the global gold and diamond mining companies including BHP, Freeport-McMoRan Copper & Gold, Homestake Mining, Placer Dome and Rio Tinto have Boards of Directors who have adopted and regularly revise their corporate environmental policies, whereas Aldanzoloto and ALROSA have not adopted formal corporate policies relating to the environment, health and safety and employee relations.

Instead, they base their environmental and social conduct on principles formulated in various corporate documents such as annual action plans of environmental, health and safety, personnel and social departments, document relating to regulatory compliance and agreement with Trade Unions. ALROSA is more prominent in the formulation of its goals and objectives which is evident in their corporate documents. In general, operations of both Aldanzoloto and ALROSA are built on same sets of principles that are outlined in the corporate environmental policies of the global gold and diamond mining companies. However, the global gold and diamond mining companies including AngloGold, Argyle, Ekati, Homestake Mining, Normandy Mining, Placer Dome and Rio Tinto (see Chapter 3) are increasingly concerned about sustainable development in relation to their operations, they commit to the principles of sustainable development and specifically emphasize their sustainability approach within their corporate environmental policies. Aldanzoloto and ALROSA have not yet developed commitments to sustainable development and do not attempt to integrate corporate social responsibility issues into one coordinated action; environment, health and safety, personnel and social issues are still considered separately and little cooperation between respective departments takes place.

The distinct feature of guiding principles of Aldanzoloto and ALROSA operations in the field of environmental protection, health and safety and employee relations is an absence of articulated willingness to go beyond regulatory requirements, which shows their reactive approach to sustainability issues as they see their responsibility only within the limits of existing legislation. Many mining companies around the world are starting to adopt corporate environmental policies which include aims to perform better than the prevailing regulatory requirements and standards. This emphasizes the voluntary commitments made by companies and contributes to the notion that environmental responsibility is an approach which should go beyond legislative requirements.

Some of the global gold and diamond mining companies apply their corporate policies on environmental protection, health and safety and employee relations to all of their operations both at home and abroad, including subsidiaries and affiliated companies such as BHP for Ekati, Placer Dome, Rio Tinto for Argyle. However, Aldanzoloto and ALROSA do not regulate any CSR issues within their daughter companies and some of their developed goals and statements are only relevant to specific sites. Furthermore, the global gold and diamond mining companies often adopt integrated policies which embrace environmental, economic, social, health and safety issues via for example, the Sustainability Policy of Placer Dome and Health, Safety, Environment and Community Policy of BHP for Ekati.

The absence of corporate policies on environmental protection, health and safety, and employee relations in Aldanzoloto and ALROSA reinforces the low profile of CSR issues within these companies. It shows an absence of important foundations upon which interactions, decision making and future policies with regard to environmental and social concerns are based and the absence of long-term visions regarding environmental and social issues. The absence of corporate policies in Aldanzoloto and ALROSA results in: difficulties for employees with environmental and social responsibilities to identify the overall approach of the companies; inefficient communication between the environmental departments within the companies; insufficient environmental education of employees; and inconsistent planning of environmental operations.

The existing situation in relation to the absence of corporate environmental and social policies in Aldanzoloto and ALROSA is rooted in the fact that the development of such documents is not developed practice in the Russian Federation as a whole. The low profile of sustainable development in the Russian Federation as a whole does not encourage business and industry to commit to the principles of sustainable development for improvement of corporate environmental and social policies.

Management of environmental and social issues A growing number of the global gold and diamond mining companies both in developed (Homestake Mining, Placer Dome and Rio Tinto) and developing countries (Freeport-McMoRan Copper & Gold, De Beers, Debswana and Namdeb) have adopted formal EMS or are certified to EMS standards, namely ISO 14001. Some also are assigned to voluntary environmental codes of conduct, for example, AngloGold, Normandy Mining, Placer Dome and Rio Tinto are assigned to the Australian Minerals Industry Code for Environmental Management, Rio Tinto is also assigned to Global Sullivan Principles and Rio Tinto and Freeport-McMoRan Copper & Gold has joined Global Mining Initiative (see Chapter 3). Aldanzoloto and ALROSA have not yet established a formal environmental management system, only ALROSA has developed a health and safety management system, and is planning to develop a quality system, which would integrate environment, health and safety issues. The general level of certification of management system and adoption of international management standards is a slow process in the Russian Federation, e.g. despite the fact that in 1998, the National Standard Organization adopted a national version of ISO 14001, only 14 industrial companies in the Russian Federation were officially certified to GOST R ISO 14001 during 1998-2002 (see Chapter 4).

Some global gold and diamond mining companies develop environmental performance indicators to assess their operations against their targets and goals, emphasized in their actions plans. Aldanzoloto and ALROSA mostly utilize performance indicators which are used for State statistical reporting. For many global gold and diamond mining companies including AngloGold, Argyle, Ashanti Goldfields, De Beers, Freeport-McMoRan Copper & Gold, Homestake Mining, Normandy Mining, Placer Dome and Rio Tinto, environmental, health and safety audits have become regular practice. ALROSA recently conducted environmental audit and is willing to pursue it regularly in the future.

An absence of formal management system in Aldanzoloto and ALROSA suggests that the companies do not clearly recognize the need for improvement of existing procedures for conducting environmental operations. Indeed, it can be argued that an absence of a formal management system in relation to environmental and health and safety operations in Aldanzoloto and ALROSA prevents the improvement of their businesses, which includes better control and increased consistency of business practices, reduced operating costs and improvement of communication with regulators and the public.

Reporting Aldanzoloto and ALROSA have developed a system for internal reporting on environmental, health and safety, and employee relations. The major form of reporting to external stakeholders on performance is the state statistical reporting that is produced at the end of each year and reporting to concerned governmental departments such as Ministry for Nature Protection, Ministry of Industries, etc. While the state agencies are regularly and fully informed about companies' environmental, health and safety performance, the public has limited access to this information from the companies or from the state agencies. Although some environmental and social information is presented in companies' Annual Reports, but no such information is presented on corporate websites and no separate Environmental and Social Reports are produced. ALROSA makes Annual Reports available to the wider public. Although Aldanzoloto includes detailed environmental and social information in its Annual Reports, they are distributed to shareholders only. The situation in relation to environmental and social reporting in Aldanzoloto and ALROSA reflects the general situation in the Russian Federation, where sustainable development has not become a recognized agenda both within the state and the industry.

The comparison between Aldanzoloto and ALROSA shows that the companies have similarities in their approaches to environmental and social issues, but the main differences relate to the fact that ALROSA is a larger, more profitable and aware company than Aldanzoloto. Whilst in comparison with global gold and diamond mining companies, the mining companies in the Republic of Sakha have many weaknesses, but also some strength that are associated with the legacy of state owned enterprise practices during Soviet Era.

Differences between environment, health and safety, and employee relations in Aldanzoloto and ALROSA:

- ALROSA has better organization of environmental services.
- ALROSA has conducted independent environmental audit.
- ALROSA has been developing quality management.
- ALROSA has adopted health and safety management system.
- ALROSA more actively cooperates with governmental agencies on environmental protection and bio-diversity conservation.
- Aldanzoloto develops plans for the entire operations of the company, while ALROSA does not draw general plans for the entire company.

Weaknesses of Aldanzoloto and ALROSA:

- Absence of formal corporate environmental policies.

- Absence of formal environmental management systems.
- No health and safety audits.
- No production of separate Environment, Health and Safety Reports and Community Reports.

Strengths of Aldanzoloto and ALROSA:

- Range of initiatives for employee reward and support for retired personnel.
- Extensive initiatives for improvement of health and well-being of employees, including the maintenance of social facilities.
- Willingness to employ and train members of the local community and indigenous people.

Notes

[1] According to the Russian official classification of waste, Aldanzoloto generates five types of waste: Class 1 waste – batteries and fluorescent lightings; Class 2 waste – petroleum products; Class 3 waste – waste that contains lead and copper; Class 4 waste – tires, slag and consumer waste; Class 5 waste – scrap metal and wood. Toxic waste is categorized into four classes: Class 1 – extremely hazardous waste; Class 2 - highly hazardous waste; Class 3- moderate hazardous waste; and Class 4 – small hazardous waste. According to the Russian legislation, on 1 February each year, all organizations that produce toxic waste should submit annual statistical form 2-ТП (toxic waste) 'Report on formation and disposal of toxic waste' to local state statistics agency; state regulatory agency for particular branch of the economy; and local division of the State Committee for Nature Protection of the Russian Federation. The report should be developed on the basis of receipts and expenditure documents. All types of toxic waste should be accounted, including defective goods that contain hazardous elements. Ready goods for further use and semi-finished goods are not accounted in the report, as well as toxic waste that are emitted to water objects and atmospheric air and are accounted according to statistical forms 2-ТП (air) – 'Report on protection of atmospheric air'; 2-ТП (water use) – 'Report on water use'. The report should contain the information on formation, elimination and storage of toxic elements, which are produced as the result of cleaning-up of released gases and sewage waters on appropriate facilities (Instruction for completion of statistical report about formation and disposal of toxic waste form N2 -ТП - toxic waste, adopted by Decision of the State Committee for Statistics of the Russian Federation on 14.09.93, N 180).

[2] According to the Law of the Russian Federation 'About enterprises and business activity', enterprises engaged in all types of economic activities which cause negative impact on the environment and engage in operations for protection and use of natural resources must submit information necessary for carrying state system of data collection and processing. There are several types of statistical reports on environmental protection: 2-ТП (air) – 'Report on protection of atmospheric air'; 2-ТП (water use) – 'Report on water use'; 3-ОС – 'Report on building of water protection facilities and stoppage of emission of contaminated sewage waters'; 2-ТП (reclamation); 1- nature protection, 4-ОС – 'Report on current expenses for environmental protection and environmental payments'; 1-ecofund 'Report on revenues and outlays of ecological fund', 2-ТП (toxic waste) – 'Report on formation and disposal of toxic waste'. All forms of reports have instructions for their completion.

Industrial associations, organizations and bodies which have stationary sources of air pollution submit annual statistical form 2-ТП (air) – 'Report on protection of atmospheric air' to local state statistics agency; state regulatory agency for particular branch of economy; and local division of the State Committee for Nature Protection of the Russian Federation. The report should be submitted by 5 December of reporting year. Annual statistical form 4-OC 'Report on current expenses for environmental protection and environmental payments' is submitted on 25 January by all organizations that carry environmental protection action to the same bodies (Protasov, V. F. and Molchanov, A.V. (1995), *Ecology, Health and Management of Nature in Russia*, Finance and Statistics, Moscow).

[3] SSBT is an abbreviation for *Sistema Standartov Bezopasnosti Truda* (System of Standards for Occupational Safety). Aldanzoloto uses the sanitary norms of industrial premises according to the following Russian health and safety standards: SSBT GOST 12.1.014-84 – Air of the working area; SSBT GOST 12.1.038-82 – Electric safety; and SSBT GOST 12.0.002-80 – Terms and definitions.

[4] Frequency rate is calculated according to the formulae: Number of injuries in the period multiplied by 100,000 and divided by total hours working during the period.

[5] Severity rate equals to total number of days lost multiplied by 1,000 and divided by total number of man hours worked.

[6] Projects of norms for maximum permissible emission (MPE) is developed according to the Law of the Russian Federation 'About protection of atmospheric air', according to which organizations which have stationary sources for emission of hazardous contaminants into atmospheric air, should conduct an inventory of hazardous contaminants released into atmospheric air, develop maximum permissible emission and maximum permissible norm for hazardous impact on atmospheric air. Maximum permissible emission (MPE) is a norm for emission of hazardous contaminant into atmospheric air, which is established for a stationary source of air pollution taking into account the technical norms for emissions and the background air pollution, following the condition that this source of pollution should not exceed hygienic and ecological norms for air quality and maximum permissible load on ecological systems. According to the project of norms for maximum permissible emission, a local environmental protection agency establishes the volume of maximum permissible emissions for a particular stationary source of air pollution and issues a permit for emission of contaminants into atmospheric air according to the project of norms of maximum permissible emission. Similarly the project of norms for maximum permissible discharge of contaminants into watercourses is developed, approved by a local environmental protection agency and permits for water use are issued by regulatory agencies according to the developed project of norms.

[7] Ecological passports in the Russian Federation were introduced in 1990. Methodology for the development of an ecological passport is presented in the state standards GOST 17.0.04-90 'Passport of industrial enterprise'. Ecological passport aims to:

- assess environmental aspects of the production such as rational use of natural resource use (use of raw materials, energy and natural resources) and discharge of contaminants per unit of production;
- assess negative environmental impacts of an enterprise by identifying total quantity of pollution and waste disposal for a period of time and relate them to volumes of production;
- assess effectiveness of environmental cleaning facilities and control the minimization of environmental impacts;
- control environmental impacts of enterprise by setting pollution payments.
- Ecological passports consist of two sections. The first section includes documents for limiting air pollution (maximum permissible emission, permit for emission of

contaminants), water pollution (maximum permissible discharge, permit for discharge to industrial sewage), permit for transportation and burial of solid waste. These documents serve as a basis for inspection of the environmental protection operations in an enterprise. The section second consists of documents relating to the issues of natural resource use. The section includes a permit for water use (defined volumes for use of water with different quality), permit for land use, etc. (Protasov, V. F. and Molchanov, A.V. (1995), *Ecology, Health and Management of Nature in Russia*, Finance and Statistics, Moscow).

[8] TEO and OVOS (Feasibility Study and Environmental Impact Assessment respectively). Environmental impacts assessment is a procedure for defining negative environmental impacts and their socio-economic consequences and development of measures for minimization and prevention of unfavorable impact on the environment. The environmental impact assessment includes information about nature and scale of impacts on the environment of a proposed activity, alternatives for development, assessment of environmental and related socio-economic consequences of these impacts, possibilities to minimizes the impacts, consideration of public opinion. Results of the environmental impact assessment are submitted to State Ecological Review.

[9] According to the Russian official classification of waste, ALROSA generates three types of hazardous waste: Class 1 hazardous waste – fluorescent lighting; Class 2 hazardous waste – petroleum products, sulfur acid and bromine; Class 3 hazardous waste – lead from exhaust batteries.

[10] Organizations and separated divisions according to the list adopted by the state statistical agency annually submit statistical form No. 7 – injury rate 'Information about injuries at workplaces and occupational diseases' to a local state statistical agency ('Instruction for completion of the form of Federal State Statistical Monitoring for injuries at workplaces, occupational diseases and related material costs (form N 7 – injuries)', adopted by Decree of State Committee of Statistics of Russia of 25 February 1997 No. 13).

Industrial enterprises, construction companies, transport and communication companies and their structural divisions annually submit statistical form No. 1 –T (labor conditions) 'Information about labor conditions and compensation for working in unfavorable working conditions' to the local state statistical agency.

Chapter 6

Models for Community Development

Introduction

Community development is becoming increasingly important for mining companies that operate in both developed and developing countries as well as in countries in transition from planned to market economy. Mining companies are increasingly required to mitigate negative impacts and deliver positive contribution to the communities in which they operate in order to respond to stakeholders concerns and expectations (Holme et al, 2000; Kolk, 2000). Relations with local communities, community development and corporate philanthropy are important parts of corporate social responsibility for a mining company (Mullen, 1997; Bennett, 1998; Buchholz et al, 1999).

Community development activities involve transferring a company's resources for the production of social goods and services (Husted, 2003). Within companies, three models for governance of community development are most common: (1) *company-led approach*, when companies internalize community development through in-house corporate programs; (2) *philanthropic corporate foundations*, when companies outsource community development through the establishment of a corporate foundation to deliver philanthropic initiatives; and (3) *partnership model*, when community concerns are addressed through a cooperation between company, state and civil society organizations.

This Chapter presents an analysis of the three models for community development drawing from case studies of gold and diamond mining companies and their corporate foundations worldwide and more specifically in the Republic of Sakha in the Russian Federation.

The first part of this Chapter analyses the company-led approach to community concerns and is supported by case studies of gold and diamond mining companies from the Republic of Sakha (previously examined in Chapter 5) and the evidence from global gold and diamond mining companies (previously discussed in Chapter 3). The second part explores the nature of corporate foundations established by international mining companies. The third part investigates partnership model for community development using a case study of community foundations linked to diamond mining in the Republic of Sakha.

Company-led Approach to Community Development

Mining companies develop various initiatives to address concerns of affected communities including: maintenance and delivery of social services to community members; charity giving; cooperation with government agencies and local authorities on issues related to community projects and state socio-economic programs; and development of corporate social programs that target community members. Community relations cover a diverse area of issues that concern the life of the local communities near mines such as:

- Consultation and communication with local communities about possible environmental and social impacts of mining operations.
- Rights of indigenous people.
- Land rights of local communities and indigenous people.
- Contribution of mining companies to community life and well-being.
- Training and employment of local community members in the mining companies.

This section draws examples from community relation strategies in Aldanzoloto, the largest gold mining company in the Republic of Sakha in the Russian Federation which is currently experiencing economic decline, and ALROSA, the largest diamond mining enterprise in the Republic of Sakha, which is a growing company. Both companies caused significant environmental and social impacts in the area of their mining operations; local communities in the adjacent areas have become economically dependent on the mining companies (see Chapter 4 and 5). During the Soviet Era, the mining industry in Russia was state-owned and the state provided for socio-economic development of communities in mining areas, while mining companies acted as state-delegated agents for development. Since the privatization of the mining industry in the 1990s, mining companies no longer receive large endowments from the state for supporting socio-economic initiatives targeted at the local community, but carry on engaging in community development due to the long history of such engagement, as well as high dependency and changing expectations of local communities.

Community Relations in Aldanzoloto

Aldanzoloto, the largest and oldest gold mining company in the Republic of Sakha (see Chapter 5), has caused a significant impact on lives of people in Aldan district, since the beginning of its mining operations in the 1920s. The development of gold mining operations in the region boosted economic development of the district. As Aldanzoloto was controlled by a state agency of Soviet Union, the state funded the company for building of local infrastructure: construction of roads between settlements in Aldan district, installation of telecommunication and electricity lines, etc. Aldan district grew economically dependent on operations of Aldanzoloto as the company was providing for economic prosperity of the region via provision of employment, economic revenues to regional and local budgets, investment in infrastructure and social sphere.

However, as soon as the financial situation in Aldanzoloto worsened, following the disintegration of Aldanzoloto from the larger state-owned Trust, Yakutzoloto Corporation, and subsequent privatization in the beginning of the 1990s, economic problems spread throughout all the settlements of Aldan district. Lack of investments, drop in gold production and financial losses led to closure of some gold mining operations and was followed by redundancies and massive migration of ex-gold miners out of the area. During the 1990s, almost 2,500 miners lost their jobs in Aldanzoloto.

In the settlement of Nizhny Kuranakh, where the Head of Office of Aldanzoloto is situated, 80 per cent of the population is still employed by the company. Even now tax payments from gold mining operations of Aldanzoloto form 50 per cent of the budget of local authority (Administration of Aldan district) which is spent on provision of public services and promotion of socio-economic development in the district. Aldanzoloto provides income and social care for the families of employees and local communities in the area. In the past, Aldanzoloto had several social care facilities such as nurseries and schools, which were maintained at company's expense. However, after facing financial crises in the 1990s, the gold mining company considerably reduced expenditure on social services and these organizations were transferred to self-management or to the state budget of the regional authorities (the Government of the Republic of Sakha). Economic crisis within the company was strongly felt by the communities in Aldan district. According to inspectors from the local environmental protection agency in Aldan district, social development in the mining area depends entirely on the activities of Aldanzoloto:

> The existence and further development of the local community; aid in social sphere and situation … are only possible in case of normal operation of Aldanzoloto with regard to its gold production.

The company still considers community relations as an important part of its operations, although they are currently hindered by financial problems. Relations with the community in Aldanzoloto are organized via the appointment of a manger with responsibilities to deal with social and community issues. In this case, the company's corporate manager on social and community issues also has a position of Deputy General Director Aldanzoloto. The company develops one-year programs with allocated resources for implementation of social and community initiatives. Aldanzoloto implements projects targeting the local community in cooperation with local and regional authorities and local community organizations. Although the company has an extensive corporate community program, the communication of results of the corporate community program to the local community and wider audiences is very limited. Aldanzoloto does not produce a Social Report, but includes some information on community initiatives such as charity giving in its Annual Reports. The company's community program generally targets employees, residential communities and indigenous communities in the areas adjacent to company's operations, and communities outside of the mining area. The corporate community program of the gold mining company has several directions: support of organizations providing social services to the community; engagement in charity giving; cooperation with government agencies; community consultation; and promotion of culture and well-being of indigenous people.

Social services Aldanzoloto makes a contribution to the life of local community via regular support of social care, education and health care institutions in different settlements of Aldan district, but primarily in the settlement of Nizhny Kuranakh, which hosts Aldanzoloto Head Office. Aldanzoloto maintains an orphanage for 98 children in Nizhny Kuranakh; the gold mining company regularly sponsors renovation and equipment update in local schools. Apart from monetary donations, the company makes gifts in kind; after the reduction in the company's size, several buildings became unused, and the company decided to give them to local hospital and schools. The gold mining company also maintains the municipal and utilities services in the settlement of Nizhny Kuranakh. As part of the corporate water collection, Aldanzoloto also provides 50 per cent of the communal water supply in the settlement. The company is regularly engaged in vegetation of streets, actions oriented towards improvement of the settlement, construction and maintenance of roads, as well as cleaning and transportation of communal waste. Moreover, the company helps municipal organizations in the neighboring settlements via assistance with fuel and construction material and also fully maintains a boiler-house in the administrative center of the Aldan district, supplying the town with household heating.

Charity giving Managers of Aldanzoloto recognize that charity giving is one of the main thrusts of Aldanzoloto activity, no significant event or project in Aldan district is conducted without financial support from Aldanzoloto. Aldanzoloto does not have a specific policy on charity giving; the Board of Directors considers appeals for assistance from individuals and organizations. Aldanzoloto usually sponsors organizations and events related to culture and sports development in the region. Aldanzoloto participated in the construction of a skiing complex; the company also supported the reconstruction of a local museum. In 2001, Aldanzoloto donated building material for the construction of museum in Megino-Khangalass district, where the company does not conduct any mining operations. In 2001, Aldanzoloto purchased equipment for a local newspaper *Aldansky Rabochy* (Aldan Worker). Aldanzoloto sponsors cultural and sporting events in the region, such as Days of friendship between Megino-Khangalass and Aldan districts, sports championships, and other celebrations and cultural events. In 2000, the costs of maintenance of the social facilities of Aldanzoloto and expenditure on charity giving and sponsorship amounted to over one million US dollars.

Cooperation with government agencies Aldanzoloto cooperates with the government agencies of the Republic of Sakha in the implementation of numerous social projects. The common approach of the regional government in the Republic of Sakha is to seek financial and resource support from the largest industrial enterprises for the implementation of important socio-economic initiatives in the region. For example, Aldanzoloto participated in the Republican movement 'Two thousand good deeds in the year 2000', initiated by the President of the Republic of Sakha. This movement promoted local authority administrations in each district of the Republic to attract resources from businesses which are operating on their territories to invest in community projects, philanthropic activities, and to support and maintain social care institutions, to develop initiatives targeting employees and neighboring communities, and to contribute to socio-economic development in the areas. As a part of the movement, the gold mining company Aldanzoloto participated in projects oriented towards children and the elderly, arts and culture, sports, education, environmental protection, health care, provision of social services, and assistance to state agencies.

Another example of Aldanzoloto's contribution to government initiated social projects is participation in rehabilitation operations after the spring floods in 2001. The Government of the Republic of Sakha appealed to several large companies in the Republic of Sakha, including Aldanzoloto to assist in rehabilitation of damaged settlements. The Government attached Aldanzoloto to Kobyai district; the gold mining company supplied construction materials and labor for reconstruction of damaged buildings. Aldanzoloto also encouraged charity donations from employees, who donated a share of their salaries to the account to help people affected by the disaster.

Aldanzoloto also regularly cooperates with individual ministries and government agencies on the range of issues. Aldanzoloto collaborates with the Ministry of Industry and on its recommendation often undertakes students and graduates for training, internships or full employment in the company. Aldanzoloto regularly cooperates with environmental agencies (i.e. Ministry for Nature Protection and its local division, the Upper Aldan Territorial Committee for Nature Protection) on environmental protection projects, research and innovations via provision of resources assistance, facilities and transport. Starting in 1992, the local environmental agency and Aldanzoloto have signed one-year agreements on environmental payments from the company.

Community consultation As part of community relations the gold mining company Aldanzoloto involves local authority administration in decision making concerning issues that could potentially affect citizens in Aldan district. Most importantly, the Administration of the Aldan district is a shareholder of Aldanzoloto and formally participates in corporate decision-making through voting at the General meetings of shareholders. Other communication and consultation with community stakeholders are limited; Aldanzoloto primarily focuses on relationships with local authority, rather than community groups, social institutions and NGOs.

> We work in friendship and agreement with local population. Of course, we consider opinion of the local population. In relation to the environmental protection activity of the Company, everything is in order. Tailings are under control. We do not enter in

confrontation with the local population (from interview with senior manager of Aldanzoloto).

Another example in the area of community consultation which indicates that Aldanzoloto tends to ignore local communities interests is the lack of corporate incentive to discuss land use issues with anyone but local authority and government agencies. No consultations with local communities and indigenous people are conducted in relation to land use.

> Lands were allocated to us. Some parts of land are not used for industrial development, and it is just taiga. The land was allocated to us long ago. In area where Aldanzoloto is operating, there are no such issues. Aldanzoloto does not enter into conflicts with local population about pastures, agricultural land. We do not have such issues arising (from interview with senior manager of Aldanzoloto).

Indeed, when the gold mining was developing in the 1920s, Aldanzoloto was given land sites to conduct mining and exploration operations, while reindeer pastures that existed in the area were moved to other sites. Thus no reindeer pastures were left in Aldan district, which constituted the main traditional activity of the indigenous Evenk people.

An important element of community consultation is the communication of corporate environmental and social impacts on livelihood of communities. Aldanzoloto mostly informs government agencies and shareholders, including local authorities about their operations, but rarely is this information related to residents in the mining area, neighboring villages and general public. Although Aldanzoloto occasionally supplies information to a local newspaper about the company's operations, this could hardly be considered as a regular provision of information to stakeholder about corporate sustainability impacts and performance. The gold mining company is not entirely secretive about its mining operations; the company organizes student visits to its facilities to introduce to industrial operations.

Indigenous people Industrial operations of Aldanzoloto have affected the life style of the local population near the mines. Traditional activities of Evenk people who live in the region which have been reindeer farming and hunting are significantly affected by negative environmental impacts of gold mining operations in the Aldan district. Mining and exploration activities disrupted forests and rivers affecting hunting grounds; no reindeer pastures are left in Aldan district. Indigenous people were deprived of their traditional economic activities or forced to move to other areas, and were not initially recruited to the gold mining industry.

> All the territory, which is between river Aldan and Yakokit, was traditionally, many years ago transferred for mining and industrial region. This means that there were no settlements here, where indigenous people would live. Pastures existed here. There were reindeer pastures. They were immediately transferred to other places. I think, in principle this territory has small feeding basis. It is not sensible to develop any traditional types of economy, mostly reindeer farming and hunting. From this point of

> view, traditional forms of economy occupy small area in this mining region. I think that there is no massive impact here. It is only local. Whereas out of 157,000 square kilometers, the total territory of Aldan district, the mining and industrial region accounts for about 80,000 hectares. Well, this means very little. The company is located within this area. This means that… I think that we have many examples, examples of nature. I mean traditional natural resource use is developing normally. This is Khatystyr and Ugoyan. From this point of view, the company does not impact negatively (from interview with senior inspector of Upper Aldan Territorial Committee for Nature Protection).

In the 1980s there were attempts to attract members of the indigenous population (Yakuts and Evenks) into gold mining, however, only a small number of indigenous people stayed to work at Aldanzoloto. Mass recruitment of indigenous people into gold mining failed, which refers not only to Aldanzoloto, but to other mining companies as well. Currently, the gold mining company recruits members from indigenous communities for training and employment at Aldanzoloto, for labor in jewelry production and wood processing, which is closer to traditional activities of indigenous people.

Nevertheless, the company regularly supports settlements of indigenous people and provides them with transport and petrol. The company itself considers that it has in no way infringed upon the interest of indigenous people.

> The company does not affect interest of the local population and does not cause damage, because the nearest settlement with the indigenous population is 60 kilometers from Nizhny Kuranakh. Before the creation of the company in Nizhny Kuranakh, the indigenous population did not constantly live in the near located areas. The company does not affect the life of the tribal communities in Kharystyr, Ulu, Ugoyan and Chagda. The company renders assistance to the indigenous population in the form of petrol, machinery, buses to these settlements, builds and maintains roads (from interview with senior manager of Aldanzoloto).

Currently, Aldanzoloto tries to support traditional forms of economy, participates in organization of cultural events such as 'Days of reindeer farmers' and contributes to private agricultural businesses in indigenous people settlements. The company sponsors holidays for children from indigenous people settlements.

Community Relations in ALROSA

ALROSA is one of the biggest diamond mining companies in the world, which operates on a vast area in the Republic of Sakha (see Chapter 5). The extent of environmental and social impacts caused by diamond mining operations in the Republic was discussed in Chapter 4. The diamond mining company, the leading enterprise in the Republic of Sakha, went through a privatization in 1992 and was in the center of disputes between the federal and regional governments concerning property, land and mineral resources rights and interests of the center and the periphery. ALROSA, via tax payments contributes up to 40 per cent of the entire budget of the Republic of Sakha, which is spent on provision of public services and socio-economic development of the region.

The diamond mining company ALROSA realizes its huge responsibility for well-being of the communities in the diamond province and in the region as a whole. Although ALROSA has not developed a corporate social or community policy or long-term plan for community issues, the company manages community issues via the following mechanisms:

- *Social program* The corporate social department of ALROSA at Head Office in Mirny develops an annual social program for ALROSA that contains corporate operations in the areas of charity giving; cooperation with local authorities-shareholders of ALROSA, cooperation with government agencies, and conservation of traditional lifestyle and culture of indigenous people of the North.
- *Regional policy* ALROSA undertakes operations oriented towards cooperation with the local community via delivery of the regional policy, which is developed and implemented by a Department for regional policy in ALROSA, based in the Head Office in Mirny. Coordination of the regional policy is assigned to the company's Vice-Presidents, based in Mirny, Moscow and Yakutsk. The regional policy deals with issues such as training of local community members for future employment in ALROSA; enforcement of cooperation with local and government authorities; social support of social projects targeting population in the diamond province; provision of corporate employees with products of the local food and agricultural industry; development of patronage relationships between structural divisions of ALROSA and local community organizations; development of sport and culture, promotion of a healthy life style; and environmental safety.
- *Lease agreement with the Government of the Republic of Sakha* According to the lease agreement, ALROSA assigns a part of its revenues, so called 'two per cent assignments' to an environmental fund, which aim is to deliver programs for socio-economic development and environmental improvement in the diamond province which has been negatively affected by the diamond mining operations. In 1993-1998, ALROSA transferred the 'two per cent assignments' to the SAPI Foundation, which delivered programs in the diamond province, and since 1999, the assignments has been forwarded to the Target Fund, managed by the Government of the Republic of Sakha.

Various corporate departments engage in delivering corporate community programs. Implementation of community projects is conducted in cooperation with local authorities and external organizations. The company reports the results of its regional policy, charity giving and other community initiatives in Annual Reports. Corporate community activity in the diamond mining ALROSA could be grouped under several themes: support in provision of social services to communities in the mining area; engagement in charity giving; cooperation with government agencies in social projects; community consultation; and maintenance of community relations via signing community agreements.

Social services ALROSA contributes to community life via provision of social services to employee and residential communities in the mining areas (see Chapter 5). ALROSA also substantially invests in municipal activities via maintaining and building roads, subsidizing culture and leisure facilities, accommodation, agricultural farms, museums, schools and hospitals. ALROSA places an emphasis on agricultural development in districts and supports the traditional economy (cattle breeding and reindeer farming). In addition, ALROSA regularly assists social care organizations (nurseries, hospitals and schools, etc). Individual divisions of ALROSA undertake patronage of certain organizations such as nurseries, clubs, schools and local voluntary organizations. Divisions of ALROSA plan and allocate resources for patronage activities on regular basis and form long-term relationships with these organizations and regularly support their operations.

Charity giving ALROSA spends a substantial amount of money on charity giving. In 1997, ALROSA established a special Corporate Committee for Charity Giving. The Board of Directors adopted special regulations, which the Committee operates under. All requests from individuals and organizations seeking sponsorship and financial assistance from ALROSA are registered and then examined at Committee's meetings, which are usually held once every two months. For instance, in 2000, the Committee held seven meetings, considered 617 written requests and satisfied 347 requests. The Board of Directors approves final decisions of the Committee. The President and the General Director of ALROSA can also grant requests up to 300 US dollars independently from the Committee. In addition, the company has established a Committee for Information Politics, which considers requests for financial assistance in publishing.

No rules apply to the process of decision-making and members of the Committee for charity giving make decisions concerning incoming requests solely at their discretion, although the Regulations stipulate that decisions on requests must be made with consideration of the financial means available in the company. Usually, the decisions of the Committee to satisfy a request rely on a degree of connection between an applicant and the company. The charity giving program of the ALROSA donates on causes such as:

- medical treatment for the poor and assistance to people in unfortunate circumstances (victims of fires and accidents);
- organization of cultural and leisure events and celebrations;
- purchase of houses, machinery and transport means;
- support in contraction development, including construction of churches;
- support of government projects, programs and initiatives;
- support of schools and orphanages.

ALROSA charity giving programs covers the Republic of Sakha and other parts of the Russian Federation, where ALROSA has structural divisions.

Cooperation with government agencies ALROSA regularly cooperates with regional and central governments on environmental protection, remediation of emergency situations, social development, support in cultural and leisure events, etc. ALROSA is closely involved in implementation of social development programs in the Republic of Sakha. When in 1999, the President of the Republic of Sakha initiated a regional movement 'Two thousand good deeds in the year 2000', the company significantly contributed to the movement via implementing community oriented programs in the diamond province such as building of industrial and social care facilities, supply of schools with equipment and computers, organization of small agricultural and industrial businesses, and sponsored holidays and medical treatment for children and their parents in corporate health complexes. In 2001, when spring floods damaged many settlements in the Republic of Sakha, the regional government approached the diamond mining company with a request to assist in restoration operations of the most significantly affected districts. ALROSA actively participated in this restoration and funded the construction of a housing development in Lensk and also helped other districts of the Republic of Sakha. The total contribution of ALROSA to the restoration of districts after flooding amounted to 130 million US dollars in 2001 (AK ALROSA, 2001).

Moreover, the diamond mining company ALROSA regularly cooperates with the Ministry for Nature Protection and its regional branch, the Mirny Territorial Committee for Nature Protection on environmental protection and bio-diversity conservation in the diamond province. In 2000, ALROSA and the Ministry for Nature Protection of the Republic of Sakha signed a three-year agreement for environmental protection, environmental rehabilitation and development of specially protected natural areas in the diamond province. As a part of the agreement, ALROSA has committed to undertake activities to support:

- Environmental communal movements.
- Introduction of environmentally clean technologies for re-cycling of waste.
- Enhancement of bio-diversity such as organization of wild animal and fish farms.
- Monitoring of wild animals in the territory of the diamond province.
- Development of specially protected areas in the territories of Anabar, Mirny and Olenek districts.
- Environmental research in the diamond province, such as environmental monitoring in the area of mining operations.

Community consultation Local authority administrations from eight districts in the diamond province (Anabar, Lensk, Mirny, Nyurba, Olenek, Suntar, Verkhenvilyuysk and Vilyuysk) are shareholders of ALROSA, each holding one per cent of shares. The shareholding of local authorities ensures their participation in decision making in ALROSA via voting at the General meeting of shareholders of ALROSA and each year a Head of administration of a district in the diamond province is elected as a member in the Supervisory Council, the strategic organ of ALROSA. Thus, the principle for involving the local community in the mining area in decision-making concerning corporate operations was set up at the establishment of privatized diamond mining company ALROSA in 1992.

Community consultation in ALROSA is dominated by building strong relationships with local authority administrations and government agencies, while the role of other community stakeholders in communication on corporate operations is diminished, as a Head of social department of ALROSA stated:

> In decision-making…Especially in our times, when everything is so thin, strict, groups of local community cannot participate in decision-making, only professionals. Only professionals can participate, this is my opinion.

The difficulty for ALROSA to consider communities as partners in decision-making is linked to the fact that the company has not yet identified interests of the local communities. Interests and concerns of the local communities are not considered and studied, because the company ignores community interest groups and voluntary organizations as entities that would reflect community interests:

> We do not have nongovernmental organizations, which would make decisions or would be somehow influential. Everyone is at business, working. Organizations, which would be demanding anything, we do not have them here (from interview with manager from ALROSA).

Nevertheless, in some areas of company's operations, where the public is more active, such as Vilyuy region, which was the focus of the environmental movement at the end of the 1980s, the company adopted more flexible approaches to community relationships. ALROSA was one of the first companies in the Republic of Sakha to conduct a consultation forum with local community members before the launch of a new mining project. ALROSA-Nyurba, a subsidiary of ALROSA established for the new diamond developments in Nyurba district, held public meetings to discuss issues on safe exploitation of Botuobinskaya and Nyurbinskaya diamond pipes and future economic and social effects of mining operations attracting local NGOs such as Public Committee for Revival of the River Vilyuy Basin, local authority administrations, government officials and scientists. Recent changes in community relations shows that ALROSA accepts that diamond mining has caused environmental damage in the past and considers the communication with local community as a way of avoiding future conflicts.

Community agreements The regional policy of ALROSA aimed at community development in districts of the diamond province is carried out through a system of one-year agreements signed between the company and local authority administration of districts-shareholders of ALROSA (Anabar, Lensk, Mirny, Nyurba, Olenek, Suntar, Verkhnevilyuysk and Vilyuysk districts); ALROSA-

Nyurba, subsidiary of ALROSA, signed an agreement for community development with Nyurba district. In 2001-2002, ALROSA spent 8.4 and 2.3 million US dollars accordingly on regional policy development in the diamond province (AK ALROSA, 2001e, 2002). The agreements stipulate the company's participation in the various initiatives for community development in the diamond province including:

- *Encouragement of business development.* The diamond mining company is willing to encourage local business development in the mining area and help setting up businesses (e.g. in diamond cutting, mining of industrial minerals, food processing and agricultural sector) to encourage economic development and employment. For example, the company invested in the development of zeolite and salt deposits in Suntar district and set up a company 'ALROSA-Suntar'.
- *Promotion of education and training.* One of the major parts of the agreement is dedicated to promotion of education and training amongst the population in the diamond province. The diamond mining company offers professional training with the opportunity of subsequent employment in ALROSA; training for teachers; and support for young people in accessing higher education. During 1992-2000, ALROSA awarded about 250 school graduates with scholarships to study for higher degrees. ALROSA participated in the establishment of the Verkhnevilyuysk regional school, Suntar polytechnic college and Mirny polytechnic institute.
- *Recruitment in the company.* The diamond mining company regularly recruits members from local communities in the diamond province to work for the company.
- *Promotion of sports, childcare, culture and recreation.* According to the agreements, the diamond mining company delivers numerous initiatives to promote social welfare, offering donations and gifts in kind (such as assistance with resources and transport) to orphanages, schools, rural communities and voluntary organizations, engaged with veterans and poor. The company sponsors sporting events, community project and offers assistance for people in need and funds medical treatments. The company also develops initiatives to improve well-being of children in the diamond province.

Comparison of Community Relations in Aldanzoloto and ALROSA

Aldanzoloto and ALROSA have developed a variety of mechanisms for addressing community relations including, involvement of local communities in decision making, training and recruitment of local and indigenous people, contribution to community development, engagement in charity giving and cooperation with local and regional authorities on various social initiatives. Between the two companies, the diamond mining company ALROSA is more forward looking in the areas of community relations having developed several ways of communicating and cooperating with stakeholders. The gold mining company Aldanzoloto on the other hand is not planning to capitalize on its community relations as the company is uncertain about its future existence, which implies that Aldanzoloto does not assume responsibility for the well-being of communities after closure of mining operations (see Table 6.1).

Table 6.1 Comparison of Aldanzoloto and ALROSA community strategies

Differences	Similarities
• ALROSA has more organized structure for community relations • ALROSA established a compensation system for local community via contribution to environmental fund (SAPI Foundation and later the Target Fund) • Relations with local communities in ALROSA are regulated via agreements with local authority administrations in the districts of diamond province • A special established committee within ALROSA supervises charity giving • ALROSA conducts consultation forums with stakeholders on effects of corporate operations	• Both companies have significant effects on communities where they operate, making mining areas highly dependent on the economic success of mining • Both companies support organizations for education, social care, health, leisure, culture and sports • Both companies contribute to socio-economic development of the area via road and town building, and infrastructure development
Strengths	**Weaknesses**
• Inclusion of the local community in decision making via shareholding • Substantial contribution to community development • Training and recruitment of local community members and indigenous people • Support of community projects within the indigenous communities • Cooperation with local and regional authorities on social projects	• No corporate community policies • No Environment and Social Reports • No corporate philanthropic foundations • Neglect of NGOs in stakeholder dialogue • No discussion land rights with local communities

The analysis of corporate strategies in the area of community relations and development shows that mining companies in the Republic of Sakha are concerned about the well-being of local communities and indigenous people and are making actions to meet their expectations and needs. However, the most common approach of the mining companies is to give priority to regional and local authorities in stakeholder relations. Center-periphery disputes and recent shifts in democratic developments in the Russian Federation significantly influenced the process of privatization of the two companies. Both companies have local authority administrations as founders and subsequent shareholders and membership of the Heads of the district administration in corporate councils, which ensures the involvement of local communities in decision-making. However, other forms of stakeholder consultation are very limited.

ALROSA organized consultation forums, attracting the local community, scientists, local and regional authorities in order to explain and advertise the company's intentions and operations in relation to the new mining developments. One of the important issues in community relations is the negotiation concerning land rights of local communities. For example, three out of the global top ten gold mining companies (Freeport-McMoRan Copper & Gold, Normandy Mining and Rio Tinto) and two out of the global top seven diamond mining companies (Argyle and Ekati) are willing to negotiate land issues with local communities. However, Aldanzoloto and ALROSA fail to acknowledge the rights of local communities for land and only consider these issues in cooperation with the state authorities.

Aldanzoloto and ALROSA undertake many attempts towards training and recruiting from local communities. Similar practices are employed by many global gold and diamond mining companies such as Argyle, Ekati, Debswana and Namdeb. Aldanzoloto and ALROSA have given important consideration to the recruitment policy of indigenous people and offer them jobs in corporate divisions with jobs closer to the traditional activities of the indigenous people such as jewelry production and wood processing that would help to stimulate them to retain the jobs. However, the percentage of indigenous people of the Republic of Sakha employed by Aldanzoloto and ALROSA does not exceed four per cent of the total number of employees.

Before privatization, when both companies were state-owned enterprises and endowed with generous state investments, they were actively involved in community development projects and were seen as agents for socio-economic development in the mining regions, as they invested in infrastructure development, built roads, schools and clinics, supported and maintained cultural, leisure and educational organizations. The newly privatized companies, although have governments as their shareholders mostly rely on private capital for investment. These mining companies now being private enterprises still carry on their responsibilities for local and regional communities, in some cases to lesser extent and in some cases improving social conducts, due to existing and changing expectations of companies' stakeholders. The existing demands to deliver good to the communities are formed by the conduct of previous state-owned enterprises that made communities highly dependent on them. New demands are formed by new economic, legal and political relations of privatization, democratic development, strengthening of regional and municipal authorities. In terms of corporate social responsibility, the mining companies form relations with the following groups of stakeholders:

- Regional authorities have more political and economic liberties than they had during the period of Soviet Union. Regional authorities are now a much stronger party in the dialogue on economic benefits from minerals development on their territory. The expectation of regional authorities from the mining companies to contribute to the socio-economic development of the region are higher then ever, as the role of state in delivering public services is diminishing due to the lack of resources.

- Local authorities find themselves in the wake of democratic development being able to demand more economic and social returns from mining operations on their territories. Local authorities want to have the control over corporate operations and participate in fair distribution of economic benefits from mining through holding shares in the companies.
- Local community organizations (schools and educational institutions, voluntary organizations, arts and culture organizations such as museums, libraries, etc) have long-term relations with the mining companies and are expecting the mining companies to supports their activities further especially as the government funding is limited.
- The mining companies do not experience considerable pressure from nongovernmental organizations or interest groups. Nevertheless, the companies should not underestimate the powers of local NGOs, as based on the example of the rise of the environmental movement in the diamond province in the 1980-90s, led by the local NGO (Public Committee Vilyuy), which resulted in significant changes in governmental and corporate policies in relation to compensation and improvement of environmental situation in the diamond province.

Corporate Foundations Established by Mining Companies

Corporate Foundations

According to the concept of CSR, corporations have an obligation to give back to the society in the form of donations for the public good. Corporate philanthropy is an essential element of many companies' community relations and CSR. Corporate philanthropy differs from commercial sponsorship in that it involves donations made without any expectation of direct commercial reward. For the past two decades, corporate charitable donations have significantly increased in Northern America and Western Europe which relates to increased corporate resources to donate to charities, successful appeals of charities to companies, government pressure on companies to contribute to community needs and consumer demands for greater social accountability. In addition to charitable donations, companies also make other charitable contributions to the community such as gifts in kind, loans of equipment and or/employees to organizations, the provision of facilities, professional advice and participation of charitable campaigns One of the crucial reasons for the rise of corporate philanthropy in the developed countries is an escalation of social problems, diminishing role and capacity of the state to offer social care and services following privatization of industries previously run by national governments (Mullen, 1997; Bennett, 1998; Buchholz et al, 1999). Corporate philanthropy can bring benefits to businesses and is used as a strategic marketing tool (Goldberg, 1997; Bennett, 1998; Marx, 1998; Business for Social Responsibility, online) to achieve business objectives such as:

- facilitation of relations with government and other key audience;
- enhancement of corporate image and increase customer loyalty;
- strengthening of employee motivation, loyalty and productivity;
- attraction of extensive media coverage;
- changing public attitudes;
- benefit communities; and
- generally support promotional campaign.

The extent of corporate philanthropy depends on several determinants such as company's size, board composition, financial performance, values of chief executive officer, institutional forces, industrial sector and organizational age (Buchholz et al, 1999). Corporate philanthropy could be organized through corporate philanthropic activities such as charity giving described in the previous section or through an establishment of a corporate foundation or trust; the latter has become a popular way for industry to respond to environmental and social concerns. Foundations (the name used in the US) or Trusts (often used in the UK) are nonprofit making organizations established for philanthropic purposes, which derive their funds primarily from the contributions of profit-making businesses, and are managed by their own wholly independent management (Norton, 1981). Foundations come in a variety of shapes – private, company sponsored, governmental and formed from several sources (Mirikitani, 1999), and sizes – local, regional, national and international (Norton, 1981) and behave in as many different ways – grant-making and operational (Porter and Kramer, 1999) and pursue various activities – community, environmental, social and charitable (Foundation Centre, 2001).

The establishment of corporate foundations as the means for delivering community development initiatives is a popular option amongst many mining companies. Companies prefer to manage their philanthropic initiatives through corporate foundations for several reasons (International Finance Corporation, 2000):

- Tax laws sometimes make it beneficial to set up an independent organization.
- Independent management structure could allow the company development program to have more community control than would be feasible for a company managed program.
- A separate organization can help the company to separate legal liability for the actions of the community development program from those of the company.
- An independent organization can outlive the company, adding to program sustainability.

Corporate foundations often maintain close ties with donor companies, whose interests and priorities are often reflected in foundations' operations; however corporate foundations are separate, legal organizations with their own endowment (Pine et al, 1998). Typically, corporate foundations are governed by Boards, which include officers from donor companies. According to the objectives and focus of a foundation, Hamann et al (2001) groups corporate foundations into three major types:

- Implementing (or community) foundations, which work directly with communities. They have their own staff that develop and implement local development initiatives together with community representatives.
- Development foundations implement development work that goes beyond the scope of the communities affected by the donor company's actions, addressing issues that affect regions, countries, or the globe.
- Partnership foundations. Their main objective is explicitly and purposefully to establish, maintain, and expand their interactions with government institutions, NGOs, donor agencies and other businesses.

The nonprofit sector has become a powerful force, which accumulates large sums of money and is able to invest significant amounts in development issues. For example, the US nonprofit sector is highly developed. According to the Council of Foundations, there are more than 2,000 corporate foundations in the US holding some 13 billion US dollars in assets. The range of activities that philanthropic foundations fund depends on type, location and size of foundations (see Table 6.2).

Table 6.2 Main areas of funding by grant-making foundations

Fields of activity for grants	UK grant-making trusts*	Main categories of grants	US corporate foundations
• Social care and development	35%	• Human services	16%
• Arts, culture and recreation	22%	• Arts and culture	13%
• Education and training	31%	• Education	24%
• Conservation and environment	23%	• Environment and animal	6%
• Health	33%	• Health	17%
• Religious activities	22%	• Religion	2%
• Infrastructure	17%	• Public and society benefit	11%
• Accommodation and housing	24%	• Science and technology	6%

* Does not add up 100% as grant-makers support more than one field.

Source: Adapted from Pharaoh and Smerdon 'Dimensions of the Voluntary Sector: Key Facts, Figures, Analysis and Trends' (1997); Foundation Centre, 'Foundation Reporting' (2000).

The nonprofit sector including corporate foundations continues to grow and provide critical social services and advocacy, creating and sustaining organizations that have become increasingly important to the society. Unlike profit making organizations, nonprofit foundations do not have a clear bottom line to determine how well they are performing and achieving their mission. Instead, performance of nonprofit organizations could be judged by using program evaluation, which assesses the program's success in achieving set goals and satisfying clients, and assessment of organizational effectiveness, which assesses how well organizational processes (e.g. leadership, board of directors, human resources) work to carry out the mission.

Corporation foundations are accountable to various stakeholders including governments, donors, clients, regulatory bodies, employees, board of directors and the communities in which they operate. Although donor companies usually define the accountability criteria, other stakeholders also expect corporate foundations to account for their operations as they are often considered to be a part of donor company philanthropic activity (Buckmaster, 1999). Responding to the public concerns to show accountability of their operations, nonprofit organizations are trying to demonstrate that they achieve specified goals using reporting. A number of foundations today produce Annual Reports and other publications designed to inform the public about their giving activities and interests, grant-making guidelines and work with their grantees.

Corporate Foundations of Mining Companies

Mining companies also establish corporate foundations to deliver community development and social initiatives. The purposes of corporate foundations established by mining companies, their operation modes, areas of funding and target population are reviewed in this section based on an analysis of publicly available documents (Annual Reports of donor companies and/or corporate foundations, other published material from donor companies and/or corporate foundations such as Community Reports and information provided on official websites of donor companies and their foundations). Corporate foundations that are established by different mining companies are selected for this review on the basis of their size in terms of market capitalization, economic performance or mineral output of their donor companies. Size has been chosen as a criterion for selection because it has been noted that company's size is a major factor in corporate philanthropy (Buchholz et al, 1999). Assets of mining companies worldwide have grown as well as their profits; therefore large mining corporations are more likely to establish corporate philanthropic foundations. Four groups of corporate foundations have been selected for the review, established by four groups of mining companies:

- *Mining foundations* Corporate foundations established by the top 50 mining companies (excluding gold and diamond mining companies as they are studied separately). Out of the top 50 mining companies, ranked according to their market capitalization in January 2001 (MJ, 2001), 37 companies provided with information on their philanthropic activities, of them 12 companies established 22 corporate foundations (see Table 6.3). BHP and Billiton merged in 2001 that is why they are viewed as one company in this study. This group of corporate foundations was chosen to provide a reference to the mining sector in general.

- *Gold mining foundations* Corporate foundations established by the top ten gold mining companies, ranked according to their annual gold production in 2000 (World Gold Analyst, 2000). Out of the top ten gold mining companies seven companies provided with information on their philanthropic activities and established 13 corporate foundations (see Table 6.4). This group of corporate foundations was chosen to provide a reference to the international gold mining industry and compare it with data on corporate community development of a case study gold mining company.
- *Diamond mining foundations* Corporate foundations established by the top seven diamond mining companies, ranked according to their annual diamond production (Rombouts, 2000). Out of the top seven diamond-mining companies three companies supplied with information on their philanthropic activities and all three established corporate foundations (see Table 6.4). This group of corporate foundations was chosen to provide a reference to the international diamond mining industry and compare it with data on corporate community development of a case study diamond mining company and the case study of community foundations in the Republic of Sakha.
- *Russian foundations* Corporate foundations established by top Russian extractive companies (except for ALROSA as it is studied separately) engaged in oil, gas, ferrous and non-ferrous metallurgy, ranked according to their economic performance in 1999 (Expert, 1999). Six nonprofit organizations were selected for this group, including one foundation, which was created by the efforts of three Russian oil and gas companies (see Table 6.4). This group of corporate foundations was chosen to provide a reference for the case study foundations and case study companies operating in the Republic of Sakha, a region of the Russian Federation.

Establishment Mining companies around the world have been establishing corporate philanthropic foundations since the 1930s. Most corporate foundations within the chosen four groups (mining, gold mining, diamond mining and Russian extractive sector) were established in the 1990s, which primarily refers to the opening of new projects or closure of old mines. The establishment of corporate foundations within the gold mining industry is a relatively new phenomenon, while the establishment of corporate foundations within the Russian extractive sector is not a very popular trend, which started only since 1993 (see Table 6.5).

Mode of operation Foundations of mining companies are often grant-making, when grants are awarded to individuals and nonprofit organizations for delivery of social programs. Grant-making foundations are very common amongst corporate foundations in general. Although some Russian foundations give away grants, all

Russian foundations refer to operating foundations, which deliver initiatives and services themselves (see Table 6.5).

Table 6.3 Mining foundations

Foundation	Donor company	Since	Location
Alcoa Foundation	Alcoa	1952	USA
Alcoa-Alumax Foundation Inc.	Alcoa-Alumax	1998	USA
AngloAmerican Chairman's Fund	AngloAmerican	1960s	South Africa
BHP Community Trust	BHP	1988	Australia
Central Highlands Community Employment Trust	Pacific Coal (Rio Tinto)	1999	Australia
Coal and Allied Community Development Trust	Coal &Allied (Rio Tinto)	1999	Australia
Development Fund financed by Privatization (FRD)	CVRD	1997	Brazil
Escondida Foundation	Minera Escondida (BHP)	1996	Chile
Fly River Development Trust	Ok Tedi (BHP)	1990	Papua New Guinea
Foundation Falconbridge Inc	Falcondo (Falconbridge)	1989	Dominican Republic
Impala Corporate Development Trust	Imapala	2000	South Africa
Iscor Foundation	Iscor	1994	South Africa
MIM Community Trust Fund	MIM Holdings	2000	Australia
Ok Tedi Development Foundation	Ok Tedi (BHP)	2000	Papua New Guinea
Outokumpu Oyj's Foundation	Outokumpu	1930s	Finland
Palabora Foundation	Palabora Mining (Rio Tinto)	1987	South Africa
Phelps Dodge Foundation	Phelps Dodge	1953	USA
Rio Tinto Coal Community	Rio Tinto Coal (Rio Tinto)	1999-2001	Singleton, Australia
Rössing Foundation	Rössing Uranium (Rio Tinto)	1979	Namibia
Tarong Coal Community Development Fund	Tarong Coal (Rio Tinto)	1999-2001	Australia
Vale do Rio Doce Foundation (FVRD)	CVRD	1968	Brazil
WMC Sir Lindesay Clark Trust Fund	WMC	1979	Australia

Target population The majority of corporate foundations primarily target local communities in the area immediately surrounding the mines of their donor mining companies. If operations of the mining company are spread geographically on a large area then its corporate foundation is likely to address local communities at the mines and communities outside the mining area and wider circles of the society. For example, amongst Russian foundations there are foundations that address the interests of specific communities such as teachers and young people that are not members of local communities near the mines (see Table 6.5).

Table 6.4 Other mining foundations

Foundation	Donor company	Since	Location
Gold mining foundation	*Gold mining company*		
AngloGold Fund and Educational Trust Fund	AngloGold	1998	South Africa
Community Development Fund	AngloGold	1998	Mali
Freeport Fund for Irian Jaya	PT-FI (Freeport-McMoRan C&G)	1996	Indonesia
Freeport McMoRan Foundation	Freeport-McMoRan Copper & Gold	1985	USA
Gold Fields Foundation	Gold Fields	1998	South Africa
Inti Raymi Foundation	Inti Raymi Mining Corporation (Newmont)	1991	Bolivia
Mercur Legacy Scholarship Fund	Barrick Gold	1989	USA
Metropolitan Arts Fund	Freeport-McMoRan C&G	1989	USA
Normandy Youth Entrepreneur Fund	Normandy Mining	1999	Australia
Rio Tinto Aboriginal Foundation	Rio Tinto	1996	Melbourne, Australia
Rio Tinto Indonesia Foundation	Rio Tinto	1995	Indonesia
Rio Tinto Zimbabwe Foundation	Rio Tinto	1974	Zimbabwe
SMEDI – Small and Medium Enterprises Developments Initiative	AngloGold	1998	South Africa
Diamond mining foundation	*Diamond mining company*		
De Beers Fund	De Beers	1998	South Africa
Debswana Donations Fund	Debswana	-	Botswana
Namdeb Social Fund	Namdeb	1977	Namibia
Russian foundation	*Russian extractive company*		
Miloserdie Charity Foundation for Social Protection	Novolipezk metallurgical complex	1999	Russia
Miloserdie Charity Foundation	Tatneft	-	Russia
Lukoil Charity Foundation	Lukoil	1993	Russia
Federation of Internet Education	YUKOS	2000	Russia
Rukhiyat Foundation	Tatneft	1997	Russia
Vernadsky Foundation	UES of Russia, Gazpom and Lukoil	1995	Russia

Often when donor companies operate only within boundaries of one continent (in the case of diamond mining foundations) or one country (in the case of Russian foundations), then corporate foundations serve communities only within that continent or country. However, when the mining company operates in several countries and on various continents, two models are available for the donor company. First, the mining company establishes several separate corporate foundations to address concerns in an individual country or in a particular mining area (Rio Tinto and its corporate foundations Tarong Coal Community Development Fund and Fly River Development Trust). Second, the mining company establishes one corporate foundation that addresses communities in all the areas of company's operations (Alcoa Foundation and De Beers Chairman's Fund).

Table 6.5 Operations of corporate foundations

	Mining foundations (22)	Gold mining foundations (13)	Diamond mining foundations (3)	Russian foundations (6)
Year of establishment				
1930-1950s	30%	0%	0%	0%
1960-1980s	20%	30%	33%	0%
1990-2000s	50%	70%	66%	100%
Mode of operation				
Grant making foundation	55%	92%	100%	16%
Operational foundation	0%	0%	0%	50%
Combined mode	5%	0%	0%	34%
Unknown	40%	8%	0%	0%
Target population				
Local communities near mines	77%	92%	0%	80%
Regional communities	1%	9%	0%	0%
National communities	1%	0%	100%	20%
International public	1%	0%	0%	0%
Financial endowment				
Less than 1 million US dollars	45%	15%	66%	-
1-5 million US dollars	23%	30%	33%	-
5-20 million US dollars	4%	8%	0%	-
More than 20 million US dollars	4%	0%	0%	-

Expenditure The majority of corporate foundations established by the international mining companies have expenditure of about one million US dollars (including Rio Tinto Coal Community and Foundation Falconbridge Inc.). Endowments of corporate foundations established by mining companies of the first group vary dramatically, starting from 70 thousand US dollars a year to 36 million US dollars as a one off donation from the mining company. Financial endowment of gold mining foundations differs from 150 thousand US dollars to 13 million US dollars. Amongst the diamond mining foundations, De Beers Fund is the largest corporate foundation, which has a budget of 3.5 million US dollars a year. Top mining companies and top gold mining companies establish the wealthiest corporate foundations (see Table 6.6).

Table 6.6 Expenditure of corporate foundations of various mining companies

Foundation	Year	US$ '000
Mining foundations		
Alcoa Foundation	1999	18,000
Alcoa-Alumax Foundation Inc.	1999	1,000
AngloAmerican Chairman's Fund	1999	2,700
Coal and Allied Community Development Trust	1999-2002	500
Development Fund financed by Privatization (FRD)	1997	36,000
Escondida Foundation	-	1,200
Fly River Development Trust	1998	500
Foundation Falconbridge Inc.	1989-1999	70
Ok Tedi Development Foundation	2000	1,000
Outokumpu Oyj's Foundation	2000	1,000
Palabora Foundation	2000	200
Phelps Dodge Foundation	1997	2,000
Rio Tinto Coal Community	1999-2002	500
Rössing Foundation	2000	2,000
Tarong Coal Community Development Fund	1999-2002	200
Vale do Rio Doce Foundation (FVRD)	1999	5,000
WMC Sir Lindesay Clark Trust Fund	2000	360
Gold mining foundations		
AngloGold Fund and Educational Trust Fund	1999	3,000
Freeport Fund for Irian Jaya	1998	13,500
Freeport McMoRan Foundation	1999	1,850
Inti Raymi Foundation	2000	1,200
Normandy Youth Entrepreneur Fund	1999	150
Rio Tinto Aboriginal Foundation	1998	600
SMEDI – Small and Medium Enterprises Developments Initiative	1999	2,000
Diamond mining foundations		
De Beers Fund	2000	3,500
Donation Fund	2000	500
Namdeb Social Fund	2000	200

Programs The majority of corporate foundations established by the top mining companies invest their resources in areas such as education and training, community development, health, arts and culture, and business development. Lesser mining foundations support environment, economic development, research, agriculture, and job creation. The selected gold mining foundations mostly implement initiatives in the areas of education and training, health, business and social development (see Table 6.7). Although mining companies cause significant environmental impacts, only a few corporate foundations donate to activities concerning environmental protection; only ten out of 44 reviewed foundations report to have programs on environmental protection.

Table 6.7 Areas of support by corporate foundations

Areas of funding (do not add up to 100%)	Mining foundations (22)	Gold mining foundations (13)	Diamond mining foundations (3)	Russian foundations (6)
Business development	36%	30%	66%	16%
Community/Civic	60%	23%	100%	50%
Culture/Arts/Sports	41%	23%	66%	83%
Economic development	18%	8%	33%	0%
Education/Training	82%	70%	66%	50%
Environment	27%	8%	33%	33%
Health	45%	62%	100%	33%
Job creation	9%	23%	33%	0%
Research	14%	15%	33%	50%
Rural/Agriculture	9%	23%	0%	0%
Social welfare	50%	30%	33%	0%
Women/Youth	18%	15%	66%	16%

The diamond mining foundations that are situated in Africa primarily invest in business development, community development, health, culture and education. In addition to corporate foundation that are nonprofit organizations, the diamond mining company Debswana set up two profit-making companies, Peo Holdings Limited and Masedi Limited, which are engaged in creation of jobs, rural and agriculture development. In the Russian Foundation popular fields for corporate foundations activities are arts and culture, education and training, community development and research; less popular topics include environment, health, women/youth programs and business development. None of the studied Russian foundations invest in economic development, job creation, agriculture and rural development. Clearly, the topics of activities on which corporate foundations focus depend on community needs and character of regional social problems.

Partnership Model for Community Development

Tri-sector Partnership for Community Development

Community development in the mining areas is recognized as an area, which requires efforts not only from the mining companies, but local and regional governments and civil society organizations. Tri-sector partnership offers a model for cooperation between the three sectors: the industry, the government and civil society for socially related projects such as development of local communities in the mining areas (see Chapter 2). Some mining companies establish corporate foundations with the view of tri-sector partnership. Hamann et al (2001) and Warhurst (2001) discuss the benefits of adopting a tri-sector partnership model in the establishment and operation of corporate foundations within the mining sector, which include:

- sustainable community development;
- enhanced transparency within governments;
- more equitable distribution of wealth;
- avoidance of local disputes;
- informal social 'license to operate' for companies;
- reduced dependency of local communities on companies.

Table 6.8 Advantages and disadvantages of adopting a tri-sector partnership model to enhance corporate foundations

Advantages and benefits	Disadvantages and constrains
• Increased buy-in from a variety of stakeholders; • Integrated needs appraisal and identification of objectives; • Effective information exchange and co-ordination of implementation activities; • Greater efficiency in the allocation of resources and less duplication of effort, with resulting cost savings; • Increased trust and capacity for conflict resolution; • Improved evaluation of programs and financial accountability; • Increased access to resources, directly through contributions from partners or through greater ability to solicit external donor support; • Ongoing capacity-building for community and government institutions; • Prolonged existence and greater continuation of development practices and impacts, leading to increased self-sufficiency and independence.	• Where there are no accountable local organizations, or where there is local conflict, a partnership approach may disable the effective implementation of foundation objectives or require disproportionate resources for capacity-building; • The potential exclusion of stakeholders not represented by any of the tri-sector partners; • The unwillingness of company representatives to give up ownership of outcomes, on the basis of the high investments tied up in a foundation – this presents obvious difficulties for the adoption of the tri-sector partnership model, particularly in foundation management.

Source: Adapted from Hamann, 'Corporate Foundations and Tri-sector Partnerships' (2001).

The major aspects to be considered when establishing and/or operating a corporate foundation on the principles of tri-sector partnership relate to design of a foundation, provision of resources, management of a foundation and program implementation. Key stakeholders should be involved in developing the objectives of a foundation and establishment of its management rules. The provision of resources for the foundation could include direct contribution from the partners or be enhancement of ability to obtain donor support. Management of a foundation should include representatives from all the partners. Places on the management boards or management team of a foundation might include representatives from communities, local government, NGOs and development agencies. Partners should be engaged in implementation of programs and delivery of the foundation's objectives. Different organizations from civil society, government could provide competencies and resources that complement those of the company and each of the partners could contribute in a coordinated fashion to the activities of a foundation (Hamann et al, 2001). The structural elements of tri-sector partnerships include legitimacy of representation; geographic boundaries and target population; responsibility, transparency and accountability; communication between partners and their constituents; capacity building; and monitoring (Davy, 2001). The adoption of a tri-sector partnership model to corporate foundations has a series of benefits and drawbacks (see Table 6.8).

In terms of delivery of a foundation's initiatives, one of the clearest ways in which the tri-sector partnership model may contribute to a foundation's objectives is by linking the activities to existing governmental or nongovernmental projects or programs. In this manner, pooling of core competencies can avoid duplication of effort, and numerous opportunities for efficient use of labor can be explored. Concerning evaluation and accountability, it is important that the foundations' operations are evaluated and its programs reviewed. This may be carried out internally by the company or by external auditors. In the tri-sector partnership approach, community consultation is an integral part of the process of defining the evaluation criteria, measurement and communication (Hamann et al, 2001).

Environmental Programs in the Diamond Province of the Republic of Sakha

This section investigates the tri-sector partnership in practice drawing from the case study of foundations established by the diamond-mining industry in the Republic of Sakha. The extent of environmental and social problems in the diamond province (see Figure 6.1) in the Republic of Sakha caused by diamond mining operations and accompanying industrial development were kept from exposure to the public almost until the late 1980s (see Chapter 4). At the end of 1980s with the start of *perestroika* and *glasnost*, the exposure of environmental and health problems in the Vilyuy region started a public movement for improvement of the environmental situation in the Vilyuy region, which was actively steered by Vilyuy Public Committee, a nonprofit organization established in 1990 in Nyurba. Along with its supporters (local authority administrations, government agencies, various associations and organizations in the region) the Vilyuy Public Committee appealed to governments of the Republic, Russia and the Soviet Union to take immediate actions for the improvement of the environmental situation in the Vilyuy region. In response to the environmental movement, the regional government of Yakutia started to develop a program for environmental improvement in the Vilyuy region in the beginning of 1990s, however, the program failed due to uncertainty in the sources of funding. In 1990-1991, the regional government acted to confer the river Vilyuy basin with the status of 'environmentally unfavorable zone' in the USSR and the RSFSR, but the Governments of Russia and the Soviet Union rejected these initiatives.

In 1992, after the declaration of sovereignty, the Government of the Republic of Sakha (Yakutia) started to formulate the official position to address environmental problems in Vilyuy region (Decree of the Presidium of the Supreme Soviet of the Republic of Sakha 'About set of primary actions for improvement of environmental situation in the river Vilyuy basin' of 20 March 1992) (Pavlov and Afanasieva, 1997). The Government developed a program for the environmental protection and socio-economic development actions in the river Vilyuy basin in the diamond province for 1992-1995, which aims were to:

- Define a state of environmental situation in the river Vilyuy basin and evaluate the damage, caused by the Yakutlamaz Corporation, the Yakutenergo (acronym for Yakut Energy) Corporation and geological exploration, to the environment and health of the population in the river Vilyuy basin.
- Develop and implement a program of compensation to the population affected by the damage using endowment from the Russian Federation, Republic of Sakha (Yakutia), the Yakutalamaz Corporation, the Yakutenergo Corporation and the Lenaneftegasgeologia (acronym for Lena Oil and Gas Geology) Corporation.
- Create a special Fund for environmental rehabilitation of the river Vilyuy basin endowed by assignments from the Government of the Russian Federation, Government of the Republic of Sakha, local authority administration of affected districts, Yakutalmaz Corporation and Yakutskenergo Corporation.

In April 1992, the Government of the Republic of Sakha established the Fund for Vilyuy Ecology with aims to rehabilitate the environmental situation in the diamond-mining region. Until the final evaluation of compensation for damage, the Yakutlamaz Corporation and the Yakutenergo Corporation were obliged to assign funds to the Fund for Vilyuy Ecology. The program for environmental and socio-economic actions in the Vilyuy region addressed six districts: Kobyai, Mirny, Nyurba, Suntar Verkhnevilyuysk and Vilyuysk. Several initiatives were included in the program such as environmental protection initiatives in the Vilyuy region, construction of water supply and sewage systems, construction of hospitals and clinics in the diamond province. The Fund for Vilyuy Ecology operated only for a year and in 1994 ceased to exist (Mironov, 1999).

In 1992, the Yakutalmaz Corporation, state-owned enterprise and the main benefactor for the Fund for Vilyuy Ecology was closed down and a new joint-stock company ALROSA was created to take over its diamond-mining operations. Chapter 5 highlighted that the diamond mining company ALROSA signed the agreement with the Government of the Republic of Sakha for 'Lease of land sites, diamond deposits and assets of Yakutalmaz Corporation' in 1993, according to which ALROSA is obliged to assign two per cent of its revenues to a special environmental fund for rehabilitation of the diamond province. In 1993, the President of the Republic of Sakha established the SAPI Financial Corporation, the state-owned enterprise to deliver the socio-economic and environmental programs in the diamond province using two per cent assignments from the diamond mining company ALROSA (Yakovleva and Alabaster, 2001).

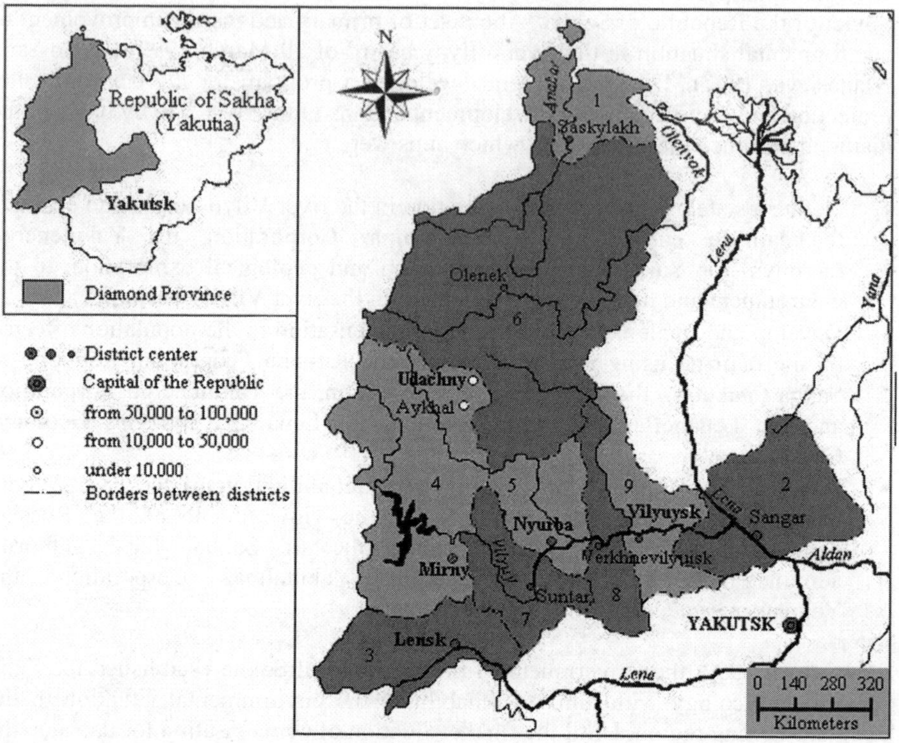

Figure 6.1 Map of the diamond province in the Republic of Sakha

Note:

No.	District	District centre	Area, sq. km	Population
1	Anabar	Saskylakh	55,600	3,900
2	Kobyai	Sangar	107,800	19,200
3	Lensk	Lensk	77,000	49,800
4	Mirny	Mirny	165,800	51,700
5	Nyurba	Nyurba	52,400	29,900
6	Olenek	Olenek	381,100	4,300
7	Suntar	Suntar	57,800	27,200
8	Verkhnevilyuysk	Verkhnevilyuysk	48,000	21,400
9	Vilyuysk	Vilyuysk	55,200	28,800
	Total		1,000,700	236,200

In 1995, the SAPI Financial Corporation was re-organized into a private nonprofit foundation and since then is known as the SAPI Foundation, which received assignments from ALROSA and delivered social and environmental programs in the diamond province. However, the Government of the Republic of Sakha was not satisfied with the situation about private management of assignments from ALROSA, which were set up by the lease agreement between the diamond mining company and the Government of the Republic of Sakha. In 1999, the Government of the Republic of Sakha decided to incorporate 'two per cent assignments' from ALROSA to the State Budget of the Republic of Sakha and manage them through a governmental Target Fund to deliver socio-economic and environmental programs in the diamond province.

SAPI Foundation

In 1993, SAPI Financial Corporation was established by the Decree of the President of the Republic of Sakha in order to create an effective mechanism for addressing socio-economic and environmental problems in the diamond province and also to increase the financial potential of the Republic of Sakha. Since then, the SAPI Financial Corporation has been designated to receive 'two per cent assignments' from ALROSA, specified in the lease agreement, and deliver environmental and socio-economic programs in eight districts (Anabar, Lensk, Mirny, Nyurba, Olenek, Suntar, Verkhnevilyuysk and Vilyuysk) of the diamond province (Yakovleva and Alabaster, 2001).[1]

Since its establishment in 1993, the SAPI Financial Corporation founded various businesses in areas such as banking, manufacturing, finance, trade, publishing, services and which have been united under the SAPI Financial Corporation. In 1995, with the introduction of the new Civic Code in the Russian Federation, the SAPI Financial Corporation (state-owned enterprise) was re-organized into the SAPI Foundation (nonprofit organization), which became the owner of the 'two per cent assignments' from diamond mining company ALROSA. In 1997, the SAPI Foundation and ten businesses, which the SAPI Foundation has set up, were formed into the SAPI Financial and Industrial Group, which became the first and the only organization of its kind in the Republic of Sakha. The chronology of the social and environmental programs in the diamond province is shown in Table 6.9.

The mission of the SAPI Foundation was to accumulate financial resources and then invest them to address socio-economic and environmental problems in the districts of the diamond province of the Republic of Sakha. The management bodies of the SAPI Foundation were: President of the SAPI Foundation, Committee of Founders and Board of Directors. The Committee of Founders was set up as a supreme management body of the SAPI Foundation, whose membership had to be approved by the President of the Republic of Sakha. The Committee included representatives of ALROSA, the Government of the Republic of Sakha, and the SAPI Foundation, and the heads of local administrations from the diamond province.

Table 6.9 Chronology of key events in the diamond province

Year	Event
1950s	• Discovery of first diamond kimberlite pipe Zarnitsa (1954) • Establishment of Yakutalmaz Corporation (1957) • Foundation of town Mirny as a centre for diamond mining industry (1959)
1960s	• Foundation of settlement Chernyshevsky (1961) • Foundation of diamond mining town Aykhal (1962) • Foundation of diamond mining town Udachny (1968) • Beginning of flooding of area for water reservoir of Vilyuy hydroelectric power station (1969)
1970s	• Completion of water reservoir of Vilyuy hydroelectric power station, covering an area of 2,170 sq.km. (1973) • Nuclear explosion with code name Kristall was carried out near town Udachny in Mirny district, creating an accidental fall-out of radioactive elements (1973)
1980s	• Level of phenol of Vilyuy river reached 0.003-0.004 milligrams per liter, 30-40 times above the maximum permissible concentration levels
1992	• Decree of Supreme Soviet of the Republic of Sakha (Yakutia) 'About set of primary actions for improvement of environmental situation in the river Vilyuy Basin' of 20 Match 1992 • Decree of the Government of the Republic of Sakha (Yakutia) 'About emergency socio-economic and environmental protection actions for rehabilitation of environmental situation in the river Vilyuy basin' of 27 April 1992 • Establishment of Fund for Vilyuy Ecology
1993	• Establishment of ALROSA Company Ltd. • ALROSA and the Government of the Republic of Sakha signed the agreement for 'Lease of land sites, diamond deposits and assets of Yakutalmaz Corporation' • SAPI Financial Corporation (Sakhaalmazproinvest) was established according to the Decree of the President of the Republic of Sakha
1995	• SAPI Financial Corporation re-organized into the SAPI Foundation, a non-for-profit organization
1997	• SAPI Foundation along with its created businesses and organizations amalgamated into the SAPI Financial Industrial Group
1999	• Inclusion of Kobyai district in socio-economic and environmental programs for the diamond province • SAPI Foundations is no longer funded by ALROSA ruled by the Decree of the President of the Republic of Sakha (Yakutia)
2000	• Establishment of the Target Fund by the Government of the Republic of Sakha to deliver socio-economic and environmental program in the diamond province

The Committee of Founders of the SAPI Foundation adopted annual budgets for the SAPI Foundation and individual budgets for the programs. During 1993-1999, the income of the SAPI Foundations comprised of 'two per cent assignments' from ALROSA and additional income from its businesses. According to the lease agreement, the diamond mining company ALROSA transferred two per cent of its revenues during 1993-1997. In 1998, the rates of lease agreement were reconsidered and thereafter, ALROSA transferred two per cent of the sales of its main production (i.e. raw and polished diamond). In 2000, the rates were reconsidered again and from 2001 ALROSA was stipulated to assign 1.8 per cent of the value of raw diamond output and sales (excluding polished diamonds); and for 2002-2003 and 2004-2005 the rates were changed to 1.5 per cent and 1.4 per cent respectively. Table 6.10 shows the assignments from ALROSA according to the lease agreement.

Table 6.10 Assignments from ALROSA to the environmental fund according to lease agreement during 1993-2000, million US dollars

	1993	1994	1995	1996	1997	1998	1999	2000
Two per cent assignments from ALROSA	10.9	13.9	23.3	22.1	23.7	28.8	27.2	29.1
Total Budget of SAPI, 100%	100	100	100	100	100	100	100	100
Transfers from ALROSA, %	92	83	86	84	96	94	64	-
Other business revenues, %	8	17	14	16	4	6	36	-

During 1993-1997, SAPI implemented annual programs in the diamond province and also participated in programs developed by the Government of the Republic of Sakha. The SAPI Foundation delivered two types of programs: social programs (concerning housing, health, environment and employment) and economic programs (supporting agriculture, diamond-cutting, gas and oil development and construction). Each year, the SAPI Foundation developed one-year programs in these areas. In 1998, the SAPI Foundation developed seven major programs for three consecutive years (1999-2001), but they were operating under the SAPI Foundation only for 1999:

1. Improvement of the environmental situation in the diamond province.
2. Development of the oil and gas industry in the Republic of Sakha.
3. Development and reconstruction of diamond cutting plants in the Vilyuy region.
4. Children of the North: Education, Culture, and Sports.
5. Introduction of special mechanisms for financing of agriculture and enforcement of internal markets for agricultural products.
6. Individual civil construction.
7. Regional center for socio-economic research in Western Yakutia.

The SAPI Foundation implemented its programs through its own established regional divisions, which the SAPI Foundation established in Anabar, Lensk, Mirny, Nyurba, Olenek, Suntar, Verkhnevilyuysk and Vilyuysk districts, and through contractors.

Target Fund

Since 1995, when the Government of the Republic of Sakha has lost the direct control over the 'two per cent assignments' from ALROSA, concerns have been raised in relation to the usage of these financial resources. During 1999-2000, several state inspections examined the financial operations of the SAPI Foundation, which revealed cases of resource misuse; these reports were then presented to the Government of the Republic of Sakha, which served as a catalyst for the creation of the Target Fund.

In 2000, the Decree of the President of the Republic of Sakha changed the process for handling the 'two per cent assignments' from ALROSA. The target budget fund for implementation of socio-economic and environmental programs in districts of the diamond province (later referred to as the Target Fund) was established within the State Budget of the Republic of Sakha and was set to receive the 'two per cent assignments' from ALROSA. The owner of the Target Fund's resources is the Government of the Republic of Sakha, and the Ministry of Finance of the Republic of Sakha carries out financial management of the Target Fund on a special budget account.

In February 2000, following the disagreement of the SAPI Foundation with the decision of the President of the Republic of Sakha, the SAPI Foundations and other organizations who were funded by the SAPI Foundation approached the Supreme Court of the Republic of Sakha with the demand to recognize the decision of the President of the Republic of Sakha as invalid. The court case proceeded till the end of the year, but the final decision was in favor of the Government of the Republic of Sakha, which carried on with the establishment of the Target Fund.[2]

In 2000, the Government of the Republic of Sakha started to deliver programs in the diamond province for rehabilitation of environmental damage caused by diamond mining developments and enhancement of socio-economic development. In addition, Kobyai district was included in the programs for the diamond province in 1999. The Presidium of the Target Fund is a collective body, which controls and manages the Target Fund. Members of the Presidium include various branch ministers and heads of government departments, a representative from ALROSA, and heads of local authority administrations in nine districts of the diamond province.

Income of the Target Fund comprises solely of the 'two per cent assignments' from ALROSA. Unlike the SAPI Foundation, the Target Fund does not invest these funds to extract additional income. The Ministry of Finance supervises the usage of resources from the Target Fund. Each year the Legislative Assembly 'Il Tumen' adopts the Law of the Republic of Sakha entitled 'State Budget of the Republic of Sakha', which sets out the main directions for usage of annual budget in the Target Fund. The mechanism of financing is as follows: the Ministry of Finance distributes the funds amongst the main recipients who are defined by the developed programs of the Target Fund (Ministry for Nature Protection, local administrations of districts in the diamond province, and contractors of the state programs such as Sakhatelecom and Sakhaneftegas). The recipients of the funds regularly report to the Ministry of Finance on the usage of resources.

Since 2000, the state agencies and state enterprises develop programs for the diamond province funded from the Target Fund, which are adopted by the Government of the Republic of Sakha. The Target Fund invests its resources into four major programs:

1. State program for oil and gas developments in the Republic of Sakha.
2. President's program for the development of new technologies and telecommunications in the Republic of Sakha for 2000-2005.
3. Regional target program for improvement of the environmental situation in the diamond province.
4. Sector programs in districts of the diamond province for 2000-2005, which includes two sub-programs: a) Program of rural development; b) Program of social and economic development.

Programs of the Target Fund are implemented through the system of government agencies, local authority administrations in the diamond province and contracted organizations.

Organization The creation of the SAPI Foundation was attributed to the rise in public concerns about the distribution of profits from diamond mining and the compensation for the damage caused to the environment and the local community in the Republic of Sakha. Most importantly, the conditions for the creation of the SAPI Foundation relate to the development of center-periphery relations, including the Declaration of Sovereignty of the Republic of Sakha and the signing of the agreement between the Russian Federation and the Republic of Sakha that distinguished the property rights and assigned the diamond deposits as the property of the Republic of Sakha. The privatization of the state-owned diamond-mining industry and the formation of the diamond-mining company ALROSA allowed consequent negotiations between the Government of the Republic of Sakha and ALROSA on the distribution of revenues from diamond mining operations through the lease agreement signed in 1993 and led to the creation of the SAPI Foundation, assigned with 'two per cent' of revenues from the diamond-mining operations of ALROSA.

Moreover, the case of the SAPI Foundation and the Target Fund demonstrates a rare example where the mining company, in agreement with the Government, decided to assign funds for social investment as a proportion of its sales. For example, Barrick Gold has undertaken a commitment to assign at least one per cent of its profits to charity. The diamond mining company ALROSA shows a great degree of commitment in comparison with international mining companies, as it agreed to assign a fixed per cent of its revenues, as oppose to its profits, to the community development and environmental protection initiatives on the top of company-led program for community development. This means that whether the company has profits or not, it will still invest in community development in the diamond province according to the lease agreement. The assignments from ALROSA for community development in the diamond province may be regarded as tax payments to the regional budget, but are regulated by the lease agreement.

Such payments are not stipulated in the legislation of the Russian Federation, which is a cause of growing fear about the uncertainty of future assignments from ALROSA and the on-going disputes between the Government of the Republic of Sakha and the Government of the Russian Federation in relation to the distribution of revenues from the mineral exploitation between the center and the periphery.

Usually, the overall management bodies of these corporate foundations include members from the Boards of Directors of the donor companies (Pine et al, 1998). Utilizing the tri-sector partnership model, management bodies of the corporate foundations also incorporate members from the local community and local or regional authorities (Hamann et al, 2001). A similar composition is found in the SAPI Foundation (Committee of Founders) and the Target Fund (Presidium), which ensures the participation of three key parties (local community, mining company and state authorities) in joint decision-making (see Table 6.11). Nevertheless, while the SAPI Foundation was a separate entity with independent structures, the Target Fund is now a part of the State Budget of the Republic of Sakha, which indicates a shift towards predominantly governmental control over the community development programs in the diamond province.

Accountability One of the major reasons for transferring the management of 'two per cent assignments' from private management under the SAPI Foundation to state control was the failure of the SAPI Foundation to demonstrate accountability and transparency in its operations with its major stakeholders in community development in the diamond province, including the Government of the Republic of Sakha, the diamond mining company ALROSA, local authorities, local communities in the diamond province and the general public in the Republic of Sakha. The Government of the Republic of Sakha strongly condemned the quality of the SAPI Foundation's financial reports and commissioned several inspections, which revealed unaccountable use of funds. Even though the SAPI Foundation produced public reports on its operations in a similar manner to other corporate foundations (such as Phelps Dodge Foundation and Rio Tinto Foundations) and other NGOs for social investment (Alaska Permanent Fund Corporation), the public questioned the effectiveness of the funds' use.

Furthermore, the SAPI Foundation never accounted for its operations to ALROSA, and the company was totally excluded from the evaluation of operations' effectiveness and, therefore, had no tools to control the social investment in the diamond province. Thus, the SAPI Foundation largely failed to meet the expectations of its stakeholders and its accountability practice resulted in great dissatisfaction from the Government, the wider public and ALROSA. This issue raises a general question as to whether the nonprofit sector in the Republic of Sakha is ready to perform and deliver social investments and be accountable for its operations. This also reflects the general mistrust in the Russian Federation of NGOs with regard to the implementation of social investment due to their accountability failures. Therefore, NGOs in the Russian Federation need to demonstrate trustworthiness by increasing their accountability standards.

Table 6.11 Evolution of the SAPI Foundation and Target Fund

	SAPI Corporation 1993-1995	SAPI Foundation 1995-1999	Target Fund 2000-2001
Legal status	State-owned enterprise	NGO	Part of State Budget
Higher management organ			
• Representatives from donor company	✓	✓	✓
• Representatives from the regional government	✓	✓	✓
• Representatives from the local authorities	✓	✓	✓
Development of annual budgets			
• Direct participation of the local authorities	✓	✓	✓
• Direct participation of regional state agencies	✓		✓
• Direct participation of donor company			
• Budget adopted by Foundation's internal management organ	✓	✓	
• Budget adopted by the regional government			✓
• Developed budget policy for Foundation		✓	✓
• Investments income of Foundation	✓	✓	
Program implementation			
• Participation of local authorities	✓	✓	✓
• Participation of regional government agencies	✓	✓	✓
• Participation of state-owned enterprises	✓	✓	✓
• Participation of nonprofit organizations	✓	✓	✓
• Participation of donor company			
Reporting to:			
• Public		✓	
• State agencies	✓	✓	✓
• Donor company			

Currently, with the Target Fund, the population of the Republic of Sakha and the Russian Federation as well as the local community in the diamond province has limited public access to information on the use of the assignments from ALROSA. The only public information concerning environmental programs in the diamond province is presented in the State Report on the Environmental Situation in the Republic of Sakha, produced by the Ministry for Nature Protection of the Republic of Sakha. The donor company ALROSA is not informed as to the progress of programs in the diamond province from the Target Fund and the Government does not recognize the need to disclose information to the local community in the diamond province and the general public about the use of funds. Government officials simply rely on the notion that the Government uses funds more purposefully. Nevertheless, the public is very concerned about the effective use of the funds and the ability of the Government to bring changes to the environmental situation and social development in the diamond province. Therefore, the Target Fund needs to be more open about its operations and disclose information on expenditure and outcomes of initiatives to the wider public via regular publicly available reports.

Programs The SAPI Foundation and the Target Fund showed a great diversity of programs undertaken, which embrace a large spectrum of social problems associated with mining areas. Few other corporate foundations established by international mining companies undertake such a vast range of initiatives. Their emphasis tends to be mainly on investment in the areas of education and training, health, arts and culture, social welfare, and business development, whereas the SAPI Foundation (see Figure 6.2) and the Target Fund (see Figure 6.3) have invested the majority of their funds in the economic development of the diamond province.

The SAPI Foundation and the Target Fund stated the aim to rehabilitate the environmental damage caused by diamond-mining developments as one of their main missions. Indeed, this has become one of the prominent initiatives in their overall activity, especially with the Target Fund. Nevertheless, the annual share of expenditure invested in environmental protection in the diamond province is significantly smaller than expenditures on socio-economic initiatives, which is a key underlying reason for the increasing dissatisfaction of regional environmentalists.

Both organizations target the local community in the mining area, which is defined as the population in the diamond province, the area affected by the diamond mining developments. Between 1992 and 2000, the notion of what area constitutes the diamond province has geographically expanded from six to nine districts (see Figure 6.1). The SAPI Foundation and the Target Fund, not strictly being corporate foundations, represent the collaboration between the Government of the Republic of Sakha and ALROSA, and serve the communities only within the boundaries of the Republic of Sakha, even though ALROSA has mining operations in other regions of the Russian Federation and abroad.

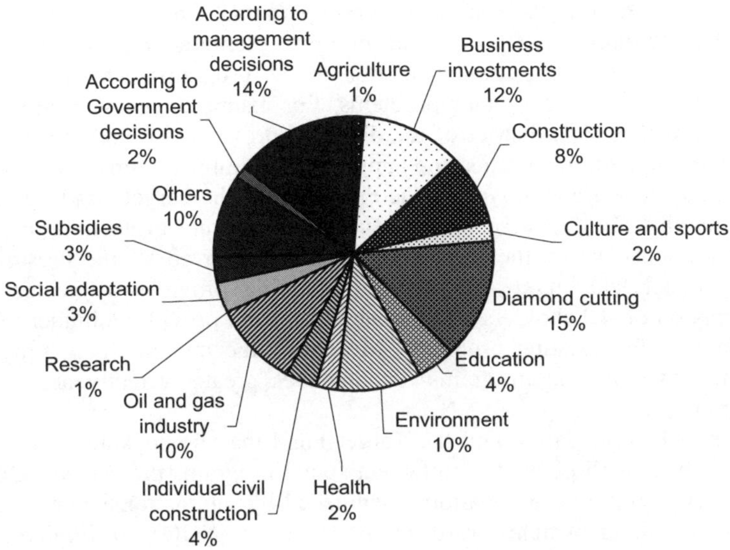

Figure 6.2 Structure of expenditure of the SAPI Foundation (1993-1999), %

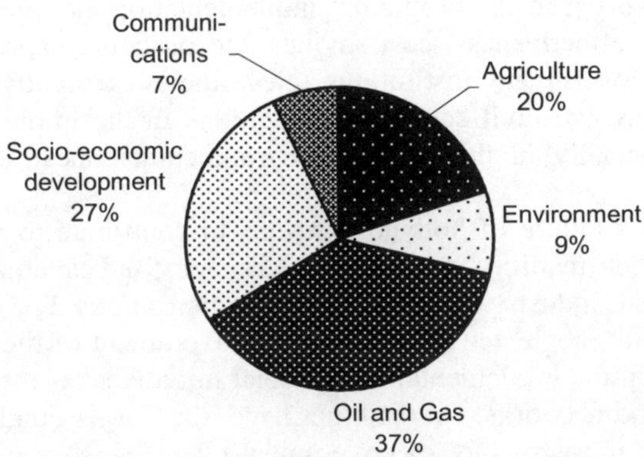

Figure 6.3 Structure of expenditure of the Target Fund (2000-2001), %

Planning and implementation The planning process in the Target Fund is different to that adopted in the SAPI Foundation. The SAPI Foundation used to develop programs in a centralized manner, with little or no participation of program implementers and other stakeholders in the decision-making process. For example, the Ministry for Nature Protection cooperated with the SAPI Foundation in the implementation of environmental protection programs in the diamond province, but was not fully integrated in the program planning. On the other hand, the Target Fund has delegated the process of program planning to stakeholders that are directly involved in the implementation of the programs, and who have greater expertise in solving particular problems. For example, social programs that are delivered within each district of the diamond province are now developed by the local authority administrations and approved by administrations' councils, and only in the final stage are approved by the Presidium of the Target Fund. Environmental programs that are delivered by the Ministry for Nature Protection, are developed by departments within the Ministry, and later approved by the Presidium of the Target Fund. This mode of program planning provides a greater degree of participation of stakeholders and implementers, and provides a higher relevance of programs to the existing problems. Moreover, the integration of programs with existing governmental programs demonstrates greater effectiveness in resource allocation.

The SAPI Foundation and the Target Fund may be considered as operational foundations, which plan and implement social programs themselves. On the other hand, many corporate foundations established by international mining companies are grant-making which award grants to other NGOs to implement certain initiatives. The key reason for choosing the operational mode of foundations in the Republic of Sakha lies in a general mistrust of civil society organizations to deliver social initiatives and the generally undeveloped nonprofit sector, which also applies to the Russian Federation as a whole. Moreover, the SAPI Foundation and the Target Fund have underestimated the advantages which the adoption of a grant-making stance would bring to the programs' implementation via integrated needs assessment, increased effectiveness, cost savings and on-going capacity building for community and government institutions. Nevertheless, recently, the Target Fund has started to involve civil society organizations in the implementation of social initiatives, especially in the area of environmental education and social protection.

The most important failure of both foundations to contribute to the tri-sector partnership model is their inability to embrace the capacity and capabilities of three sectors: the government, industry and civil society organizations. For example, the SAPI Foundation mainly neglected to involve the Government of the Republic of Sakha and ALROSA in the implementation of social initiatives, as it mainly relied on its own structure and networks. On the other hand, the Target Fund delivers the initiatives through the infrastructure of government agencies (sectoral Ministries, such as the Ministry for Nature Protection) and the local authority administrations and extensively uses their resources and capacity. However, yet again the Target Fund has neglected to incorporate complementary resources from the SAPI Foundation, which has now attained sufficient expertise in delivering social investments in the diamond province and has built substantial capacity and appropriate infrastructure because the situation resulted in the breaking of ties with the SAPI Foundation. Nevertheless, sometimes the local authority administrations in the diamond province, which are implementers of some social initiatives, still maintain cooperation with the SAPI Foundation, which is used as a facilitator, such as Saidy-Tour of the SAPI Foundation for education programs in Mirny district.

Integration of initiatives with existing governmental programs is advocated as a great advantage of tri-sector partnerships (Hamann et al, 2000). In the case of the Target Fund, the integration of environmental initiatives for the diamond province with wider environmental programs in the Republic of Sakha and the Russian Federation underlines the usefulness of this approach and has brought obvious improvements to the environmental initiatives. Firstly, as environmental programs for the diamond province have been incorporated into environmental programs of the government, the duplication of initiatives is avoided and economy of financial resources is achieved. Secondly, the environmental programs have been developed utilizing professional staff and delivery is monitored by the system of environmental protection agencies. Thirdly, the needs assessment for environmental initiatives is conducted by the environmental protection agencies, which have greater expertise in issues of environmental protection. Fourthly, the environmental programs have now started to operate on a more regular basis, systematically, and on a larger scale.

Selection of the target population for some social programs in the Target Fund is mainly based on personal experiences and observations of program planners, as opposed to being based on substantial investigation, which requires regular investigation of social problems in the diamond province. There is an absence of thorough monitoring of program implementation, which results in slow and incomplete implementation of initiatives. Social investments of the Target Fund are, in essence, not assessed and no evaluation of program outcomes is conducted; this makes it difficult to evaluate the effectiveness of resource allocation and outline improvements for future programs.

Comparison of the SAPI Foundation and Target Fund

The case of the SAPI Foundation and the Target Fund in the Republic of Sakha shows that true partnership between the government, industry and civil community organizations is difficult to sustain. The SAPI Foundation and Target Fund represent an innovative mechanism for compensation to local communities for the negative impacts caused by the diamond mining industry to the environment and socio-economic structure in the diamond province. The creation of the SAPI Foundation was the result of negotiations between the diamond-mining company ALROSA and the Government of the Republic of Sakha, which is a remarkably proactive approach to be adopted in the mining industry. The operations of the SAPI Foundation and Target Fund have clearly brought many benefits for the development of the diamond province. For the first time, social, environmental, economic and health problems present in the diamond province were formally addressed and the stability of substantial funding was ensured in the long-term Lease agreement. Key stakeholders (ALROSA, local community and the regional government) have been involved in decision making via ensuring their position in management bodies of the SAPI Foundation and Target Fund.

However, disagreement in the usage of funds resulted in major conflicts: the transfer of funds' control from the nonprofit organization (the SAPI Foundation) to the Government of the Republic of Sakha (the Target Fund); the court case concerning the possession of assignments from ALROSA; the breakage of effective dialogue between the SAPI Foundation and the Government of the Republic of Sakha. Transition from the SAPI Foundation (potential tri-sector partnership) to the Target Fund (bilateral partnership) happened for several reasons that reveal inability of partners to handle fundamental issues of supervision, accountability and promotion of initiatives funded by the assignments from AROSA. They include:

- Insufficient communication and inability to maintain partnership requirements between the nonprofit organization and public bodies, which led to the exclusion of the Government from the important decision-making process in relation to the usage of the assignments.
- Concerns relating to the purposeful use of the assignment from ALROSA in the SAPI Foundation and state revisions of the financial accounts of the SAPI Foundation highlighted the inappropriate usage of funds.
- Disputes on the property rights of the assignments from ALROSA caused by legal difficulties following the introduction of a new Civil Code of the Russian Federation, which led to the transfer of state assets into private property.
- ALROSA never participated in the development and implementation of initiatives in the diamond province, apart from stipulating the general mission of the assignments in the lease agreement. ALROSA was never fully integrated into the decision-making process and the initiatives in the diamond province were never accountable to ALROSA.

The Target Fund is a step backwards from a potential tri-sector partnership to a bilateral partnership between the Government and ALROSA. The Government of the Republic of Sakha now has a crucial control over the operations the Target Fund as it owns and controls distribution of assignments from ALROSA via a system of government agencies. Although, the Government ensured targeted and accountable use of funds via a range of procedures, the Target Fund still fails to effectively engage other interested parties. The Target Fund fails to attract complementary resources from ALROSA for the implementation of social investments in the diamond province; this raises a dilemma as to whether ALROSA should play a more important role that a mere source of funding for implementation of initiatives. Both partnerships (SAPI Foundation and Target Fund) have similar problems concerning management of the partnership such as:

- Failure to incorporate key stakeholders (ALROSA, the local community, the Administrations of districts in the diamond province and the Government of the Republic of Sakha) in decision making related to the initial design of the Foundations. It was the decision of the President of the Republic of Sakha to give the legal form to the SAPI Foundation as a state enterprise. Even at the time of reorganization of the SAPI Foundation into a nonprofit organization, the decision apparently involved all parties, but later revealed the conflict of interests. Finally, the establishment of the Target Fund according to the Decree of the President of the Republic of Sakha was not preceded with any consultation process with other stakeholders.

- Low level of involvement of NGOs with the mining industry in the implementation of social investment initiatives and ignoring the potential use of complementary resources from the other sectors which might contribute to the effectiveness and efficiency of operations.
- Failure to demonstrate accountability and transparency via external disclosure on operations and failure to evaluate and assess the outcomes of the programs in the diamond province.

For the Target Fund to continue its operations for the benefit of the population in the diamond province and to sustain effective partnership with the diamond mining company and local communities, the Government of the Republic of Sakha needs to strengthen its operations in several areas. First, setting guidelines and allocation of resources for needs assessment could improve stage of program planning. Second, the establishment of formal procedures for program monitoring and increasing the capacities of program implementers could enforce program implementation. Third, evaluation of program outcomes and distribution of information on program results to a range of important stakeholders and the general public will increase accountability of the Target Fund and contribute to stability of the partnership.

Summary

- Company-led approaches to community development usually include charitable giving, support of social care facilities, cooperation with government agencies in social and economic programs in the mining areas, cooperation with NGOs, community and educational organizations for the benefit of local communities. An important aspect of community relations of the mining companies is on-going consultation with local communities, compensation for damage caused, discussion of land rights and contribution to lives of indigenous communities.
- Corporate foundations are a common approach chosen by mining companies to deliver community development initiatives in mining areas. Commonly corporate foundations established by the mining companies target local communities near mines, and invest their resources into education and training, community development, social development, health, arts and culture, and business development.
- Tri-sector partnership, which involves cooperation between the government, industry and civil society organization for community development in mining areas is a promising approach, however, using the case study of foundations in

the Republic of Sakha proves to be difficult to sustain with unequal distribution of interests and powers between the parties.

Notes

[1] SAPI is an acronym for *Sakhalmazprovinvest*, which stands for Sakha Diamond Province Investment.

[2] The SAPI Foundation and other organizations filed a case to the Supreme Court of the Republic of Sakha with the demand to recognize the Decree of the President of the Republic of Sakha of 17 January 2000 No. 948 as invalid, which took 'two per cent assignments' from the SAPI Foundation to the Government. In April 2000, the decision of the Supreme Court restored the rights of the SAPI Foundation as an agent for the use of the two per cent assignments from ALROSA. After the appeal in July 2000, the decision was overturned and re-directed for another hearing. On the final hearing of the case in October 2000, the Supreme Court declared Article One of the disputed Decree to be invalid; however, the establishment of the Target Fund remained in force, ending the battle over the two per cent assignments from ALROSA in favor of the Government of the Republic of Sakha.

Conclusions

Organizational Structure for CSR Management

There is a growing need for mining companies to reassess their corporate strategies in relation to environmental and social issues in order to demonstrate greater corporate social responsibility and undertake a more effective and prominent role in progress towards sustainable development (see Chapter 1 and 2). Global mining companies as shown in the analysis of corporate disclosure of the top gold and diamond mining companies are increasingly concerned about their environmental, social and economic impacts on the society (see Chapter 3). They have started to incorporate concepts of corporate social responsibility in their business strategies in the areas of environmental protection, employee and health and safety management, stakeholder and community relations (see Chapter 5).

The analysis of the gold and diamond mining companies shows that there are three key models of organizational structure for managing corporate social responsibility in a company:

- Centralized integrated structure.
- Centralized disintegrated structure.
- Decentralized disintegrated structure.

A centralized integrated structure for managing corporate social responsibility in a company suggests that the corporate social responsibility agenda is taken to the highest levels of corporate management starting from the Board of Director down to division level, with allocation of responsibilities to manage corporate social responsibility activities. If a company has the centralized integrated organizational structure of corporate social responsibility, a company is likely to have a specially formed committee within the Board of Directors, which promotes, monitors and assesses corporate social responsibility within the company. For example, some global top gold and diamond mining companies have established committees within their Boards of Directors such as Placer Dome Sustainability Committee and Rio Tinto Committee for Social and Environmental Accountability.

A centralized integrated structure often implies that the company is likely to have a corporate social responsibility manager at a corporate level, who would steer or cooperate with other corporate managers which are responsible for issue-based management (such as environmental protection, personnel, health and safety, community relations, supply chain and etc.) to promote a unified approach to corporate social responsibility in the company.

Issue based corporate managers in turn coordinate the operations of managers at division and site levels. This type of organizational structure leads to greater penetration of corporate social responsibility visions and policies from the top management to operating divisions and sites and can improve planning, implementation and monitoring of corporate social operations with clear distribution of responsibilities between managers of different levels. Centralized integrated structure could also contribute to greater efficiency in collecting, processing and distributing information for evaluation and reporting on corporate social and environmental performance.

A centralized disintegrated organizational structure for corporate social responsibility is characterized by the establishment of separate management streams in different areas of corporate social responsibility (e.g. environmental protection, employee policy, community relations) without necessarily integrating them under a single management body at the highest management level. In companies with a centralized disintegrated organizational structure, the Board of Directors may have an established committee for one or two relevant areas of corporate social responsibility such as environmental protection or community relations. For example, Homestake Mining has established Environment, Health and Safety Committee within its Board of Directors.

According to the centralized disintegrated structure, the company is likely to have separate corporate managers, which coordinate different issues relating to corporate social responsibility, e.g. corporate environmental manager, corporate community relations manager, corporate health and safety manager, corporate employee manager, etc. For example, Freeport-McMoRan Copper & Gold appointed a Vice President for Environmental Affairs. Each corporate social responsibility issue is dealt with in its own right, leading to separate management processes, measurement and reporting practices and limited unification of operations with regard to corporate social responsibility. For example, ALROSA has a centralized disintegrated structure, whereby it has a corporate environmental manager, a corporate regional policy manager, a corporate personnel and social issues manager, who direct operations of the entire company in their specific areas and coordinate appropriate managers at division and site levels. However, little collaborative effort is conducted between important CSR issues.

A decentralized disintegrated organizational structure is likely to be present in companies with a low level of corporate social responsibility. This organizational structure implies an absence of a separate line of management for core CSR issues (such as environmental protection and health and safety) at the highest management level as they are incorporated in the production related management. Usually, companies with the decentralized disintegrated organizational structure do not have committees within their Board of Directors specifically dedicated to corporate social responsibility issues. Furthermore, such companies often do not have corporate environmental, health and safety, and community relations managers, as corporate production and personnel managers regulate these issues. Only at division level there could be employees assigned with responsibilities for managing CSR issues and some could share such responsibilities with other production related responsibilities. For example, Aldanzoloto does not have an environmental manager at corporate level; instead the production manager at corporate level supervises environmental managers that are operating at division levels.

The disadvantage of the decentralized disintegrated structure is that environmental managers in divisions have less authority to be directly involved in decision-making on important issues concerning environmental protection at corporate level. This could reflect a low level of consideration of CSR issues within the company and high dependency of environmental and social related decisions on production and economic priorities and unwillingness to consider corporate social responsibility that goes beyond economic and technical requirements of the business. The abovementioned models of organizational structures for corporate social responsibility are summarized in Appendix 1.

Corporate Social Responsibility Policies

Along with the classification of models for the formation of organizational structures for managing corporate social responsibility in a company, mining companies demonstrate different types of formulation of policies related to CSR such as:

1. *Corporate policies oriented towards single area of corporate social responsibility* At the beginning and middle of the 1990s, mining companies have started to formulate corporate policies where they expressed commitments for responsible conduct of operations relating to the natural environment. For example, Placer Dome adopted a formal Environmental Policy in 1990, where the company expressed its corporate goals and objectives related to the natural environment and established a Placer Pacific Environmental Management Review Committee. Later gold and diamond mining companies started to develop similar policies in other areas of CSR such as health and safety and community relations. For example, Barrick Gold adopted a Safety and Occupational Health Policy Statement in 1996.
2. *Integrated policies oriented towards two or three areas of corporate social responsibility* In the middle and end of the 1990s mining companies started to combine several environmental and social issues under single corporate policies such as Corporate Environment, Health and Safety Policies or Corporate Environment and Community Policies. For example, in 1998-1999, BHP operated under a single Environmental Policy and in 2000 has adopted integrated Health, Safety, Environment and Community Policy.
3. *Corporate social responsibility and sustainable development policies* Towards the end of the 1990s, beginning of 2000, with the rise of the sustainable development agenda and corporate social responsibility concept within the mining industry, mining companies started to view various social and environmental issues as being part of wider CSR or sustainable development. In the end of the 1990s mining companies started to develop policies that would address issues of sustainable development or corporate social responsibility issues in general. For example, Placer Dome established a

position of Vice-President for Sustainable Development in 1997 and adopted Sustainable Policy in 1998.

Not all mining companies went through such evolution of policy development. Some mining companies, omitted the second stage of policy development, for example, Newmont first developed their Mission Statement of Environmental Compliance in 1991 and in 2003 adopted a Social Responsibility Policy, which expressed corporate commitments to sustainable development, protection of human life, health, the environment and contribution to communities in which the company operates. Some companies have been slower in setting their businesses towards socially responsible conduct, for example AngloGold first adopted formal Environmental Policy, Employment Equity Philosophy and Health and Safety Policy, approved by its Board of Directors, only in 1999. Other mining companies opted for reorganization of their corporate policies into policies dedicated to separate areas of CSR as well as adopting Sustainable Development Policy. For example, in 2005, Rio Tinto has been operating under a range of corporate policies devoted to Communities, Employment, Environment, Human Rights, Land Access, Political Involvement, Safety and Sustainable Development.

Mining companies usually develop their corporate policies, which apply to their entire global operations as well as operations of their subsidiaries. This is particularly important for dissemination of business strategies concerning corporate social responsibility from the developed countries to developing countries, as the top gold mining companies usually originate in developed countries and operate their mines in developing countries of Asia, Africa and Latin America.

Moreover, when mining companies with different levels of corporate social responsibility management unify their business strategies under a single approach as a result of mergers and acquisitions it is an interesting subject, which requires further investigation. This book covers an analysis of the top gold and diamond mining companies during 1998 and 2000, however, several companies under investigation have since merged or been acquired by the other companies, such as BHP and Billion, AngloGold and Ashanti Goldfields, Newmont and Normandy Mining, Barrick Gold and Homestake Mining.

The top gold and diamond mining companies formulate corporate policies in important areas of CSR, which usually have the following statements:

Environment

- Commitment to meet regulatory requirements and standards.
- Commitment to minimize environmental impacts and effects.
- Commitment to continuous improvement of environmental performance.
- Commitment to reclaim and return land to beneficial users.
- Commitment to conserve resources and materials.
- Commitment to ensure management systems for control and monitoring of environmental impacts.
- Commitment to conduct research.
- Commitment to protect the environment.

- Commitment to consult with stakeholders.
- Commitment to communicate with government and the community on environmental issues.
- Commitment to report on environmental performance.

Employee welfare, health and safety

- Commitment to improve health and safety performance.
- Commitment to health and safety standards.
- Commitment to provide resources to maintain a safe and healthy working environment.
- Commitment to investigate the causes of accidents and develop immediate preventative or remedial actions.
- Commitment to train employees concerning health and safety.

Community

- Commitment to comply with regulatory requirements and standards.
- Commitment to conduct business based on mutual trust.
- Commitment to consult with stakeholders.
- Commitment to communicate and consult with traditional land owners on land management.
- Commitment to implement cross-cultural awareness programs.
- Commitment to respect cultural heritage.
- Commitment to recognize and respect the traditional rights of indigenous peoples.
- Commitment to cooperate with indigenous people for their long-term benefit.
- Commitment to commission social and community research.
- Commitment to report on community issues.
- Commitment to promote employment from the local areas through recruitment and training programs.
- Commitment to engage employees, business partners, contractors, suppliers and visitors in a shared responsibility.

Although gold and diamond mining companies operate in areas with a strong presence of indigenous people, derived from the analysis of corporate disclosure during 1998-2000, only one gold mining company, i.e. Normandy Mining has developed a specific Indigenous People Policy, other companies embrace commitments towards welfare and communication with indigenous people in their Community Policy.

In each area of corporate social responsibility (e.g. environment, health and safety, employee relations and community, and the list is not exhaustive), mining companies configure their approaches following a similar pattern starting from compliance with regulatory requirements; committing to operate comprehensive management systems and to allocating sufficient resources for regulating CSR issues; committing to principles of ethics, sustainable development, voluntarism, proactive behavior, responsiveness and accountability; and finally committing to communicate and report on corporate operations as well as performance to external and internal stakeholders and the general public.

Environmental and Social Reporting

Environmental and social reporting reflects the openness of companies and their willingness to communicate their environmental and social policies and performance with a range of stakeholders. Repeating the pattern of evolution of corporate policy development in the area of CSR, environmental and social reporting practices of gold and diamond mining companies went through the same three stages of evolution in approaches to reporting:

1. *Disintegrated approach to environmental and social reporting* Mining companies started with the production of reports in a single area of corporate social responsibility, such as the BHP Environment Report 1998.
2. *Integrated approach to environmental and social reporting* Mining companies started to publish reports that would cover several environmental and social issues such as the BHP Environment and Community Report 1999 and Rio Tinto Social and Environment Report 1998.
3. *Sustainability reporting or corporate social responsibility reporting* Coinciding with the development of sustainability policies in mining companies, mining companies started to produce highly integrated policies which would have appeal to wider corporate social responsibility and sustainable development issues. For example, Placer Dome produced its first Sustainability Report in 1999 and Barrick released Environmental, Health, Safety and Social Responsibility Report in 2002.

The analysis of environmental and social disclosure by the top gold and diamond mining companies during 1998-2000 reveals that the majority of companies report on their global operations and some produce site-specific reports. The analysis of corporate reporting practices in gold and diamond mining companies during 1998-2000 revealed several patterns of reporting:

- *Advanced reporting* is characterized by a relatively long history of engagement in reporting practices; early development of stand-alone reports; greater coverage of corporate social responsibility issues; wide use of media (Annual Reports, corporate websites, produce stand alone Environment and Social Reports, and produce other publications focusing on CSR issues); more detailed coverage of site specific information (production of site specific reports or inclusion of site specific data in Environmental and Social Reports); use of environmental and social performance indicators in reporting; and the use of third party verification statements in stand alone Environmental and Social Reports. Rio Tinto, Normandy Mining and Placer Dome could be categorized as advanced reporters.

- *Intermediate reporting* is characterized by later engagement in environmental and social reporting practices; inclusion of extensive environmental information in Annual Reports and corporate websites; later development of stand-alone Environmental and Social Reports; absence of site specific data; absence of comprehensive quantitative environmental and social data in reports; limited use of other media for corporate social disclosure (such as production of separate publication on corporate policies or community development actions); absence of third party verification statements in stand alone Environmental and Social Reports. AngloGold, De Beers and Freeport-McMoRan Copper & Gold and could be categorized as intermediate reporters.
- *Beginner reporting* is characterized by minimum disclosure of environmental and social information; use of limited media for environmental and social disclosure (use of Annual Reports and corporate websites only); absence of stand-alone Environmental and Social Reports; absence of site specific information; absence of quantitative environmental and social performance data. SouthernEra and Rex Mining could be categorized as beginner reporters.

The analysis of social disclosure of top gold and diamond mining companies provide that larger companies disclosure more environmental and social information than smaller companies, which has been previously noticed in other sectors (Meek et al, 1995; Niskala and Pretes, 1995; Deegan and Rankin, 1996; Frost, 2000). Moreover, the disclosure analysis shows that companies originated in developed countries (Australia, Canada, Europe and USA) tend to disclose more environmental and social information than their counterparts in developing countries and countries in transition.

Although both gold and diamond mining industries are dominated by large mining companies, it is evident that the top gold mining companies are more prominently engaged in environmental and social reporting than the diamond mining companies. This could be explained with several reasons:

1. *Lack of leadership* There is a lack of leadership in promoting CSR in the diamond mining industry. De Beers, the leader in diamond mining production and a company which has control over the majority of the world's diamond mining production has a potential to influence adoption of CSR approach in the diamond mining industry through its participation in diamond mining projects in Botswana and Namibia (Debswana and Namdeb) and supply chain pressure for improvement of environmental and social performance, when acquiring diamonds through CSO from other diamond producers. However, the analysis of corporate disclosure in 1998-2000 shows that De Beers lags behind some proactive gold mining companies and produced its first Safety, Health and Environment Review in 2000, several years later than some top gold mining companies such as Rio Tinto and Homestake Mining, who started Environmental Reports in 1996.
2. *Closed sector vs. open sector* World gold mining is controlled by diversified companies, which extract not only gold but also other minerals such as Rio

Tinto and Freeport-McMoRan Copper & Gold. Contrary to this, the diamond industry is a fairly closed industry, which comprises of mining companies with sole focus on the extraction of diamonds. In each mineral sector different forces form the pressures for promotion of CSR, such as industry association and NGOs. As mining companies adopt similar CSR approaches to all minerals extraction operations, experiences shared across different minerals divisions.

3. *Limited scale of operations* Diamond mining companies focus their mining operations on a single country or region, usually within the developing world, such as Debswana operating in Botswana and Numbed operating in Namibia. Developing countries as it has been mentioned lack effective government regulation of environmental and social issues (Poulin and Sinding, 1993; Acquah, 1994; Bomsel et al, 1996), which leads to limited exposure from stakeholder base on diamond mining companies. The top global gold mining companies operate mines in a number of countries situated on different continents, and are more exposed to regional and international developments concerning CSR issues from national governments, local communities, employees, local and international NGOs.

Almost all top global gold and diamond companies disclose environmental and social information in Annual Reports and corporate websites. In 1998-2000, especially Ekati Diamond Mine, Freeport-McMoRan Copper & Gold, Homestake Mining, Normandy Mining, Placer Dome and Rio Tinto were extensively using their corporate websites to disclose environmental information. These transnational mining companies disclose environmental and social information about their worldwide operations as well as operations of their subsidiaries and affiliated companies.

Factors that contribute to greater engagement in reporting practices in gold and diamond mining companies include: adherence to industry code of conduct; adopted corporate policies relating to CSR issues; and establishment management systems of CSR issues.

Environmental Protection

Environmental protection plays one of the key roles in corporate social responsibility conduct in the mining industry. Impacts of the mining operations on the natural environment are very important to communities in which mining companies operate and responsible environmental protection of mining operations is a necessary condition for the establishment of effective relationships and communication between mining companies and their stakeholders. Corporate attitude to the natural environment in the context of the mining industry is key factor for maintaining 'license to operate'.

The analysis of global gold and diamond mining companies shows that mining companies are increasingly concerned about environmental impacts they are causing. Proactive gold and diamond mining companies adopted important elements of environmental management systems outlined by Dias and Begg (1994), Watkins (1998) and Welford (1998), including:

1. Development of corporate policies with commitments to the protection of the environment.
2. Establishment of appropriate organizational structures and allocation of resources for the delivery of these policies.
3. Establishment of environmental management systems and certification of EMS according to international standards such as ISO 14001.
4. Monitoring of environmental operations and environmental impacts and research into environmental impacts and ecological risk assessment.
5. Assessment of environmental performance and conduct of independent environmental audits.
6. Reporting on environmental operations and performance, production of stand-alone Environmental and Social Reports and use of other corporate media such as Annual Reports, corporate websites and other corporate publications.
7. Update of environmental policies.

Nevertheless, the absence of formal EMS does not mean that the companies do not have procedures for environmental management in place, as gold and diamond mining companies still report to have procedures for planning, implementation, measurement and reporting of environmental operations and performance. It is evident that in order to exhibit responsible stewardship of operations related to the natural environment, mining companies adopt common approaches including:

- Establishment of specific measures for safe management of tailings and overburden.
- Waste recycling and disposal.
- Efficient water use.
- Efficient use of energy resources.
- Efficient protection from air pollution.
- Rehabilitation of disturbed land involving re-vegetation.

A variety of tools are available for mining companies in the area of environmental protection, which could enhance corporate responsible conduct and which demonstrate important CSR principles of accountability, proactive behavior, responsiveness and voluntarism, they include: adoption of industry codes of conduct and voluntary guidelines; voluntary agreements with governmental agencies; cooperation with stakeholder on wildlife conservation and environmental research; and stakeholder consultation on environmental issues.

Industry codes of conduct AngloGold, Normandy Mining, Placer Dome and Rio Tinto are assigned to the Australian Minerals Industry Code for Environmental Management, Rio Tinto is also assigned to the Global Sullivan Principles; and Rio Tinto and Freeport-McMoRan Copper & Gold have joined Global Mining

Initiative. Moreover, the Global Reporting Initiative has recently developed a mining sector supplement to its core GRI Sustainability Reporting Guidelines to which 16 leading mining and metals companies comprising the International Council on Mining and Metals (ICMM) have agreed to report on their sustainable development performance (Global Reporting Initiative, online). From the gold and diamond mining sectors, ICMM includes Anglo American plc, AngloGold Ashanti, BHP Billiton, Freeport-McMoRan Copper & Gold Inc., Newmont Mining Corporation, Placer Dome Inc., Rio Tinto plc and WMC Limited (ICMM, online).

Voluntary agreements with government agencies In the area of bio-diversity conservation, ALROSA shows a remarkably progressive approach by adopting the conservation concept and conducting environmental protection initiatives in cooperation with the Ministry for Nature Protection of the Republic of Sakha. This voluntary approach of ALROSA is linked to a growing movement in bio-diversity conservation, which started in the 1990s in the Republic of Sakha. Some global top gold and diamond mining companies undertake similar voluntary initiatives on bio-diversity conservation including Ekati, which conducts bio-diversity monitoring programs in cooperation with indigenous peoples groups and regional authorities.

Cooperation for environmental conservation and research It has become a common approach amongst the mining companies to invest in environmental research and conservation, protection of wildlife and assisting in establishment of nature reserves. For example, De Beers is participating in the establishment of Venetia Limpopo Nature Reserve, which aims to conserve wildlife and bio-diversity of the region as well as contribute to regional economy and eco-tourism. In the case of the diamond mining company ALROSA, the company established several animal farms and also cooperates with regional environmental protection agencies for the establishment of specially protected natural areas.

Stakeholder consultation on environmental issues Proactive corporate behavior in the area of environmental protection includes consultation with stakeholders on a range of environmental issues related to mining operations. As mining operations cause a significant impact on natural environment, which is of a great concern to local communities, mining companies try to explain their positions with mining operations and listen to concerns about sensitive environmental issues through:

- Conducting consultation forums with members of local communities, environmental NGOs, government agencies and general public on issues concerning new mine development (e.g. ALROSA and De Beers), mine closure (e.g. Argyle).
- Signing negotiated agreements with local communities, which include statements on environmental issues. For example, Musselwhite General Agreement signed by Placer Dome with First Nations, federal and provincial

governments addresses environmental protection as one issues covered in the agreement (ICME, 1999).

Employee Welfare, Health and Safety

Mining companies around the world employ people from within local communities of the mining areas. Human resource management in mining companies faces a number of common challenges related to:

- Provision of housing, health and childcare services and recreation for employees as often mining projects start to operate in undeveloped areas lacking town building, communal services, infrastructure and in areas where there is a lack of state welfare support.
- Adjustment of employees in new mining areas and relationships with local and indigenous communities. Often migrant labor working for mining companies are willing to relocate from the mining areas after their employment in the mining companies is finished.
- Sufficient reward to employees, concerns about employee career and development, provision of education and training for employees, guarantees of financial security after retirement.
- Communication with employees on issues concerning impact of corporate operations on employee welfare, education and training, health and safety, community involvement and impacts on the natural environment.

Voluntary initiatives in the area of human resource management in the context of the mining industry that go beyond employee related legislation could be in the following areas:

- Shareholding for employees in mining projects. In both case study companies, Aldanzoloto and ALROSA, employees hold shares in the mining businesses that allow them to have greater influence on corporate decision-making; it creates shared responsibility for corporate actions and contributes to employee motivation and ownership of corporate profits.
- Support for education of employee families via allocation of grants and fellowships. The majority of gold and diamond mining companies provide such initiatives for their employees including AngloGold, Argyle Diamond, Barrick Gold, De Beers, Freeport-McMoRan Copper & Gold, Namdeb, Normandy Mining, Placer Dome and Rio Tinto.
- Support for future retirement via establishment of corporate pensions in addition to state pension schemes. For example, ALROSA has established a Pension Fund for its employees like some of the other diamond mining companies (e.g. Argyle, De Beers, Debswana, Ekati and Rex Mining).
- Support for relocation of employees. For example, mining companies in the Republic of Sakha not only help in relocation of employees after their

retirement, but also buy or build houses for their retired employees outside the mining regions.

- Building housing for employees and provision of interest-free credits to buy houses (e.g. ALROSA and Barrick).
- Maintenance of nurseries, sports and recreation centers for employees. Aldanzoloto and ALROSA invest substantial resource in maintaining such facilities for use of mine employees, their families and local community members.
- Healthcare programs designed for employee and their families, often extended to members of local communities, could be delivered using resources of mining companies themselves (e.g. healthcare programs within Argyle, Ekati, Debswana and Namdeb) and via establishment of tri-sector partnerships between interested NGOs and government health agencies (e.g. health care initiative adopted by MINCA, Placer Dome operation in Venezuela). Mining companies provide health care through creation of company clinics and hospitals as well as company contributions to healthcare and medical aid costs.

Although health and safety is usually strictly regulated by national regulations and standards, the corporate social responsibility approach to health and safety could be demonstrated via engagement in initiatives, which demonstrate principles of CSR such as accountability, proactive behavior, responsiveness and voluntarism. Effective management of health and safety in mining companies includes several common elements such as development of corporate health and safety policies; establishment of health and safety management systems; establishment of appropriately resourced health and safety services; allocation of appropriate resources; development of regular action plans; regular monitoring of health and safety operations at divisions and sites; training and improvement of health and safety awareness among employees; employee health monitoring; assessment of health and safety performance with set targets and standards involving health and safety audits; reporting on health and safety performance.

Proactive and voluntary initiatives in the area of health and safety within gold and diamond mining companies include the following elements:

- Proactive mining companies adopt a zero injury tolerance approach via set the highest standard for ensuring health and safety of their employees and undertaking risk assessment and analysis of safety incidents with consequent improvement of preventive measures (e.g. Argyle and Normandy).
- Negotiations with Trade Unions on the issues concerning health and safety is an indicator of responsive corporate behavior, where concerns and demands of employees as important stakeholders of corporate operations are taken into account and incorporated in decision making concerning workplace issues.
- The companies, usually with operations in Africa, are developing appropriate programs to target health issues amongst their employees such as HIV and AIDS to prevent the spread of the virus and protect welfare of HIV-positive

employees (e.g. AngloGold, Ashanti, De Beers, Debswana, GoldFields, Namdeb, Placer Dome and Rio Tinto).

Models for Community Development

Community development deals with a number of important issues that mining projects face in their interaction with communities such as:

- Employment of local community members and indigenous people.
- Training for local community members.
- Negotiation over land rights with local communities and indigenous people.
- Conduct of social and community research.
- Consultation with stakeholders.
- Consideration community concerns on regular basis.
- Contribution to local community life.
- Compensation to local community for negative impacts.
- Charity giving and establishment of philanthropic foundations.
- Cooperation with government, NGOs and industry associations on social issues.

Mining companies can use models for community development, discussed in Chapter 6, i.e. company-led approach, corporate foundations and partnership approach on their own, but some mining companies are pursuing different models of community development at the same time (e.g. Placer Dome and Rio Tinto). However, for mining companies, which cause significant impact on the fiber of community life, the adoption of in-house or a company-led approach to community development is perhaps essential in the context of corporate social responsibility.

Each approach has its benefits and challenges; and the choice of approach for a particular mining project depends on circumstances including: the extent of mining operations; size of the affected community; severity of community problems; capabilities of mining companies; willingness of mining companies, government and civil society organizations to cooperate. The models for community development offer different ways of organizational structures, procedures for funding, planning, execution, evaluation and reporting of initiatives (Appendix 2).

The company-led approach characterized by planning, execution and evaluation of social projects using only resources of the company. The company-led approach for mining companies could be a useful way of improving communication between mining companies and communities; managing direct impacts of mining operations on communities; addressing needs and concerns of communities immediately near mining operations. Moreover, mining companies often maintain company-led community programs by integrating them with initiatives targeted at employees and their families, which offers possibilities of cost-saving.

Although the company-led approach could be an effective tool for addressing community concerns via developed community development programs, this model requires substantial internal resources for the company to deliver extensive initiatives for community development. The company-led approach has a danger to become a one-way activity, when the community is considered to be a passive recipient of company's initiatives. Unless the company is willing to consult and communicate with communities with regard to their needs, demands and expectations, there is a danger that the company might misinterpret and ignore vital community issues and direct its efforts to less effective courses. In order to avoid a one-sided approach to the community, mining companies try to cooperate with local community groups, NGOs, federal and local governments and international organizations to understand the issues that the mining project could contribute to community life. A promising tool that could ensure communication and consideration of interests between mining companies and local communities is the signing of negotiated agreements or memorandums of understanding with local communities and indigenous peoples (e.g. Placer Dome and Freeport-McMoRan Copper & Gold).

The choice to externalize community development programs via establishment of a corporate foundation is a popular trend among mining companies (see Chapter 6). Creation of corporate foundations is more effective in addressing community concerns that are outside the direct impact of mining operations. By externalizing processes of planning, execution, monitoring and evaluation of initiatives mining companies could save their resources and delegate these tasks to professionals with greater expertise in delivering social programs and independent judgment. Although mining companies have less opportunity to steer specific initiatives of corporate foundations, operations of corporate foundations could enhance community development via engaging nongovernmental organizations in implementation of initiatives and address greater range of societal problems such as combating poverty, improving education and reviving cultural development.

Partnerships for community development within the mining industry could have different forms and could include various number and combination of parties (mining company, either government agency or nongovernmental organization) (see Chapter 2). Partnership approach to community development has several potential benefits in establishing effective relationship between the mining industry and communities in which it operates, including: involvement of key stakeholders (mining company, government and community) in decision making concerning community initiatives which could lead to a higher degree of considerations of community concerns and demands; input of resources from different parties and involvement of different organizations in implementation of community initiatives. However, partnerships being a more complicated mechanism for addressing community concerns pose a number of challenges in relation to management of community initiatives. They include:

- Sustaining effective communication between parties in a partnership and ensuring fair involvement in decision-making process.
- Effective distribution of responsibilities and effective balance of inputs from the parties.

- Demonstration of accountability and transparency of operations within the partnership, ensuring appropriate and efficient usage of funds, involving reporting on partnership operations to concerned stakeholders.
- Securing appropriate distribution of property rights within a partnership and distribution of benefits from the partnership.
- Evaluation of outcomes of partnership's operations and continuous improvement of the partnership.

Case Study Companies

The evidence from the case study companies on Aldanzoloto ad ALROSA demonstrates that the precious metals and minerals industry plays a key role in determining sustainability of the Republic of Sakha. These mining companies form the basis for the regional economy via contributing tax payments to the state budget of the Republic, export revenues, investing in infrastructure development, and promoting trade and business development in the areas of mining operations. Aldanzoloto and ALROSA cause substantial impacts on communities of the region via providing employment, attracting migrant labor, maintaining mining towns and settlements, providing social care to employees and other residents in the mining areas. However, in the past, state management of gold and diamond mining development in the Republic of Sakha has negatively affected regional communities: relocated indigenous people, affected traditional life styles of indigenous communities and even contributed to health deterioration amongst the rural population of Vilyuy region linked to environmental pollution from the diamond mines.

Of all the industrial sectors in the Republic of Sakha, the gold and diamond mining industries have caused and still are causing the most significant impact the natural environment, disturbing land for exploration and mine development; impacting on wildlife and vegetation; releasing polluted wastewaters into natural watercourses; and producing air emissions from mines, factories and mining transport. Given the history of negative impacts from gold and diamond mining on local communities and the natural environment and the growing role of the precious metals and minerals industry in contributing towards economic growth of the region, it is evident that there is a clear need for the mining industry to take a more responsible approach to minimization of negative environmental and social impacts and improvement of well-being of the communities in the mining areas.

The transition from planned economy to market economy in the Russian Federation, privatization of formerly state owned industrial sectors, legal and political changes in the 1990s have led to reorganization of the mining industry in the Republic of Sakha. Public concerns formed from the interactions of the state owned mining enterprises with regional and local communities and the environment in the Soviet Era are now directed towards newly privatized mining companies, Aldanzoloto and ALROSA, they include:

- Increased public awareness about environmental and social impacts and risks of mining development (use of nuclear explosions, pollution of rivers and impacts on human health in the diamond province), particularly since the late 1980s. Activist groups lobbied the regional government for improvement of environmental situation in the Vilyuy region.
- Increased concerns about compensation for damage caused by mining development to local communities and concerns about fair distribution of economic revenues from mineral exploitation between the local, regional and national levels.
- Concerns about long-term employment in the mining industry, welfare of employees and their families, access to health and social case services. Mining companies in the Republic of Sakha currently provide employment to c. 45 thousand people, which comprise ten per cent of the total employment in the region. Concerns about employment in the mining industry include employment of local community members and indigenous people, provision for retired personnel and employment after mine closure.
- Expectations from local authorities for mining companies to contribute to community development via enhancement of infrastructure, provision of services for health and education for the local population.
- Promotion of environmental, social and economic improvements initiated by the regional and national governments, which were reflected in the growth of environmental protection initiatives in the region, promotion of employment of local community members in mining, changes in regulatory standards and enforcement of the regulatory structure.

Political and economic changes concerning land rights and rights for mineral resource exploitation in the Russian Federation and associated shifts in center-periphery relations in the 1990s between the Governments of the Republic of Sakha and the Russian Federation have influenced the course of mining industry formation. The Government of the Republic of Sakha negotiated a right for retaining a share of mineral production within the region, the concerns about economic contributions from gold and diamond production to local, regional and federal levels have been solved from the onset of privatization through participation of regional and federal governments and local authorities in the establishment of Aldanzoloto and ALROSA, and subsequent shareholding in the mining businesses.

Privatization of gold and diamond mining sectors led to an important shift from 'state owned enterprise, government and community' relationship to a new form of relationship 'private company, government and community'. Before the privatization, the mining companies were seen as agents of development assigned by the state and were seen as part of the state, while after privatization, the mining companies are assigned not only by the state, but also by other shareholders (which include local authorities) to conduct their operations. The Lease agreement between the Government of the Republic of Sakha and ALROSA, that stipulated payments for use of land and mineral resources, has acted as a key moment in the new phase of the relationship between the mining industry and the Government. Inclusion of an Article for the assignments from ALROSA sales for rehabilitation of environmental damage caused by previous diamond mining operations in the diamond province put the newly formed company responsible for actions if its predecessor, but most importantly shows a remarkable voluntary approach of ALROSA to the management of environmental and social impacts of mining operations. Such cooperation and voluntary negotiation between regional and national governments and the mining industry is also seen in global gold and diamond mining companies (similar to Impact and Benefits Agreements signed between Ekati diamond mine and indigenous groups) and demonstrates a progressive way of accepting responsibility for environmental and social impacts of business operations.

Aldanzoloto and ALROSA increasingly realize their role in the socio-economic development of the region, their role in contributing towards welfare of local communities and the need to improve the environmental performance of mining operations. The analysis of the case study companies during 1993-2000 shows that there is a distinct change in behavior in comparison with previously state-owned mining companies in the following areas:

- *Environmental protection* Aldanzoloto and ALROSA are clearly underway with environmental improvements; evidence shows that environmental impacts of their mining operations have decreased and the companies are starting to undertake some proactive approaches to environmental management, including improved management of environmental pollution, greater cooperation with local environmental protection authorities for compliance with regulatory requirements, conduct of environmental audits (ALROSA) and voluntary initiatives for bio-diversity conservation (ALROSA).
- *Employee health and safety* Aldanzoloto and ALROSA have clearly improved health and safety performance over the period 1993-2000 due to continuous investment in health and safety operations and employee training, accreditation of workplaces with national health and safety standards and close cooperation with Trade Unions for the promotion of health and safety standards. ALROSA adopted health and safety management systems before similar actions were taken in other areas of corporate social responsibility due to significant pressure from corporate employees and Trade Unions.
- *Employee management* Aldanzoloto and ALROSA have adopted a far more proactive approach to employee relations than in other areas of corporate social responsibility. Aldanzoloto and ALROSA undertake a range of voluntary initiatives in the area of employee relations, which include development of flexible employee reward schemes, initiatives for the improvement of health and well-being of employees and retired personnel (including maintenance of nurseries, sport and leisure centers and employee accommodation), and creation of a private pension scheme (ALROSA). Aldanzoloto and ALROSA provide health services and education and training

opportunities to employees and members of their families in a similar fashion to some global gold and diamond mining companies (e.g. Argyle, Debswana and Namdeb). Aldanzoloto and ALROSA have undertaken voluntary initiatives to provide corporate employees with housing and financial credits. Current practices of case study companies on employee issues are highly relevant to those developed by the top global gold and diamond mining companies and once again emphasize the importance of employee relations in the context of the mining industry.

- *Community relations* Aldanzoloto and ALROSA undertake a wide range of voluntary initiatives in the area of community relations including contribution to community life (supporting schools, hospitals, cultural and other local organizations); willingness to employ and train local and indigenous people; charity giving; development of community consultation initiatives (ALROSA); development of compensation systems (ALROSA); and building community development issues on the basis of formal agreements (ALROSA). These practices demonstrate that mining companies in the Republic of Sakha are, in many ways, similar to the top global gold and diamond mining companies by acknowledging their responsibility for the improvement of the well-being of local communities and undertaking initiatives which go beyond the normal economic framework of business operations. However, financial problems and loss of substantial financial support from the Government forced Aldanzoloto and ALROSA to reassess their community strategies, decrease financial assistance to schools and in some cases discharge the maintenance of several social care facilities (such as nurseries) to local and regional authorities.

The analysis of corporate operations of the case study mining companies shows that the mining industry significantly contributes to the fabric of community life in mining areas and reveals heavy reliance of communities on the success of mining operations in the conditions of declining state provision of welfare services and lack of non-mining business development. This creates dependency and jeopardizes long-term sustainability of communities in the mining areas, for which the mining companies and the Government have not yet considered a long-term strategy.

In general, ALROSA has shown a more progressive approach to corporate social responsibility issues than Aldanzoloto, demonstrating greater integration of environmental and social concerns in its business operations. As a growing mining company, ALROSA was able to adopt a more proactive and consistent approach to environmental protection by establishing a cohesive structure of environmental protection services; conducting environmental audits; cooperating with regional government on a number of wildlife conservation projects; and disclosing a greater amount of environmentally related information to stakeholders. ALROSA shows more standardized health and safety management and adopted a formal health and safety management system and has aspiration for further improvement through certification to international health and safety standards, whereas Aldanzoloto has not done so. With regard to employee relations, ALROSA invests more resources in the well-being of its employees, has a proactive scheme for corporate pension, and extensively invests in employee's health and education. ALROSA has a more structured and systematic approach to community development than Aldanzoloto by using mechanisms of negotiated agreements with communities and formal committees and procedures for charity giving.

Even though Aldanzoloto produces less environmental pollution by volume, which directly relates to a drop in gold production, it is evident that ALROSA being a larger enterprise is more prominent in advancing its environmental and social responsibilities. ALROSA due to its larger resource base, larger areas of operations and greater business involvement is more exposed to influences from its stakeholders. Aldanzoloto being a smaller enterprise often fails to deliver progressive environmental and social policies due to a lack of financial resources, economic problems and a lack of organizational incentive.

The analysis of the environmental and social performance of case study companies in the Republic of Sakha, Aldanzoloto and ALROSA shows that several factors affect corporate conduct in managing CSR issues such as:

- Shortage of appropriate financial and technological resources hampers the implementation of environmental and social programs. The difference in environmental performance between Aldanzoloto and ALROSA is highly relevant to the budgets of environmental protection operations. ALROSA being a more profitable company invests more resources in environmental protection, while Aldanzoloto finds it hard to fund its land rehabilitation and facilities for environmental protection due to the lack of financial and technological resources.
- Expansion of mining operations often relates to the expansion of environmental protection operations. Many substantial investments in environmental protection facilities are linked with the expansion of mining operations, such as opening of new mining facilities. For example, investment in tailings facilities, which is viewed as an important part of environmental protection within a mining enterprise, is also an important technological facility upon which the modern mining processes are based, which enables efficiency of water use within the mining enrichment process, providing turnover of water flows.

The transition from planned economy to market economy has affected the mining industry in the region in both negative and positive ways. Different approaches to privatization (remaining centralization of the diamond mining industry and decentralization of the gold mining industry), characteristic features of minerals markets (secured sales of diamond production with secured prices and free gold market with fluctuating prices) and the ability to export mineral commodities have clearly impacted on the economic performance of mining sectors in the Russian Federation. The gold mining industry has found itself in a very unfavorable economic position and is struggling to improve its gold output, while the diamond mining industry has shown remarkable growth in diamond output and economic performance, a rare exception amongst industries in the Russian Federation as a whole. Financial problems and the slow pace in attracting foreign investment have led to continued use of obsolete technology and consequently a worsening of environmental performance in Aldanzoloto.

In general, directions of policies in the area of environmental protection, health and safety, employee welfare and community development in Aldanzoloto and ALROSA are similar to policies of the global gold and diamond mining companies. Nevertheless, the mining companies in the Republic of Sakha lag behind international companies in a number of areas:

- Unlike many global gold and diamond mining companies, Aldanzoloto and ALROSA have been slow in integrating the notion of sustainable development and corporate social responsibility in their business practices. Although Aldanzoloto and ALROSA demonstrate some elements of accepting their social responsibilities, Aldanzoloto and ALROSA have not developed formal cohesive corporate policies in relation to their conduct in the areas of corporate social responsibility. The absence of strategic vision leads to the lack of a voluntary, proactive and accountable approach to environmental and social aspects of business operations.
- Aldanzoloto and ALROSA lack an integrated approach to corporate social responsibility. Each area relating to corporate social responsibility, i.e. environmental protection, human resources, health and safety and community development and stakeholder communication, is dealt in its own right.
- Aldanzoloto and ALROSA demonstrate a lack of general management excellence, a lack of management systems, poor performance analysis and insufficient dissemination of performance information within and outside the companies.
- Aldanzoloto and ALROSA tend to have a reactive approach to some environmental issues mainly focusing on compliance with environmental protection regulations and lack a voluntary approach to environmental protection, not committing themselves to perform beyond regulatory requirements.
- Stakeholder consultation processes in Aldanzoloto and ALROSA rarely go beyond engagement with corporate employees, local authorities and governmental agencies, all of which have economic ties with the companies. The companies tend to neglect the concerns of local communities and NGOs and have not yet developed initiatives for effective communication with community stakeholders. There is a clear lack of continuous corporate involvement in community consultation concerning impacts of mining operations on communities and the environment. Nevertheless, ALROSA attempted to conduct a consultation forum involving various stakeholder groups for the new diamond mining developments in Nyurba district.
- Aldanzoloto and ALROSA demonstrate less environmental and social accountability than the top global gold and diamond mining companies and have less developed auditing and reporting systems. Aldanzoloto and

ALROSA disclose limited environmental and social information in their Annual Reports and have a limited use of other media. Factors that drive greater environmental and social disclosure within the global mining industry such as pressure from financial institutions and peer pressure have not demonstrated their significance in the case of Aldanzoloto and ALROSA due to limited trading of their shares.

- Aldanzoloto and ALROSA as corporations with a higher level of organization in the region are not yet prone to share their expertise with business peers and other organizations to promote environmental and social values. These mining companies find themselves in a situation when local communities and local businesses are dependent on mining operations. However, these companies are not involved in initiatives oriented towards empowering their stakeholders by sharing organizational knowledge, steering partnerships and demonstrating leadership.

The improvement of environmental and social operations in Aldanzoloto and ALROSA is clearly related to increased public concerns (Khanna, 1999), enforcement of environmental regulation and cooperation with the government, which are considered to be drivers for corporate social responsibility (Otto, 1997). Most importantly, public pressure for improvement of environmental and social performance within the case study mining companies relates to environmental disasters (use of nuclear explosions, pollution of rivers and impacts on human health in the diamond province), which is similar to the cause of environmental movements within the global mining industry, emphasized by Shrivastava (1995) and Warhurst (2001). However, unlike global mining companies, which experience a wide range of pressures to improve environmental and social performance in the mining industry, in the case of the precious metals and minerals industry in the Republic of Sakha these factors are less developed, including:

Low level of promotion of principles of sustainable development One of the major obstacles in relation to the promotion of environmental standards by the governmental agencies is the low profile of the sustainable development in the Republic of Sakha and the Russian Federation as a whole. Although the Government of the Russian Federation has signed the Rio Declaration for Sustainable Development, national policy development has been slow and the Government does not put necessary pressure on the industry to improve its contribution to sustainable development by improving performance in three areas economic, social and environmental.

Low level of promotion of standards by governmental agencies Environmental protection agencies in the Republic of Sakha are not fully using their potential to influence environmental change in the mining companies. There is a great degree of loyalty with state officers of the environmental protection agencies towards corporate managers from the mining companies. One reason for this is the absence of effective tools to regulate environmental protection in the mining enterprises via pollution payments and fines, as they do not act as substantial determinants for mining companies and have little impact on mining operation costs. Another reason is that some of the decisions made by the state environmental protection officers are influenced by the general need to promote economic development in the region, which at the bottom line reflects the significant dependency of regional development upon the mining companies. Finally, environmental policing is also hampered by insufficient provision of financial and technological resources for environmental protection agencies.

Low level of pressure from the local, national and international NGOs It seems to be that the regional NGOs, especially the Public Committee Vilyuy which started the environmental movement in the diamond province and contributed to the formation of public opinion on the consequences of mining operations in the Republic of Sakha, are currently experiencing a stage of stagnation in their activity. Currently, the case study mining companies do not really recognize the influence from NGOs, and national and international NGOs do not appear to target mining companies in the Republic of Sakha or steer public attitudes.

Low level of pressure from employees Although demands especially for improved health and safety at workplaces and employee welfare come from employees of the mining companies, and Aldanzoloto and ALROSA effectively negotiates these demands with Trade Unions via agreements and reporting arrangements, these negotiations pay very little attention to improvement of environmental and community issues.

Low level of financial incentives and drivers to improve environmental and social performance Financial pressure that some global mining companies experience from financial institutions, which relate to trading shares on the stock exchanges, does not fully apply to the mining companies in the Republic of Sakha. ALROSA does not trade its shares on the stock exchange and therefore its share prices are not dependent upon the views of shareholders on environmental and social performance of the mining company. The companies are fearful of increasing costs of production through the installation of new systems for management of environment, health and safety, and community issues, especially in the case of the gold mining industry, which is suffering in the current financial crisis.

Low level of peer pressure and supply chain influence Low level of integration of Aldanzoloto in the wider international market leads to absence of significant peer pressure from global competitors to improve environmental performance. In the case of ALROSA, although it is integrated into international diamond market, the monopolization of the diamond mining industry, lack of open competition and lack of leadership on issues corporate social responsibility in the global diamond mining industry does not form sufficient drivers for corporate change with regards to environmental and social issues. Aldanzoloto and ALROSA are not experiencing pressure from their business partners, suppliers and customers to improve their environmental and social performance due to limited business competition within the region.

The case of the precious metals and minerals sector in the Republic of Sakha clearly demonstrates that industry and government actively cooperate in order to support social development of local communities, help ensure growth of the local economies and promote traditional lifestyles and cultures of the indigenous people of the North. Since the separation of areas of engagement between the center and periphery, when regions of the Russian Federation were given more freedom in managing issues of environmental protection, the Republic of Sakha undertook a number of proactive measures to improve regulation of environmental protection in the region (see Chapter 4). However, notwithstanding many positive actions and developments undertaken by the regional authorities in the Republic of Sakha in the field of environmental protection, the improvement of the environmental situation in the region is less likely to advance unless programs for sustainable development are initiated at the federal level and changes are made to environmental protection legislation.

Governments should not require companies and nongovernmental organizations to operate in a more accountable and socially responsible manner, but provide observance of standards and provide overall framework aiming at promotion of CSR practices via developing broad principles, approaches and tools and supporting evaluation and independent verification of CSR practices (such as environmental management, auditing and reporting).

Moreover, in order to promote activities civil society organizations, the Russian Federation needs to enhance its legislation that promotes and supports activities of NGOs, which play a crucial role in forming demands for CSR of businesses (Warhurst and Mitchell, 2000; McWilliams and Siegel, 2001). In order to boost philanthropic activities within businesses, the Russian Federation, perhaps, should consider further development of legislation concerning charity giving by enterprises, granting tax benefits, and providing legal conditions for creation of partnerships between industry, government and nonprofit sectors for social development.

In a sense, Aldanzoloto and ALROSA are undergoing a similar evolution of corporate social responsibility development observed in the western business community by Kolk (2000). This is demonstrated by increased focus on employee relations and impacts of Trade Unions on the formation of voluntary approaches to human resource management ahead of environmental and community issues, the pattern which has been also noticed in the evolution of social disclosure by Niskala and Pretes (1995). The case studies of Aldanzoloto and ALROSA demonstrate that although some elements of corporate social responsibility, such as adoption of voluntary initiates in environmental protection, health and safety, employee relations and community development are present, stakeholder management as a key of the corporate social responsibility concept (Frederick et al, 1992; Buchholz, 1998) has not significantly expanded beyond shareholders (who in the case study companies include employees, local authorities, regional and federal governments).

Companies in the Republic of Sakha need to recognize the importance of establishing effective relationships with their stakeholders, moving towards greater sustainability, and integrating corporate social responsibilities into mainstream business strategies and management, and enhance their proactive and voluntary approaches to environmental and social issues.

Recommendations for Further Research

This book has explored only selected issues concerning corporate social responsibility within the mining industry, including environmental protection, health and safety, employee policy and community development and reporting. However, many important CSR issues have been uncovered by this book such as supply chain responsibilities, fair trade, human rights, employee volunteerism, land rights, illegal mining, business ethics and corruption. Further research into CSR in the mining industry could be undertaken in several key areas, they include: ethical dimension of mining industry; land right of indigenous people; impacts of mining projects on the social and cultural structure of local communities; economic contribution of mining projects to local economic development; communication between mining companies and stakeholders; and evaluation of community programs.

Although the analysis of social disclosure by gold and diamond mining companies shows that mining companies are willing to discuss land rights of local communities and indigenous people, further research is required to investigate the mechanisms for such discussions and their effect on local communities. Due to continuous changes in legislation concerning rights of indigenous people and land property further research could explore the changes in the relationships between mining companies and local communities and their interactions concerning land within mining areas.

Aspects of community relations within the mining companies, which have not been fully studied, are cultural and social impacts of the mining developments on the local communities. Some evidence has been presented on relocation of the local indigenous population caused by infrastructure development and loss of traditional skills, language and culture. In order to understand the full picture of social and cultural impacts of the mining developments on the cultural values, effects on traditional lifestyles of communities and changes in economic livelihood, further research should be undertaken into communities surrounding mining developments in both developed and developing countries using participatory approaches.

Communication with stakeholders is an important area of corporate social responsibility, and many initiatives within the field of communications such as consultation forums, negotiated agreements and reporting could improve understanding and build essential trust between mining companies and stakeholders. Further research into communication between mining companies and NGOs could explore the nature of community concerns and expectations in relation to mining operations and the ways in which they could be addressed using various communication tools and techniques, and how communities could be engaged in decision-making concerning vital impacts on their livelihood.

In order to assess the full extent contribution of the mining industry towards sustainability of nations, assessment of corporate social responsibility within the mining industry should include the evaluation of outcomes of programs delivered by or on behalf of mining companies for local, regional, national and international communities. It is important to understand whether corporate community programs are effective, accommodate the needs and bring long-term benefits for the communities.

Appendix 1

Organizational structures for corporate social responsibility

Model of organizational structure	Board of Directors	Corporate level	Division level
Centralized integrated	Committee dealing with wide range of issues relating to CSR	CSR manager who often coordinates corporate managers dealing with separate CSR issues (environmental manager, community relations, etc)	Managers who deal with separate CSR issues (e.g. environmental manager, social relations manager, etc)
Centralized disintegrated	Several committees each dealing with separate CSR issues, e.g. health and safety committee, environmental committee	Managers for various CSR issues (e.g. corporate environmental manager, corporate personnel manager, etc)	Managers who deal with separate CSR issues (e.g. environmental manager, personnel manager, etc)
Decentralized disintegrated	No CSR related committee	Manager for various business issues incorporating CSR as one of the aspects (e.g. corporate production manager, corporate personnel manager, etc)	Managers for various CSR issues (e.g. environmental manager, personnel manager, health and safety manager etc)

Appendix 2

Models for community development

	Company-led approach	Corporate foundation	Tri-sector partnership
Organizational structure	Establishment of organizational structures within the company and allocation of resources within the company. Involves only corporate resources	Establishment of separate organizational structure outside the company, but with the inclusion of company representatives on foundation's board. Involves resources of corporate foundation assigned from donor company and perhaps external fund-raising	Establishment of formal or informal organizational structures for partnerships which usually involve representatives from mining company and other parties Involves resources from a company and its partners
Planning of initiatives	Conducted by company management. Involves views of company management with possible external consultation (communities or NGOs)	Conducted by independent management of corporate foundation, which usually include company representatives and possibly community representatives to consider their views	Conducted in partnership and involves views of company and its partners (government and NGOs)
Implementation of initiatives	Involves company with possible engagement of external organizations	Involves corporate foundations and usually grant-bearing external organizations	Involves inputs from company and its partners (government and NGOs)
Program evaluation	Involves company resources	Involves resources of corporate foundation	Involves resources from parties in a partnership
Reporting	Incorporated in company's reporting systems (Social or Community Reports, Annual Reports and websites)	Involves reporting systems of corporate foundation (Annual Reports and websites) and a donor company	Involves reporting systems of company and partners

Bibliography

Abramov, V.N. (1994), *Predpriyatie po Obrabotke Yuvelirnykh Almazov: Ekonomika, Organizatsiya, Upravlenie* [Enterprise for Cutting Jewellery Quality Diamonds: Economics, Organisation and Management], Kiev.

Acquah, P.C. (1994), 'Government Environmental Regulations and Implementation in the Mining Sector', *Natural Resources Forum*, Vol. 18 (3), pp. 193-206.

Acutt, N., Hamann, R., Carter, A. and Kapelus, P. (2001), *Towards Evidence of the Costs and Benefits of Tri-sector Partnerships*, Business Partners for Development Natural Resources Cluster, Working Paper No. 10, URL: http://bpd-naturalresources.org (accessed on 25/10/01).

Adams, C.A., Hill, W.Y., and Roberts, C.B., Association of Chartered & Certified Accountants (1995), *Environmental, Employee and Ethical Reporting in Europe*, ACCA Research Report No. 41, Certified Accountants Educational Trust, London.

AK Aldanzoloto [Aldanzoloto Company] (1996a), *Poyasnitel'naya Zapiska k Godovomu Bukhgalterskomu Balansu za 1996 God* [Explanation Note to Annual Balance Sheet for 1996], Unpublished Report, Nizhny Kuranakh.

AK Aldanzoloto [Aldanzoloto Company] (1996b), *Vozdeistvie Deyatel'nosti AK Aldanzoloto na Prirodnuyu Sredu* [Impact of Aldanzoloto Operations on the Natural Environment in 1996], Unpublished Report, Nizhny Kuranakh.

AK Aldanzoloto [Aldanzoloto Company] (1996c), *Ustav AK Aldanzoloto* [Charter of Aldanzoloto Company], Unpublished Document, Nizhny Kuranakh.

AK Aldanzoloto [Aldanzoloto Company] (1997a), *Poyasnitel'naya Zapiska k Godovomu Bukhgalterskomu Balansu za 1997 God* [Explanation Note to Annual Balance Sheet for 1997], Unpublished Report, Nizhny Kuranakh.

AK Aldanzoloto [Aldanzoloto Company] (1997b), *Vozdeistvie Deyatel'nosti AK Aldanzoloto na Prirodnuyu Sredu* [Impact of Aldanzoloto Operations on the Natural Environment in 1997], Unpublished Report, Nizhny Kuranakh.

AK Aldanzoloto [Aldanzoloto Company] (1998a), *Poyasnitel'naya Zapiska k Godovomu Bukhgalterskomu Balansu za 1998 God* [Explanation note to annual balance sheet for 1998], Unpublished Report, Nizhny Kuranakh.

AK Aldanzoloto [Aldanzoloto Company] (1998b), *Vozdeistvie Deyatel'nosti AK Aldanzoloto na Prirodnuyu Sredu* [Impact of Aldanzoloto Operations on the Natural Environment in 1998], Unpublished Report, Nizhny Kuranakh.

AK Aldanzoloto [Aldanzoloto Company] (1999a), *Poyasnitel'naya Zapiska k Godovomu Bukhgalterskomu Balansu za 1999 God* [Explanation Note to Annual Balance Sheet for 1999], Unpublished Report, Nizhny Kuranakh.

AK Aldanzoloto [Aldanzoloto Company] (1999b), *Vozdeistvie Deyatel'nosti AK Aldanzoloto na Prirodnuyu Sredu* [Impact of Aldanzoloto Operations on the Natural Environment in 1999], Unpublished Report, Nizhny Kuranakh.

AK Aldanzoloto [Aldanzoloto Company] (2000a), *Poyasnitel'naya Zapiska k Godovomu Bukhgalterskomu Balansu za 2000 God* [Explanation Note to Annual Balance Sheet for 2000], Unpublished Report, Nizhny Kuranakh.

AK Aldanzoloto [Aldanzoloto Company] (2000b), *Vozdeistvie Deyatel'nosti AK Aldanzoloto na Prirodnuyu Sredu* [Impact of Aldanzoloto Operations on the Natural Environment in 2000], Unpublished Report, Nizhny Kuranakh.

AK Aldanzoloto [Aldanzoloto Company] (2001), *Kollektivny Dogovor mezhdu AK Aldanzoloto i Profsoyuzom* [Collective Agreement between Aldanzoloto Company and Trade Union], Unpublished Paper, Nizhny Kuranakh.

AK ALROSA [ALROSA Company] (1993), *Godovoi Otchet 1993* [Annual Report 1993], AK ALROSA, Mirny.

AK ALROSA [ALROSA Company] (1994a), *Okhrana Prirody 1994* [Nature Protection 1994], Unpublished Report, Mirny.

AK ALROSA [ALROSA Company] (1994b), *Godovoi Otchet 1994* [Annual Report 1994], AK ALROSA, Mirny.

AK ALROSA [ALROSA Company] (1995a), *Okhrana Prirody 1995* [Nature Protection 1995], Unpublished Report, Mirny.

AK ALROSA [ALROSA Company] (1995b), *Godovoi Otchet 1995* [Annual Report 1995], AK ALROSA, Mirny.

AK ALROSA [ALROSA Company] (1996a), *Ustav AK Almazy Rossii Sakha* [Charter of ALROSA Company], Unpublished Document, Mirny.

AK ALROSA [ALROSA Company] (1996b), *Okhrana Prirody i Razional'noe Isponzovanie Prirodnykh Resursov 1996* [Nature Protection and Rational Use of Natural Resources in 1996], Unpublished Report, Mirny.

AK ALROSA [ALROSA Company] (1996c), *Godovoi Otchet 1996* [Annual Report 1996], AK ALROSA, Mirny.

AK ALROSA [ALROSA Company] (1997a), *Okhrana Prirody i Razional'noe Isponzovanie Prirodnykh Resursov 1997* [Nature Protection and Rational Use of Natural Resources in 1997], Unpublished Report, Mirny.

AK ALROSA [ALROSA Company] (1997b), *Godovoi Otchet 1997* [Annual Report 1997], AK ALROSA, Mirny.

AK ALROSA [ALROSA Company] (1997c), *Polozhenie o Poryadke Rassmotreniya Obrashenii Grazhdan po Voprosam Material'noi i Sponsorskoi Pomoschi* [Regulation about Order for Consideration of Appeals from Citizens for Resource and Sponsorship Aid], Unpublished Document, Mirny.

AK ALROSA [ALROSA Company] (1998a), *Godovoi Otchet 1998* [Annual Report 1998], AK ALROSA, Mirny.

AK ALROSA [ALROSA Company] (1998b), *Okhrana Prirody i Razional'noe Isponzovanie Prirodnykh Resursov 1998* [Nature Protection and Rational Use of Natural Resources in 1998], Unpublished Report, Mirny.

AK ALROSA [ALROSA Company] (1998c), *Tarifnoe Soglashenie mezhdu AK ALROSA i Profsoyuzom Rabotnikov AK ALROSA na 1998 god* [Tariff Agreement between ALROSA and Trade Union of ALROSA Employees for 1998], AK ALROSA, Mirny.

AK ALROSA [ALROSA Company] (1999a), *Godovoi Otchet 1999* [Annual Report 1999], AK ALROSA, Mirny.

AK ALROSA [ALROSA Company] (1999b), *Poyasnitel'naya Zapiska o Vozdeistvii Khozyaistvennoi Deyatel'nosti AK ALROSA na Prirodnuyu Sredu v 1999 Godu (po Materialam Statotchetnosti)* [Explanation Note about Impact of Industrial Operations of ALROSA on the Natural Environment in 1999 (According to Data for Statistical Reporting)], Unpublished Report, Mirny.

AK ALROSA [ALROSA Company] (1999c), *Tarifnoe Soglashenie mezhdu AK ALROSA i Profsoyuzom Rabotnikov AK ALROSA na 1999-2000 Gody* [Tariff Agreement between ALROSA and Trade Union of ALROSA Employees for 1999-2000], AK ALROSA, Mirny.

AK ALROSA [ALROSA Company] (1999d), *Okhrana Truda v 1999* [Labor Protection in 1999], Unpublished Report, Mirny.

AK ALROSA [ALROSA Company] (1999e), *Plan Proizvodstvenno-khozyaistvennoi Deyatel'nosti Udachninskogo GOKa v Sfere Ekologii na 2000 God* [Plan of Industrial Operations of Udachny Mining Complex in the Field of Environment for 2000], Unpublished Report, Mirny.

AK ALROSA [ALROSA Company] (2000a), *Godovoi Otchet 2000* [Annual Report 2000], AK ALROSA, Mirny.

AK ALROSA [ALROSA Company] (2000b), *Poyasnitel'naya Zapiska o Vozdeistvii Khozyaistvennoi Deyatel'nosti AK ALROSA na Prirodnuyu Sredu v 2000 Godu (po Materialam Statotchetnosti)* [Explanation Note about Impact of Industrial Operations of ALROSA on the Natural Environment in 2000 (According to Data for Statistical Reporting)], Unpublished Report, Mirny.

AK ALROSA [ALROSA Company] (2000c), *Okhrana Prirody i Razional'noe Ispol'zovanie Prirodnykh Resursov v Zone Deyatel'nosti AK ALROSA (Materially k Chetvertomu Zasedaniyu Nabludatel'nogo Soveta)* [Nature Protection and Rational Use of Natural Resources in the Area of ALROSA Operations (Data for Forth Meeting of the Supervisory Council)], Unpublished Report, Mirny.

AK ALROSA [ALROSA Company] (2000d), *Usloviya i Okhrana Truda* [Conditions and Orotection of Labor], Unpublished Report, Mirny.

AK ALROSA [ALROSA Company] (2000e), *Vypolnenie Meropriyatii Vnutrifirmennoi Programmy Pervoocherednykh Mer po Uluchsheniyu Uslovii i Okhrany Truda v AK ALROSA za 2000 God* [Implementation of Actions of Corporate Program for Primary Measures to Improve Labor Conditions and Labor Protection in ALROSA in 2000], Unpublished Document, Mirny.

AK ALROSA [ALROSA Company] (2000f), *Spravka-informaziya o Sostoyanii i Perspektivakh Regional'noi Politiki v AK ALROSA (ZAO)* [Note-information about Situation and Perspectives of the Regional Policy in ALROSA], Unpublished Document, Mirny.

AK ALROSA [ALROSA Company] (2001a), *O Prirodookhrannoi Deyatelnosti v AK ALROSA* [About Environmental Protection Activity in ALROSA], Unpublished Report, Mirny.

AK ALROSA [ALROSA Company] (2001b), *Tarifnoe Soglashenie mezhdu AK ALROSA i Profsoyuzom Rabotnikov AK ALROSA na 2001-2002 Gody* [Tariff Agreement between ALROSA and Trade Union of ALROSA Employees for 2001-2002], AK ALROSA, Mirny.

AK ALROSA [ALROSA Company] (2001c), *Konzepziya Sokhraneniya i Vosstanovleniya Resursov Zhovotnogo Mira v Zone Deyatel'nosti AK Almazy Rossii-Sakha* [Conception for Conservation and Rehabilitation of Wild Animals in the Area of ALROSA Operations], Unpublished Document, Mirny.

AK ALROSA [ALROSA Company] (2001d), *O Sostoyanii Okhrany Truda i Promyshlennoi Bezopasnosti Podrazdelenii Kompanii za Pervy Kvartal 2001 Goda* [Situation on Labor Protection and Industrial Safety in Divisions of the Company in 1st Quarter of 2001], Unpublished Report, Mirny.

AK ALROSA [ALROSA Company] (2001e), *Godovoi Otchet 2001* [Annual Report 2001], AK ALROSA, Mirny.

AK ALROSA [ALROSA Company] (2002), *Godovoi Otchet 2002* [Annual Report 2002], AK ALROSA, Mirny.

AK ALROSA [ALROSA Company] (2003), *Godovoi Otchet 2003* [Annual Report 2003], AK ALROSA, Mirny.

AK ALROSA and Yakutniproalmaz [ALROSA Company and Yakutniproalmaz Institute] (2001), *Otchet o Nauchno-issledovatel'skoi Rabote: Sovershenstvovanie Sozial'no-ekonomicheskikh Otnosheniy na Predpriyatii AK ALROSA (23-96-624). Korrektirovka Sozial'nogo Pasporta AK ALROSA* [Research Report: Improvement of Socio-economic Relationships in ALROSA Company Ltd. (23-96-624). Correction of Social Passport of ALROSA Company Ltd.], Unpublished Report, Mirny.

AK ALROSA i Ministerstvo Okhrany Prirody Respubliki Sakha (Yakutia) [ALROSA Company and the Ministry for Nature Protection of the Republic of Sakha (Yakutia)] (2001), *Programma 'Sozdanie Pitomnika Losei v Mirninskom Uluse'* [Program 'Creation of Elk Farm in Mirny District'], Unpublished Document, Yakutsk.

AK ALROSA i Ministerstvo Okhrany Prirody Respubliki Sakha (Yakutia) [ALROSA Company and the Ministry for Nature Protection of the Republic of Sakha (Yakutia)] (2000), *Soglashenie mezhdu Ministerstvom Okhrany Prirody RS(Y) i Akzionernoi Kompaniei ALROSA (ZAO) v Oblasti Okhrany Prirody i Vosstanovlenie Okruzhayuschei Sredy, Razvitie Sistemy Osobo Okhranyaemykh Prirodnykh Territorii (Ytyk Kere Sirder)* [Agreement between the Ministry for Nature Protection of the Republic of Sakha and ALROSA Company in the Field of Nature Protection, Environmental Rehabilitation and Development of System of Specially Protected Natural Areas (Ytyk Kere Syrder)], Unpublished Document, Mirny.

Aldanzoloto (1994), *70 Let Akzionernoi Kompanii Aldanzoloto 1924-1994* [70 Years of Aldanzoloto Company 1924-1994], Aldanzoloto, Moscow.

Alekseev, V. (1999), 'Vozmozhno li Razumnoe Prirodopol'zovanie pri Rynochnoi Ekonomike?' [Is Reasonable Natural Resource Use Possible in the Conditions of Market Economy?], *Yakutia,* Vol. 55 (29081), 26 March 1999, URL: http://www.yakutia.ru/ (accessed on 30/04/99).

Ali, S. and Behrendt, L. (2001), 'Mining and Indigenous Rights: the Emergence of a Global Social Movement', *Cultural Survival Quarterly*, Vol. 25(1), URL: http://www.cs.org/publications/CSQ (accessed on 29/10/01).

Amey, E.B. (1999), *Gold, Minerals Yearbook 1998*, United States Geological Survey, URL: http://minerals.usgs.gov/ (accessed on 1/03/01).

Amey, E.B. (2000), *Gold, Minerals Yearbook 1999*, United States Geological Survey, URL: http://minerals.usgs.gov/ (accessed on 1/03/01).

Amey, E.B. (2001), *Gold, Minerals Yearbook 2000*, United States Geological Survey, URL: http://minerals.usgs.gov/ (accessed on 30/04/02).

AngloGold Limited [online] URL: http://www.anglogold.com (accessed 01/05/01).

AngloGold Limited (1999), *AngloGold Annual Report*, AngloGold Limited, Johannesburg.

AngloGold Limited (2000), *AngloGold Limited Annual Report*, AngloGold Limited, Johannesburg.

Anosov, S. (1999), *Mineral'no-syrievoi Sektor Rossii* [Minerals Sector of Russia], Institut Ekonomicheskikh Problem Perekhodnogo Perioda [Institute of Economic Problems of Transition Period].

Apparat Pravitelstva Republiki Sakha (Yakutia) [Government of the Republic of Sakha (Yakutia)](1998), *Informaziya o Rabote FPG SAPI v Ulusakh Respubliki Sakha (Yakutia) za 1998 God* [Information on Operations of the SAPI Financial and Industrial Group in Districts of the Republic of Sakha (Yakutia)], Unpublished Report, Yakutsk.

Argyle Diamonds (1999), *Annual Social and Environmental Report*, Argyle Diamond Mines PTY Limited, West Perth.

Argyle Diamonds [online] URL: http://www.argylediamonds.com.au (accessed 18/04/01).

Arrow, K.J. (1985), 'The Economics of Agency', in Pratt, J.W. and Zeckhauser, R.L. (eds), *Principal and Agents: the Structure of Business*, Harvard Business School Press, Boston.

Ascher, W. (1994), 'Survey of Issues in Government Policy and Public Management of State and Mining Companies in Developing Countries', *Natural Resources Forum*, Vol. 18(1), pp. 3-11.

Ashanti Goldfields Company Limited [online] URL: http://www.ashantigold.com (accessed 01/05/01).

Ashanti Goldfields Company Limited (1998), *Annual Report*, Royal Print Limited, England.

Ashanti Goldfields Company Limited (1999), *Annual Report*, Royal Print Limited, England.

Ashanti Goldfields Company Limited (2000), *Annual Report*, Royal Corporate Print, England.

Ashton Mining (1998), *Annual Report*, John Herrod & Associates.

Ashton Mining (1999), *Annual Report*, John Herrod & Associates.

Assoziazia Menedgerov [Association of Managers] (2001), *Analitichesky Otchet po Rezultatam Issledovaniya Assoziazii Menedgerov 'Sozialnaya Rol' Biznesa v Obschestvennom Razvitii: Korporotivnaya Blagotvoritel'nost i Sponsorstvo'* [Analytical Report on Results of the Study of the Association of Managers 'Social Role of Business in Societal Development: Corporate Philanthropy and Sponsorship'], URL: http://www.amr.ru/sociology/charity.zip (accessed on 04/06/02).

Atkinson, J. (1999), 'Industry Codes and Community Rights', *Mining Environmental Management*, Vol. 7(6), pp. 13-14.

Auty, R.M. (1993), *Sustaining Development in Mineral Economies: the Resource Curse Thesis*, Routledge, London.

Avocat, R., Milian, J.F. and Bertoni, C. (2001), 'The Impacts of Artisanal Gold Mining on Surface Water Quality', *Mining Environmental Management*, Vol. 9(1), pp. 18-22.

Azapagic, A., and S. Perdan (2000), 'Indicators of Sustainable Development for Industry: a General Framework, *Trans IChemE* (Proc. Safety Envir. Prot.), Vol. 78(B4), Part B, pp. 243-261.

Azzone G., Manzini, R. and Noci, G. (1996), 'Evolutionary trends in environmental reporting', in J.P. Ulhøi and H. Madsen (eds), *Industry and the Environment: Practical Applications of Environmental Management Approaches in Business*, The Aarhus School of Business, pp. 15-28.

Azzone, G., Brophy, M., Noci, G., Welford, R. and Young, W. (1997), 'A Stakeholder's Views of Environmental Reporting', *Long Range Planning*, Vol. 30(5), pp. 699-709.

Balabanis, G., Phillips, H.C. and Lyall, J. (1998), 'Corporate Social Responsibility and Economic Performance in the Top British Companies: Are They Linked?' *European Business Review*, Vol. 98(1), pp. 25-44.

Balazik, R.F. (1999), *Diamonds, industrial*, US Geological Survey Publications, URL: http://minerals.usgs.gov/ (accessed on 14/04/01).

Balkau, F. (1993), *Pollution Prevention and Abatement Guidelines for the Mining Industry*, UNEP/IE/PAC, Paris.

Balkau, F. and Parsons, A. (1999), 'Emerging Environmental Issues for Mining in the PECC Region', *1st Pacific Economic Co-operation Committee Minerals Forum in Lima, Peru, 22 April 1999*, URL: http://www.natural-resources.org/minerals/ (accessed on 15/05/00).

Banks, G. and Ballard, C. (eds) (1997) *The Ok Tedi Settlement: Issues, Outcomes and Implications*, National Centre for Development Studies.

Barnes, J.W. (1995), *Ores and Minerals: Introducing to Economic Geography*, John Wily and Sons Ltd, Chichester.

Barrick Gold Corporation [online] URL: http://www.barrick.com (accessed 08/04/02).

Barrick Gold Corporation (1998), *Barrick Gold Corporation Annual Report*, Barrick Gold Corporation, Toronto.

Barrick Gold Corporation (1999), *Barrick Gold Corporation Annual Report*, Barrick Gold Corporation, Toronto.

Barrick Gold Corporation (2000), *Annual Report*, Barrick Gold Corporation, Toronto.

Bartelmus, P. (1994), *Environment, Growth and Development the Concepts and Strategies of Sustainability*, Routledge, London.

Belykh, V.E. and Vybornov, M.I. (eds) (1984), *Lyudi i Almazy* [People and Diamonds], Yakutsk.

Belykh, Z.P. (1993), *Novye Grani Almaznogo Kraya* [New Facets of Diamond Province], Soyuzreklamkultura, Moscow.

Benevolsky, B.I. and Ivanov, V.N. (1999), 'Mineral'no-syr'evaya Baza Zolota Rossii na Rubezhe XXI v' [The Gold Resource Base in Russia at the Turn of 21 Century], *Mineral'nye Resursy Rossii: Ekonomika i Upravlenie* [Mineral Resources of Russia: Economics and Managamenet], (1), pp. 9-18.

Bennett, R. (1998), 'Corporate Philanthropy in France, Germany and the UK', *International Marketing Review*, Vol. 15(6), pp. 458-475.

Bergenstock, D.J. and Maskulka, J.M. (2001), 'The De Beers Story: are Diamonds Forever?' *Business Horizons*, Vol. 44(3), pp. 37-44.

Bew, D. (1995), 'Diamonds and Trade', *TED Case Studies*, Vol. 4(1), URL: http://www.american.edu/TED/class/all.htm (accessed on 29/05/02).

BHP (1998), *Environment Report*, The Broken Hill Propriety Company Limited, Melbourne.

BHP (1998), *Report to Shareholders*, The Broken Hill Propriety Company Limited, Melbourne.

BHP (1999), *BHP Environment and Community Report*, The Broken Hill Propriety Company Limited, Melbourne.

BHP (1999), *Report to Shareholders*, The Broken Hill Propriety Company Limited, Melbourne.

BHP (2000), *Annual Report*, The Broken Hill Proprietary Limited, Melbourne.

BHP (2000), *BHP Environment and Community Report*, The Broken Hill Proprietary Company Limited, Melbourne.

BHP Diamond Inc [CD-ROM] (2000), *Environmental Agreement and Water Licence Annual Report 1999*, BHP Diamonds Inc, Yellowknife.

BHP Diamonds Inc (2000), *Annual Report: EKATI Diamond Mine Environmental Agreement*, BHP Diamonds Inc, Yellowknife.

BHP Diamonds Inc (2000), *Summary Report: Environmental Agreement Annual Report 1999, Environmental Impact Report 2000*, BHP Diamonds Inc, Yellowknife.

BHP Diamonds Inc (2001), *Environmental Agreement Annual Report 2000*, BHP Diamonds Inc, Yellowknife.

BHP Diamonds Inc [CD-ROM] (2001), *Environmental Agreement and Water Licence Annual Report 2000*, BHP Diamonds Inc.

Bomsel, O., Börkey, P., Glachant, M. and Lévêque, F. (1996), 'Is There Room for Environmental Self-regulation in the Mining Sector?' *Resources Policy*, Vol. 22(1/2), pp. 79-86.

Bond, A.R. (1984), 'Air Pollution in Noril'sk: a Soviet Worst Case?' *Soviet Geography,* Vol. 25(9), pp. 665-680.

Bond, A.R. (1994), 'The Non-energy Mineral Industries of Post-Soviet Russia', *Post-Soviet Geography*, Vol. 35(9), pp. 543-565.

Bowen, H.R. (1953), *Social Responsibilities of the Businessmen*, Harper and Row, New York.

Brack, D. (ed.) (1998), *Trade and Environment: Conflict or Compatibility?* Proceedings of the Royal Institute of International Affairs Conference, Chatham House, London, April 1997, Royal Institute of International Affairs, London.

Bradshaw, M.J. (1995), 'The Russian North in Transition: General Information', *Post-Soviet Geography*, Vol. 36(4), pp. 195-203.

Brophy, M. (1998), 'Environmental Policies', in R. Welford (ed.), *Corporate Environmental Management: Systems and Strategies*, 2nd ed., Earthscan Publications Ltd, London.

Brummer, J.J. (1991), *Corporate Responsibility and Legitimacy: an Interdisciplinary Analysis*, Greenwood Press, New York.

Buchholz, A.K., Amason, A.C. and Rutherford, M.A. (1999), 'Beyond Resources: the Mediating Effect of Top Management Discretion and Values on Corporate Philanthropy', *Business and Society*, Vol. 38(2), pp. 167-188.

Buchholz, R.A. (1998), *Principles of Environmental Management: the Greening of Business*, 2nd ed., Prentice-Hall Inc, New Jersey.

Buckley, C. and International Network for Environmental Management (1999), *Case Studies of Environmental Management in Central and Eastern Europe*, Casebook Series, Vol. II, Greenleaf Publishing, Sheffield.

Buckley, R. (1992), 'Sustainability in Mining', *Environmental Management and Health*, Vol. 3(3), pp. 23-26.

Buckmaster, N. (1999), 'Associations between Outcome Measurement, Accountability and Learning for Non-profit Organisations', *The International Journal of Public Sector Management*, Vol. 12 (2), pp. 186-197.

Bullough, M. and Johnson, D. (1995), 'Corporate Environmental Reporting in Practice', *Business Strategy and the Environment*, Vol. 4(1), pp. 36-39.

Burtsev, I.S. and Kolodeznikova, E.N. (1997), *Raioazionnaya Obstanovka v Almazonosnykh Raionakh Yakutii* [Radioactive Situation in the Diamond Regions of Yakutia], YNZ SO RAN [Yakutsk Scientific Centre, Siberian Branch, Russian Academy of Sciences], Yakutsk.

Burtseva, E.I. (1993), 'Prirodopol'zonabie v Proshlom i Nastoyaschem: Tradizii Korennogo Naseleniya Yakutii i Gornodobyvayuschaya Promyshlennost' [Natural Resource Use in Past and Present: Traditions of Indigenous People in Yakutia and the Mining Industry], in D.D. Savvinov (ed.), *Ekologicheskie Tradizii Aborigenov Severa* [Ecological Traditions of Indigenous People of the North], Institut Prikladnoi Ekologii Severa SO RAN [Institute of Applied Ecology of the North, Siberian Branch, Russian Academy of Sciences], Yakutsk.

Business for Social Responsibility [online], *Introduction to Corporate Social Responsibility*, White Paper, URL: http://www.bsr.org (accessed on 19/05/02).

Business Impact [online], URL: http://www.business-impact.org (accessed on 12/06/02).

Business Partners for Development Natural Resources Cluster [online], URL: http:// http://www.bpd-naturalresources.org/ (accessed on 12/06/02).

Cai, J. Chaeng, Y.L. and Wong, M.C.S. (2001), 'What Moves the Gold Market?' *The Journal of Future Markets*, Vol. 21(3), pp. 257-278.

Carroll, A.B. (1979), 'A Three-dimentional Conceptual Model of Corporate Performance, *Academy of Management Review*, Vol. 4(4), pp. 497-505.

Carroll, A.B. (1999), 'Corporate Social Responsibility: Evolution of a Definitional Construct', *Business and Society*, Vol. 39 (3), pp. 268-295.

Carroll, A.B. and Buchgolz, A.K. (1999), *Business and Society: Ethics and Stakeholder Management*, 4th ed., South-Western College Publishing, Cincinnati.

Carter, A.S. (2000), 'Delivering Social Investment Programmes through Corporate Foundations', *MERN Research Bulletin & Newsletter*, No. 15, pp. 200-210.

CECGRF (Centre for Economic Conjuncture under the Government of the Russian Federation) (2003), 'Russian Foreign Trade in 2002', URL: http://www.cea.gov.ru/ (29/04/04).

CERES (1998), *Report Standard Form*, Coalition for Environmentally Responsible Economies, URL: http://www.ceres.org (accessed on 19/02/00).

Chernogrivov, P.N. and Bareisha, V.M. (2001), 'Ekologicheskii Menedjment i Audit v Tomskoj Oblasti' [Environmental Management and Audit in Tomsk Oblast], in -, *Ekologicheskii Menedjement i Obshestvenny Dialog: Perspektivy v Rossii, Materialy Mezhdunarodnoi Konferenzii, Moskva, 15-16 Noyabrya 2001* [Environmental Management and Public Dialogue: Perspectives for Russia, Proceedings of International Conference, Moscow, 15-16 November 2001], GEOPLAN-Ecoline, Moscow, URL: http://www.ecoline.ru/ (accessed on 02/07/02).

Chidambaran, N.K., Fernando, C.S., Spindt, P.A. (1999), *Credit Enhancement through Financial Engineering: Freeport-McMoRan's Gold-denominated Depository Shares*, The Warton School, University of Pennsylvania, URL: http://fic.wharton.upenn.edu/fic/papers/99/9935.pdf (accessed on 3/11/04).

Chomchoev, A.I., Mikulenko, K.I. and Protopopov, Ju.Kh. (1993), 'Geologicheskaya Obstanobka i Vosmozhnie Ekologicheskie Posledstviya Podzemnykh Yadernykh Vzryvov: na Primere 'Kraton-4' [Geological Situation and Possible Environmental Consequences of Underground Nuclear Explosions: a Case of 'Kraton-4'], in Goskomitet po Delam Grazdanskoj Oborony, Chrezvychainym Situaziyam i Likvidazii Posledstvii Stikhiinykh Bedstvii Respubliki Sakha (Yakutia) i Ministerstvo Zdravookhranenia Respubliki Sakha (Yakutia) [State Committee for Emergency Situation of the Republic of Sakha (Yakutia) and the Ministry of Health of the Republic of Sakha (Yakutia)], *Radioazionnoe Zagryaznenie Territorii Respubliki Sakha (Yakutia): Problemy Radiazionnoi Bezopasnosti* [Radioactive Contamination on the Territory of the Republic of Sakha (Yakutia): Problems of Radioactive Safety], Yakutsk, pp. 18-28.

Clark, A. (1992), 'Socio-cultural Impacts', in Department for Technical Co-operation for Development, United Nations, and the Development Policy Forum of the German Foundation for International Development, *Mining and the Environment: the Berlin Guidelines: a Study Based on International Round Table in June 1991, Organised by the Department for Technical Co-operation for Development, United Nations, and the Development Policy Forum of the German Foundation for International Development*, Mining Journal Books Ltd, London, pp. 106-110.

Clark, A.L. and Clark, J.C. (1999), 'The New Reality of Mineral Development: Social and Cultural Issues in Asia and Pacific Nations', *Resources Policy*, Vol. 25(3), pp. 189-196.

Clarke, A. and Dawson, R. (1999), *Evaluation Research: an Introduction to Principles, Methods and Practice*, Sage Publications, London.

Clinch, J.P. (2001), 'Towards a More Integrated Approach to Environmental and Health and Safety Policy', *Eco Management and Auditing*, Vol. 8(1), pp. 12-24.

Connolly, P. (2001), *Creating a Blueprint for the Future, Building to Last: a Grantmakers' Guide to Strengthening Nonprofit Organisation*, Conservation Company, URL: http://www.tccgrp.com/ (accessed on 16/05/01).

Connors, T.D. (2001), *Nonprofit Handbook: Management*, 3rd ed., John Wiley and Sons Ltd, Chichester.

Cormier, D. and Gordon, J.M. (2001), 'An Examination of Social and Environmental Reporting Strategies', *Accounting, Auditing & Accountability Journal*, Vol. 14(5), pp. 587-616.

Corporate Citizenship Company [online], URL: http://www.corporate-citizneship.co.uk (accessed on 1/07/02).

Cottrell, G. and Rankin, L. (1998), *Developments in Environmental Reporting in the Resources Industry: Future Directions in Australia*, PriceWaterHouseCoopers, URL: http://www.pwcglobal.com (accessed 1/07/01).

Cottrell, G. and Rankin, L. (2000), *Creating Business Value through Corporate Sustainability: Sustainability Strategies and Reporting for the Gold Industry*, PriceWaterHouseCooppers, URL: http://www.pwcglobal.com/ (accessed on 22/07/02).

Council of Foundations [online], URL: http://www.cof.org (accessed 1/07/02).

Cowell, S.J., Wehrmeyer, W., Argust, P.W. and Robertson, G.S. (1999), 'Sustainability and the Primary Extraction Industries: Theories and Practices', *Resources Policy*, Vol. 25(4), pp. 277-286.

Crotty, J. and Crane, A. (2000), *Transition in Environmental Risk in a Transitional Economy: Management Capability and Stakeholder Trust in Russia*, Liverpool Research Papers in Economics, Finance and Accounting, No. 0025.

Crotty, J., Crane, A., Matten, D. (2001), *Environmental Management in Russia: Transition in Corporate Citizenship and Regulation from a Risk Society Perspective*, Liverpool Research Papers in Economics, Finance and Accounting, No. 0106.

Crowson, P. (1997), 'Mining During the Next 25 Years: Issues and Challenges', *Natural Resources Forum*, Vol. 21(4), pp. 231-238.

Crowson, P. (1998), 'Environmental and Community Issues and the Mining Industry', *Natural Resources Forum*, Vol. 22(2), pp. 127-130.

Danilov-Danilyan, V.I. (1999), 'Prirodookhrannoe Delo v Rossii i Perekhod k Ustoichivomu Razvitiyu' [Environmental Protection in Russia and Transition to Sustainable Development], *Vtoroi Vserossiskii S'ezd po Okhrane Prirody* [Second Russian National Congress on Environmental Protection], URL:http://www.ecocom.ru/arhiv/ecocom/2th Congress/3-3.htm(accessed on 3/11/04).

Davis-Walling, P. and Batterman, S. (1997), 'Environmental Reporting by Fortune 50 Firms', *Environmental Management*, Vol. 21(6), pp. 865-875.

Davy, A. (1999), 'Social Responsibility and Governments', *Mining Environmental Management*, Vol. 7(1), January, pp. 16-17.

Davy, A. (2001), *Managing Community Expectations through Tri-sector Partnerships*, Business Partners for Development Natural Resources Cluster, Working paper No. 8, URL: http://bpd-naturalresources.org (accessed on 25/10/01).

De Beers (2000b), *Safety, Health and Environment Review*.

De Beers [online] URL: http://www.debeersgroup.com (accessed 01/05/01).

De Beers Consolidated Mined Limited & De Beers Contenary AG (1998), *Annual Report*, Creda Communications.

De Beers Consolidated Mines Limited & De Beers Centenary AG (1999), *Annual Report*, De Beers Corporate Communications Department.

De Beers Consolidated Mines Limited & De Beers Centenary AG (2000a), *Annual Report*, De Beers Corporate Communications Department.

Debswana Diamond Company [online] URL: www.debswama.com (accessed 01/05/01).

Debswana Diamond Company (1998), *Annual Report*, Sukelela Communications.

Debswana Diamond Company (1999), *Annual Report*, Debswana House, Gaborone.

Debswana Diamond Company (2000), *Annual Report*, Debswana Diamond Company, Gaborone.

Deegan, C. and Rankin, M. (1996), 'Do Australian Companies Report Environmental News Objectively? An Analysis of Environmental Disclosures by Firms Prosecuted Successfully by the Environmental Protection Authority', *Accounting, Auditing & Accountability Journal*, Vol. 9(2), pp. 50-67.

Department for Technical Co-operation for Development, United Nations, and the Development Policy Forum of the German Foundation for International Development (1992), *Mining and the Environment: the Berlin Guidelines: a Study Based on International Round Table in June 1991, Organised by the Department for Technical Co-operation for Development, United Nations, and the Development Policy Forum of the German Foundation for International Development*, Mining Journal Books Ltd, London.

Desyatkin, T.G. (1998), *Zoloto, Almazy i Moya Zhizn'* [Gold, Diamonds and My Life], Yana, Moscow.

Dias, A.K. and Begg, M. (1994), 'Environmental Policy for Sustainable Development of Natural Resources: Mechanisms for Implementation and Enforcement', *Natural Resources Forum*, Vol. 18(4), pp. 276-286.

Dibley, D. and Newson, M. (1998), 'What the Readers Want to See', *Groundwork*, Vol. 1(3), pp. 16-17.

Dixon, R.A. and Dillon, M.C. (eds) (1990), *Aborigines and Diamond Mining: the Politics of Resource Development in the East Kimberley, Western Australia*, University of Western Australia Press, Nedlands, Western Australia.

Donaldson, T. (1983), 'Constructing a Social Contruct for Businesses', in Donaldson, T. and Werhane, P. (eds), *Ethical Issues in Business*, Prentice-Hall, Englewood Cliffs, pp. 153-165.

Dorian, J.P. and Humphreys, H.B. (1994), 'Economic Impacts of Mining: a Changing Role in the Transitional Economics', *Natural Resources Forum*, Vol. 18(1), pp. 17-29.

Down, C.G. and Stocks, J. (1977), *Environmental Impacts of Mining*, Applied Science Publishers Ltd, London.

Edwards, M. and Hulme, D. (eds) (1998), *Non-governmental Organisations-Performance and Accountability*, Earthscan Publications Ltd, London.

Eggert, R.G. (ed.) (1996), *Mining and the Environment: International Perspective on Public Policy*, Resources for the Future, Washington.

Egorov, E.G. (1990), 'Ekonomicheskie Aspekty Ekologii Basseina Reki Vilyuy' [Economic Aspects of Ecology in the Basin of the River Vilyuy] in I.Ya. Egorov and V.F. Chernyavsky (eds), *Vorposy Regional'noj Gigieny, Sanitarii i Epidemiologii* [Questions of Regional Hygiene, Sanitation and Epidemiology], Minzdrav Yakutskoi ASSR [Ministry of Health of the Yakut Autonomous Soviet Socialist Republic], Yakutsk, pp. 22-23.

Egorov, N.K., Novikov, F.L.P. and Popov, V.K. (eds) (2000), *The Radiation Legacy of the Soviet Nuclear Complex: an Analytical Overview*, Earthscan Publications Ltd, London.

EIU (1998), *Russia Country Report*, 4th Quarter 1998, Economist Intelligence Unit.

EIU (1999a), *Russia Country Report*, 2nd Quarter 1999, Economist Intelligence Unit.

EIU (1999b), *Russia Country Report*, 4th Quarter 1999, Economist Intelligence Unit.

EIU (2000a), *Russia Country Report*, 1st Quarter 2000, Economist Intelligence Unit.

EIU (2000b), *Russia Country Report*, September 2000, Economist Intelligence Unit.

EIU (2000c), *Russia Country Report*, December 2000, Economist Intelligence Unit.

EIU (2001), *Russia Country Report*, June 2001, Economist Intelligence Unit.

EIU (2002), *Russia Country Report*, January 2002, Economist Intelligence Unit.

Ekati Diamond Mine [online] URL: http://ekati.bhpbilliton.com (accessed 07/05/01).

Ellman, M. (2000), The Russian Economy Under El'tsin, *Europe-Asia Studies*, Vol. 5(8), pp. 1417-1432.

Energia: Ekonomika, Technika, Ekologia (2000), 'Zolotodobyvayushaya Promyshlennost' Rossii' [Gold Mining Industry of Russia], *Energia: Ekonomika, Technika, Ekologia*, No. 5, pp. 24-28.

Epps, J.M. (1996), 'Environmental Management in Mining: an International Perspective of an Increasing Global Industry', *The Journal of South African Institute of Mining and Metallurgy*, Vol. 96(2), pp. 67-90.

Ericsson, M. and Tegen, A. (1992), 'Dynamics of State Mining Enterprises During the 1980s and the Outlook for the 1990s', *Natural Resources Forum*, Vol. 16(3), pp. 178-191.

European Commission (2001), Promoting a European Framework for Corporate Social Responsibility: Green Paper. Luxemburg: Office for Official Publications of the European Communities.

Expert (1999), 'Expert-200, Reiting Krupneishikh Kompanii Rosii po Ob'emu Realizatsii Produkzii v 1998 Godu' [Expert-200: the Rating of Biggest Russian Companies According to Sales Volumes in 1998', *Expert*.

Fitz-Gibbon, C.T. and Morris, L.L. (1987), *How to Design a Program Evaluation*, 2nd ed., Program Evaluation Kit Series 3, Sage Publications, London.

Foundation Centre (2000), *Foundation Reporting*, Foundation Today Series, URL: http://www.fdncentre.org (accessed on 20/04/01).

Foundation Centre (2001), *Foundation Giving Trends*, Foundation Today Series, URL: http://www.fdncentre.org (accessed on 20/04/01).

Fowler, A. (1998), 'Assessing NGO Performance: Difficulties, Dilemmas and Way Ahead', in M. Edwards and D. Hulme (eds), *Non-governmental Organisations: Performance and Accountability*, Earthscan Publications Ltd, London, pp. 143-156.

Frederick, W.C., Post, J.E. and Davis, K. (1992), *Business and Society: Corporate Strategy, Public Policy and Ethics*, 7th ed., McGraw-Hill Inc, New York.

Freeport-McMoRan Copper & Gold Inc (1998), *Annual Report*, Freeport-McMoRan Copper & Gold Inc, New Orleans.

Freeport-McMoRan Copper & Gold Inc (1999), *Annual Report*, Freeport-McMoRan Copper & Gold Inc, New Orleans.

Freeport-McMoRan Copper & Gold Inc (2000), *Annual Report*, Freeport-McMoRan Copper & Gold Inc, New Orleans.

Freeport-McMoRan Copper & Gold Inc [online] URL: http://www.fcx.com (accessed 6/03/01).

Friedman, M. (1970), 'The Social Responsibility of Business Is To Increase Its Profits', *New York Times Magazine*, 13 September, p. 33.

Frost, G. (2000), 'Environmental Reporting Practices by Australian Companies', *12th Asian Pacific Conference on International Accounting Issues, Beijing, October 21-24, 2000.*

Gibbons, M.J. and Wickens, P.A. (1999), 'Impacts of Deep Sea Mining', *Mining Environmental Management*, Vol. 7(4), pp. 22-23.

Global Mining Initiative [online], URL: http://www.globalmining.com (accessed on 12/06/02).

Global Reporting Initiative [online], URL: http://www.globalreporting.org (accessed on 11/02/02).

Gold Fields (1998), *Annual Report*, Ince (Pty) Ltd, Johannesburg.

Gold Fields (1999), *Annual Report*, Ince (Pty) Ltd, Johannesburg.

Gold Fields (2000), *Annual Report*, Russell & Associates, Johannesburg.

Gold Fields [online] URL: http://www.gold-fields.com (accessed 01/05/01).

Gold Fields Minerals Services (1997), 'Gold', *Metals & Minerals Annual Review 1996, Supplement to Mining Journal,* pp. 17-23.

Gold Fields Minerals Services Ltd (1999), 'Gold', *Precious Metals and Minerals: Annual Review, Supplement to Mining Journal*, pp. 15-21.

Goldberg, B. (1997), 'Corporate Philanthropy Seeks Returns', *Colorado Business*, Vol. 27(1), pp. 22-26.

Goldman, M.I. (1972), *The Spoils of Progress: Environmental Pollution in the Soviet Union*, MIT Press, London.

Gorokhov, K.V. (1989), 'Gde Pasti Severnykh Olenei' [Where to Pasture the Rain-deer], *Socialist Yakutia*, 13 August 1989.

Goskomekologia (1998a), *Gosudarstvenny Doklad o Sostoyanii Okruzhayuschei Sredy v Rossiiskoi Federazii v 1997 Godu* [State Report on the Environmental Situation in the

Russian Federation in 1997], Gosudarstvenny Komitet po Ekologii [State Committee for the Environment], Moscow.

Goskomekologia (1998b), 'Osnovnye Napravleniya Razvitiya Deyatel'nosti v Oblasti Ekologicheskogo Audirovaniya i Menedjmenta v Rossiiskoi Federazii' [General Trends in the Development in the Area of Environmental Auditing and Management in the Russian Federation], URL: http://www.ecoline.ru/mc/management/directions.html (accessed on 5/07/02).

Goskomekologia (1999), *Gosudarstvenny Dolkad o Sostoyanii Prirodnoi Sredy v Rossiiskoi Federazii v 1998 Godu* [State Report on the Environmental Situation in the Russian Federation in 1998], Gosudarstvenny Komitet po Ekologii [State Committee for the Environment], Moscow.

Goskomekologia (2000), *Gosudarstvenny Dolkad o Sostoyanii Prirodnoi Sredy v Rossiiskoi Federazii v 1999 Godu* [State Report on the Environmental Situation in the Russian Federation in 1999], Gosudarstvenny Komitet po Ekologii [State Committee for the Environment], Moscow.

Goskomekologia (2001), *Gosudarstvenny Doklad o Sostoyanii Prirodnoi Sredy v Rossiiskoi Federazii v 2000 Godu* [State Report on the Environmental Situation in the Russian Federation in 2000], Gosudarstvenny Komitet po Ekologii [State Committee for the Environment], Moscow, URL: http://www.eco-net.ru (accessed on 20/04/04).

Goskomitet po Delam Grazdanskoi Oborony, Chrezvychajnym Situaziyam i Likvidazii Posledstvii Stikhiinykh Bedstvii Respubliki Sakha (Yakutia), Ministerstvo Zdravookhranenia Respubliki Sakha (Yakutia) [State Committee for Emergency Situation of the Republic of Sakha (Yakutia) and the Ministry of Health of the Republic of Sakha (Yakutia)] (1993), *Radioazionnoe Zagryaznenie Territorii Respubliki Sakha (Yakutia): Problemy Radiazionnoi Bezopasnosti* [Radioactive Contamination in the Territory of the Republic of Sakha (Yakutia): Problems of Radioactive Safety], Yakutsk.

Goskomstat RS(Y) (1994), *Sozial'no-ekonomicheskoe Polozhenie v Respublike Sakha (Yakutia) za Yanvar-Dekabr 1993 Goda* [Socio-economic Situation in the Republic of Sakha (Yakutia) in January-December 1993], Gosudarstvennii Komitet Respubliki Sakha (Yakutia) po Statistike [State Committee for Statistics of the Republic of Sakha (Yakutia)], Yakutsk.

Goskomstat RS(Y) (1995), *Goroda i Ulusy Respubliki Sakha (Yakutia) za Yanvar-Dekabr 1994 Goda, Kompleksnii Statbulleten 11/3633* [Cities and Districts of the Republic of Sakha (Yakutia) in January-December 1994, Complex Statistical Bulletin 11/3633], Gosudarstvennii Komitet Respubliki Sakha (Yakutia) po Statistike [State Committee for Statistics of the Republic of Sakha (Yakutia)], Yakutsk.

Goskomstat RS(Y) (1996), *Sozial'no-ekonomicheskoe Polozhenie za Yanvar-Dekabr 1995 Goda v Dvukh Tomakh* [Socio-economic Situation in January-December 1995 in Two Volumes], Gosudarstvennii Komitet Respubliki Sakha (Yakutia) po Statistike [State Committee for Statistics of the Republic of Sakha (Yakutia)], Yakutsk.

Goskomstat RS(Y) (1997), *Sozial'no-ekonomicheskoe Polozhenie za Yanvar'-Dekabr' 1996 Goda v Dvukh Tomakh* [Socio-economic Situation in January-December 1996 in Two Volumes], Gosudarstvennii Komitet Respubliki Sakha (Yakutia) po Statistike [State Committee for Statistics of the Republic of Sakha (Yakutia)], Yakutsk.

Goskomstat RS(Y) (1998), *Soizial'no-ekonomicheskoe Polozhenie Respubliki Sakha (Yakutia) za 1992-1997 gg. i Yanvar'-Avgust 1998 Goda* [Socio-economic Situation in the Republic of Sakha (Yakutia) in 1992-1997 and in January-August 1998], Gosudarstvennii Komitet Respubliki Sakha (Yakutia) po Statistike [State Committee for Statistics of the Republic of Sakha (Yakutia)], Yakutsk.

Goskomstat RS(Y) (1999), *Soizial'no-ekonomicheskoe Polozhenie Respubliki Sakha (Yakutia): Yanvar-Dekabr 1998 Goda v Drukh Tomakh* [Socio-economic Situation in the Republic of Sakha (Yakutia): January-December 1998 in Two Volumes], Gosudarstvennii Komitet Respubliki Sakha (Yakutia) po Statistike [State Committee for Statistics of the Republic of Sakha (Yakutia)], Yakutsk.

Goskomstat RS(Y) (2000), *Sozial'no-ekonomicheskoe Polozhenie v Respublike Sakha (Yakutia): Yanvar-Dekabr 1999 Goda v Dvukh Tomakh* [Socio-economic Situation in the Republic of Sakha (Yakutia): Janurary-December 1999 in Two Volumes], Gosudarstvennii Komitet Respubliki Sakha (Yakutia) po Statistike [State Committee for Statistics of the Republic of Sakha (Yakutia)], Yakutsk.

Government of Alberta [online], URL: http://www.gov.ab.ca/ (accessed 20/04/01).

Grafé-Buckens, A. and Hinton, A.F. (1998), 'Engaging the Stakeholder: Corporate Views and Current Trends', *Business Strategy and the Environment*, Vol. 7(3), pp. 124-133.

Gray, R., Kouhy, R. and Lavers, S. (1995), 'Corporate Social and Environmental Reporting: a Review of the Literature and a Longitudinal Study of UK Disclosure', *Accounting, Auditing &Accountability Journal*, Vol. 8(2), pp. 47-77.

Gray, R.H. (1990), *The Greening of Accounting: the Profession After Pearce*, The Chartered Association of Certified Accountants, Certified Research Report, 17.

Green, T. (1993), *Sovremenny Mir Almazov* [Modern World of Diamonds], Progress, Moscow.

Guerin, T.E. (2000), 'Self-regulation as an Opportunity for Sustainability', *Mining Environmental Management*, Vol. 8(2), pp. 16-18.

Gurov, V. (2001), 'Press-sluzhba AK ALROSA Soobshaet' [Press-service of ALROSA Company Informs], *Yakutia*, 15 December 2001, URL: http://www.gazetayakutia.ru/ (accessed on 16/12/01).

Gush, E.P. (1994), 'Diamonds', *Journal of South African Institute of Mining and Metallurgy*, Vol. 94(11-12), pp. 337-340.

Guthrie, J. and Parker, L.D. (1990), 'Corporate Social Disclosure Practice: a Comparative International Analysis', *Advances in Public Interest Accounting*, No. 3, pp. 159-175.

Halal, W.E. (2001), 'The Collaborative Enterprise: a Stakeholder Model Uniting Profitability and Responsibility', *Journal of Corporate Citizenship*, Vol. 1(2), pp. 27-42.

Halme, M. and Huse, M. (1997), 'The Influence of Corporate Governance, Industry and Country Factors on Environmental Reporting', *Scandinavian Journal of Management*, Vol. 13(2), pp. 137-157.

Hamann, R. (2003), 'Mining Companies' Role in Sustainable Development: the 'Why' and 'How' of Corporate Social Responsibility from a Business Perspective', *Development Southern Africa*, Vol. 20, pp. 237-254.

Hamann, R., Acutt, N. and Carter, A. (2001), *Corporate Foundations and Tri-sector Partnerships*, Business Partners for Development Natural Resources Cluster, Working Paper No. 11, May 2001, URL: http://bpd-naturalresources.org (accessed on 25/10/01).

Harte, G. and Owen, D. (1991), 'Environmental Disclosure in the Annual Reports of British Companies: a Research Note', *Accounting, Auditing & Accountability Journal*, Vol. 4(3), pp. 51-61.

Harvey, B. (2000), 'Development, Mining and Local Community', *2000 Annual Warwick Corporate Citizenship Conference, University of Warwick, 10-17 July 2000*, URL: http://www.warwick.ac.uk/ (accessed on 15/11/01).

Heemskerk, M. (2001), 'Maroon Gold Miners and Mining Risks in the Suriname Amazon', *Cultural Survival Quarterly*, Vol. 25(1), URL: http://www.cs.org/publications/CSQ (accessed on 29/10/01).

Henriques, I. and Sadorsky, P. (1999), 'The Role of Codes of Conduct in the Advertising Self-Regulation Framework', *Journal of Macromarketing*, Vol. 20(2), pp. 155-166.

Herbst, T. (1998), 'Environmental Disclosure: Corporate Use of the World Wide Web', *Corporate Environmental Strategy*, Vol. 5(2), pp. 81-90.

Herrero, S.G., Saldana, M.A.M, del Campo, M.A.M. and Ritzel, D.O. (2002), 'From Traditional Concept of Safety Management to Safety Integrated with Quality', *Journal of Safety Research*, Vol. 33(1), pp. 1-20.

Hester, R.E. and Harrison, R.N. (eds) (1994), *Mining and its Environmental Impact*, Royal Society of Chemistry Series: Issues in Environmental Science and Technology, Vol. 1.

Heuer, M. and Innonet (1999), *Nonprofit Organisational Effectiveness*, URL: http://www.inetwork.org (accessed on 20/04/01).

Hill, R. (1999), 'Overview' in ICME (1999), *Mining and Indigenous Peoples: Case Studies*, International Council on Metals and the Environment, Ottawa, pp. v-ix, URL: http://www.icme.com (accessed on 15/03/02).

Hilson, G. (2000), 'Pollution Prevention and Cleaner Production in the Mining Industry: an Analysis of Current Issues', *Journal of Cleaner Production*, Vol. 8(2), pp. 116-126.

Hilson, G. (2002), 'An Overview of Land Use Conflicts in Mining Communities', *Land Use Policy*, Vol, 19 (1), pp. 65-73.

Hilson, G. and Murck, B. (2001), 'Progress towards Pollution Prevention and Waste Minimization in the North America Gold Mining Industry', *Journal of Cleaner Production*, Vol. 9(5), pp. 405-415.

Hine, D.W., Lewko, J. and Blanco, J. (1999), 'Alignment to Workplace Safety Principles: an Application to Mining', *Journal of Safety Research*, Vol. 30(3), pp. 173-185.

Holme, R., Watts, P., WBCSD (2000), *Corporate Social Responsibility: Making Good Business Sense*, World Business Council for Sustainable Development, URL: http://www.wbcsd.org (accessed on 30/05/02).

Homestake Mining Company [online] URL: http://www.homestake.com (accessed 01/05/01).

Homestake Mining Company (1998a), *Progress Report*, Homestake Mining Company, San Francisco.

Homestake Mining Company (1998b), *Environment, Health & Safety Report*, Homestake Mining Company, San Francisco.

Homestake Mining Company (1999a), *Annual Report*, Percieve LLC, Long Beach.

Homestake Mining Company (1999b), *Environment, Health & Safety Report*, Percieve LLC, Long Beach.

Homestake Mining Company (2000a), *Annual Report*, Homestake Mining Company, Canada.

Homestake Mining Company (2000b), *Environment, Health and Safety*, Perceive LLC, Long Beach.

Homestake Mining Company (ud.), *Environmental, Health, Safety and Government Affairs Department*, Homestake Mining Company, San Francisco.

Honkasalo, A. (2000), 'Occupational Health and Safety and Environmental Management Systems', *Environmental Science and Policy*, Vol. 3(1), pp. 39-45.

Hopkinson, P. and Whitaker, M. (1999), 'The Relationship between Company Environmental Reports and Their Environmental Performance: a Study of the UK Water Industry', in M. Bennett and P. James (eds), *Sustainable Measures: Evaluation and Reporting of Environmental and Social Performance*, Greenleaf Publishing, Sheffield, pp. 392-411.

Hore-Lacy, I. (1992), *Mining ... and the Environment*, 4th ed., Australian Mining Industry Council, URL: http://www.minerals.org.au/ (accessed on 09/02/99).

Husted, B.W. (2003), 'Goverbnance Choices for Corporate Social Responsibility: to Contribute, Collaborate or Internalize?', *Long Range Planning*, Vol. 36, pp. 481-498.

ICME (International Council on Metals and Environment) (1999), *Mining and Indigenous Peoples: Case Studies*, ICME, Ontario, URL: http://www.icme.com (accessed on 15/03/02).

ICMM (International Council on Mining and Metals) [online] URL: http://www.icmm.com (accessed 3/02/05).

IIED (International Institute for Environment & Development) (2002), *Mining Minerals and Sustainable Development: Project Draft Report for Comment 4 March 2002*, IIED, London.

Il'ina, L. (1995), 'Almazobrilliantovy Kompleks: Kakov on Est, Kakim Emu Byt' [Diamond Complex: What Is It and What Will It Be], *Delovoy Mir*, 30 Janury-5 February 1995, pp. 14-15.

IMF (1998), *Press Information Notice*, No. 98/51, 21 July 1998. International Monetary Fund [online], URL: http://www.imf.org/ (accessed on 7/06/02).

Interfax-M&CN (1998), 'Russia', *Mining Annual Review 1998, Supplement to Mining Journal,* pp. 240-249.

International Finance Corporation (2000), *Investing in People: Sustainable Communities through Improved Business Practice: a Community Development Resource Guide for Companies*, International Finance Corporation, URL: http://www.Mineralresources forum.org (accessed on 30/04/02).

Janse (Bram), A.J.A. and Sheahan, P. (1995), 'Catalogue of World Wide Diamond and Kimberlite Occurrence: a Selective and Innovative Approach', *Journal of Geochemical Exploration*, Vol. 53(1-3), pp. 73-111.

Jennings, C.M.H. (1995), 'The Exploration Context for Diamonds', *Journal of Geochemical Exploration*, Vol. 53(1-3), pp. 113-124.

Kempton, D.R. (1996), 'The Republic of Sakha (Yakutia): the Evolution of Centre-periphery Relations in the Russian Federation', *Europe-Asia Studies*, Vol. 48(4), pp. 587-613.

Khan, A.S. (1999), 'O Konzepzii Razvitiya Almazo-brilliantovogo Kompleksa Rossii' [About the Development of Diamond Complex in Russia], *Mineral'nye Resursy Rossii: Ekonomika i Upravlenie*, No. 6, URL: http://www.geoinform.ru/ (accessed on 1/10/01).

Khanna, T. (1999), 'Mining and the Environmental Agenda', *Mining Magazine*, Vol. 181(3), pp. 158-163.

Kirrilov, A.F. (1990), 'O Vliyanii Zaregulorovannogo Rechnogo Stoka Vilyua na Ekosystemu Pritoplyaemogo Uchastka' [About the Impact of Artificial Flow Regulation on the River Vilyuy to the Flooded Area], in I.Ya. Egorov and V.F. Chernyavsky (eds), *Vorposy Regional'noi Gigieny, Sanitarii i Epidemiologii* [Questions of Regional Hygiene, Sanitation and Epidemiology], Minzdrav Yakutskoi ASSR [Ministry of Health of the Yakut Autonomous Soviet Socialist Republic], Yakutsk, p. 223.

Kiseleva, G. (1999), 'Put' k Almazam' [Road to Diamonds], *Yakutia*, No. 29 (29055), 17 February 1999, URL: http://www.yakutia.ru/ (accessed on 10/03/99).

Klein, L.C. and Connolly, P. (2000), 'Getting in Shape: Fitness Tips for Established Nonprofits', *The National Nonprofit Leadership and Management Journal*, Vol. 18(1), January/February. URL: http://www.consco.com/ (accessed on 20/06/02).

Knaus, E. and Klarer, J. (1997), *Russian Federation: Russia's Pollution Charge System Needs Extensive Reform - But How to Do It?* URL: http://www.rec.org/ (accessed on 27/11/99).

Kolk, A. (2000), *Economics of Environmental Management*, Pearson Education Ltd, Harlow, England.

Kolmogorov, N.K. (2000), 'Zolotodobyvayushaya Promyshlennost Rossii: Promlemy i Perspektivy' [Gold Mining Industry in Russia: Problems and Perspectives], *Mineral'nye Resursy Rossii: Ekonomika i Upravlenie* [Journal of Mineral Resources of Russia: Economics and Management], No. 2, URL: http://geoinform.ru/ (accessed on 02/02/01).

Komarov, B. (1978), *The Destruction of Nature in the Soviet Union*, Pluto Press, London.

Kommersant-daily (1993), 'Brilliantovaya Promyshlennost v Rossii i v Mire' [Diamond Industry in Russia and in the World], *Kommersant-daily*, No. 42, 9 March 1993, p. 2.

Kopylov, R. N., Marshintsev, V. K. and Tyaptirgyanov, M. M. (1993), 'Obschaya Ekologisheskaya Situatsia Territorii Yakutii' [General Ecological Situation on the Territory of Yakutia)], in Goskomitet po Delam Grazdanskoi Oborony, Chrezvychainym Situaziyam i Likvidazii Posledstvii Stikhiinykh Bedstvii Respubliki Sakha (Yakutia), Ministerstvo Zdravookhranenia Respubliki Sakha (Yakutia) [State Committee for Emergency Situation of the Republic of Sakha (Yakutia), the Ministry of Health of the Republic of Sakha (Yakutia)], *Radioazionnoe Zagryaznenie Territorii Respubliki Sakha (Yakutia): Problemy Radiazionnoi Bezopasnosti* [Radioactive Contamination in the Territory of the Republic of Sakha (Yakutia): Problems of Radioactive Safety], Yakutsk, pp. 3-10.

Kotlyakov, V.M. and Agranat, G.A. (1999), 'Rossiyskiy Sever – Krai Bolshikh Vozmozhnostei' [Russian North – a Land of Big Opportunities], *Vestnik Rossiyskoi Akademii Nauk* [Journal of Russian Academy of Sciences], No.1, URL: http://vivovoco.nns.ru/VV/JOURNAL/VRAN/RUSNORD.HTM (accessed on 31/01/01).

Koutaisoff, E. (1987), 'Survey of Soviet Material on Environmental Problems', in F. Singleton (ed.), *Environmental Problems in the Soviet Union and Eastern Europe*, Lynne Rienner Publishers, London, pp. 11-42.

Kriuchkov, V.V. (1976), *Delicate Sub-Arctic* [Chutkaya Subarktika], Moscow.

Krivoshapkin, V.G., Neustroeva, G.A., Timofeev, G.A. (1990), 'Promyshlennoe Zagryaznenie Basseina Reki Vilyuy i Sostoyanie Zdorovya Naseleniya v Gruppe Vilyuyskikh Rayonov', in I.Ya. Egorov and V.F. Chernyavsky (eds), *Vorposy Regional'noi Gigieny, Sanitarii i Epidemiologii* [Questions of Regional Hygiene, Sanitation and Epidemiology], Minzdrav Yakutskoi ASSR [Ministry of Health of the Yakut Autonomous Soviet Socialist Republic], Yakutsk, pp. 91-92.

Krut, R. and Gleckman, H. (1998), *ISO 14001: a Missed Opportunity for Sustainable Global Industrial Development*, Earthscan Publications Ltd, London.

Krut, R. and Moretz, A. (2000), 'The State of Global Environmental Reporting: Lessons from the Global 100', *Corporate Environmental Strategy*, Vol. 7(1), pp. 85-91.

Kurlyandsky, B.A. and Sidorov, K.K. (eds) (1998), *Predel'no-dopustimye Konzentrazii (PDK) Khimicheskykh Veschestv v Vode Vodnykh Objektov Khozyaistvenno-pitjevogo i Kulturno-bytovogo Vodopol'zovaniya: Gigienicheskie Normativy* [Maximum Permissible Concentrations (MPC) of Chemical Compounds in Water of Watercourses for Industrial and Drinking, and Cultural and Domestic Water Use: Hygienic Standards], Rossiski Registr Potenzial'no Opasnykh Khimicheskikh i Biologicheskikh Veschestv Minzdrava Rossii [Russian Register of Potentially Hazardous Chemical and Biological Substances of Ministry of Health of Russia], Moscow.

Kuwait Information Office [online], URL: http://www.kuwait-info.org/ (accessed on 10/04/01).

Labutina, T.M. and Smolenko, L.I. (1990), 'Vliyanie Promyshlennogo Kompleksa na Khimicheskii Sostav Prirodnykh Vod' [The Impact of Industrial Complex on Chemical Composition of Natural Waters], in I.Ya. Egorov and V.F. Chernyavsky (eds), *Vorposy Regional'noi Gigieny, Sanitarii i Epidemiologii* [Questions of Regional Hygiene, Sanitation and Epidemiology], Minzdrav Yakutskoi ASSR [Ministry of Health of the Yakut Autonomous Soviet Socialist Republic], Yakutsk, pp. 221-222.

Lambert, J. and Elix, J. (2001), *Assessing Corporate Social Performance: a Brief Overview of the State of Play*, URL: http://www.tec.nccnsw.org.au/ (accessed on 4/07/02).

Laulajainen, R. (1990), 'Gold Price Round the Clock: Technical and Fundamental Issues', *Resources Policy*, Vol. 16(2), pp. 143-152.

Lehman, C. (1995), 'A Legitimate Concern for Environmental Accounting', *Critical Perspectives on Accounting*, Vol. 6, pp. 393-413.

Levine, R.M. (1994), *The Mineral Industry of Russia*, US Geological Survey-Minerals Information – 1994, URL: http://minerals.usgs.gov/ (accessed on 10/12/01).

Levine, R.M. (1995), *The Mineral Industry of Russia*, US Geological Survey-Minerals Information – 1995, URL: (accessed on 10/12/01).

Levine, R.M. (1996), *The Mineral Industry of Russia*, US Geological Survey-Minerals Information – 1996, URL: http://minerals.usgs.gov/ (accessed on 12/06/02).

Levine, R.M. (1997), *The Mineral Industry of Russia*, US Geological Survey-Minerals Information – 1997, URL: http://minerals.usgs.gov/ (accessed on 10/12/01).

Levine, R.M. (1998), *The Mineral Industry of Russia*, US Geological Survey-Minerals Information – 1998, URL: http://minerals.usgs.gov/ (accessed on 10/12/01).

Levine, R.M. (1999), *The Mineral Industry of Russia*, US Geological Survey-Minerals Information – 1999, URL: http://minerals.usgs.gov/ (accessed on 10/12/01).

Levine, R.M. and Steblez, W.G. (1990), 'Soviet Union', *Mining Annual Review 1990, Supplement to Mining Journal*, pp. 156-168.

Levine, R.M. and Wallace, G.J. (2000), The Mineral Industry of the Commonwealth of Independent States, United States Geological Survey, URL: http://minerals.usgs.gov/ (accessed 29/04/04).

Line, M., Hawley, H. and Krut, R. (2002) 'The Development of Global Environmental and Social Reporting', *Corporate Environmental Strategy*, Vol. 9(1), pp. 69-78.

Lluardo, J.C. (1995), 'Highlights: Opportunities for Improving the Russian Environment', *Environmental Management and Health*, Vol. 6(5), pp. 4-6.

Logan, D. (1998), 'Corporate Citizenship: Defining Terms and Scoping Key Issues', The Corporate Citizenship Company, URL: www.corporate-citizenship.co.uk (accessed on 15/01/05).

Lvovsky, K. (1999), 'Health and Environment', *Environment Matters*, No. 5, Annual Review, pp. 44-47, URL: http://www-esd.worldbank.org/envmat/home.html (accessed on 12/06/02).

Lynn, N.J. (1999), 'Constructing 'Federal Democracy' in Russia: Debates Over Structure of Power in the Regions', *Presented on Annul Conference of the Royal Geographical Society, Leicester, January 1999*, URL: http://www.geo.ed.ac.uk/~njl (accessed on 3/12/99).

Lynn, N.J. and Fryer, P. (1998), 'National-territorial Change in the Republics of the Russian North', *Political Geography*, Vol. 17(5), pp. 567-584.

MacLean, C. and Warhurst, A. (2000), 'Conceptual Approaches to Health and Well-being in Minerals Development: Case Study in South Africa', *MERN Research Bulletin & Newsletter*, No. 15, 1999/2000, pp. 47-55.

Madeley, J. (1999), *Big Business, Poor People: the Impact of Transnational Corporations on the World's Poor*, Zed Books, London.

Maignan, I. and Ferrell, O.C. (2004), 'Corporate Social Responsibility and Marketing: an Integrative Framework', *Journal of the Academy of Marketing Sciences*, Vol. 32(1), pp. 3-19.

Makarov, S.V. (2001), *Ekologichesky Menedgment: Nesupolzuemye Vozmozhnosti* [Environmental Management: Unused Opportunities], Ecoline, URL: http://www.ecoline.ru/ (accessed on 5/07/02).

Makarov, S.V., Tarasov V.V., Guseva T. V. (1995), 'Environmental Auditing: Opportunities in Russia at Local and Territorial Levels', *Trans IChemE*, Vol. 73(B), August, pp. 223-226.

Makarov, V.N. and Shats, M.M. (2000), 'Masshctabnye Izmeneniya Sredy Yakutii Svyazannya s Promyshlennoi Deyatel'nostyu' [Large Scale Changes in the Environment of Yakutia Linked to Industrial Operations], *Nauka i Obrazovanie* [Journal of Science and Education], No. 1, URL: http://sitim.sitc.ru/E-books/Journals/sin_edu/ (accessed on 20/04/01).

Marples, D. (1999), 'Environmental and Health Problems in the Sakha Republic', *Post Soviet-Geography and Economics*, Vol. 40(1), pp. 62-77.

Martynov, P.N. (1997), 'Osnovnye Voprosy Vosstanovleniya Ekologii Vilyuyskogo Regiona' [Main Issues of Remediation of the Environment in Vilyuy Region], in N.P. Pavlov and B.M. Afanasieva (eds), *Tragediya i Bol' Sedogo Vilyuya* [Tragedy and Pain of Grey Vilyuy], SAPI Togr-Kniga, Yakutsk, pp. 44-45.

Marx, J.D. (1998), 'Corporate Philanthropy: Implications for Social Work', *Social Work*, Vol. 43(1), pp. 34-42.

Maryland Association of Nonprofit Organizations (1998), *Standards for Excellence: an Ethics and Accountability Code for Nonprofit Sector*, URL: http://www.mdnonprofit.org/ (accessed on 19/03/01).

Mathews, M.R. (1993), *Socially Responsible Accounting*, Chapman & Hall, London.

Matveev, A. (1998), 'Almazy Rossii-Sakha: Sostoyanie, Perspektiby, Problemy Kompanii' [Almazy Rossii-Sakha: Situation, Perspectives and Problems of the Company], *Problemy Teorii i Praktiki Upravleniya* [Journal of Problems of Theories and Practice of Management], No.3, URL: http://www.ptpu.ru/ (accessed on 1/02/01).

Matveev, A.S. and Cherny E.D. (2000), 'AK ALROSA Strategiya Razvitiya' [ALROSA Company: Strategy for the Development], *Mineral'no-syr'evye Resursy Rossii* [Journal of Mineral Resources of Russia], No.5-6, URL: http://www.geoinform.ru/ (accessed on 1/02/01).

McMahon, G. and Strongman, J. (1999), 'The Rise of the Community', *Mining Environmental Management*, Vol. 7(1), pp. 8-11.

McMahon, G. and Remy, F. (eds) (2001) *Large Mines and the Community: Socioeconomic and Environmental Effects in Latin America, Canada and Spain*, World Bank, Washington, DC and International Development Research Centre, Ottawa.

McPhail, K. and Davy, A. (1998), *Integrating Social Concerns into Private Sector Decisionmaking: a Review of Corporate Practices in the Mining, Oil and Gas Sectors*, World Bank Discussion Paper No. 384, The World Bank, Washington.

McWilliams, A. and Siegel, D. (2001), 'Corporate Social Responsibility: a Theory of the Firm Perspective', *Academy of Management Review*, Vol. 26(1), pp. 117-127.

Meadows, D.H., Meadows, D.L., Randers, J., Behrens III, W.W. (1972), *The Limits to Growth - A Report for the Club of Rome's Project on the Predicament of Mankind*, Pan Books Ltd, London.

Meek, G.K., Roberts, C.B. and Gray, S.J. (1995), 'Factors Influencing Voluntary Annual Report Disclosure by US, UK and Continental European Multinational Corporations', *Journal of International Business Studies*, Vol. 26(3), pp. 555-572.

MEM (2002), 'Business, Government and Society Working Together', *Mining Environmental Management*, Vol. 10(2), pp. 20-22.

Menshikov, N. (1999), 'Dragozennyye Metally v Rossii' [Precious Metals in Russia], *Dragozennyye Metally, Dragozennyye Kamni* [Journal of Precious Metals and Precious Stones], No. 9.

Miller, R.O. (1999), 'Mining, Environmental Protection, and Sustainable Development in Indonesia', in N.J. Vig and R.S. Axelrod (eds), *The Global Environment: Institutions, Low and Policy*, Earthscan Publications Ltd, London, pp. 317-332.

Ministerstvo Okhrany Prirody Respubliki Sakha (Yakutia) [Ministry for Nature Protection of the Republic of Sakha (Yakutia)] (2001), *Respublikanskaya Tselevaya Programma 'Ozdorovlenie Ekologicheskoi Obstanovki Almaznoi Provinzii v 2002-2005 gg.'*

[Regional Target Programme 'Improvement of the Environmental Situation in the Diamond Province in 2002-2005'], Unpublished Document, Yakutsk.

Minnesota Council of Nonprofits (1998), *Principles and Practices for Nonproft Excellence*, URL: http://www.mncm.org/ (accessed on 4/05/01).

Mirikitani, C. (1999), 'The role of Philanthropy in the Intersection between Cultures and Commerce', *Journal of Arts Management, Law & Society*, Vol. 29(2), pp. 128-132.

Mironov, V. (1999), 'Pyat Let Suschestvovaniya SAPI: No Nelzya Vse Spikhivat' na Nego' [Five Years of the SAPI Operations: Should Not Be Entirely Blamed], *Yakutia*, No. 29066, 4 March 1999, URL: http://www.yakutia.ru/ (accessed on 5/05/99).

Mitchell, A. (1999), 'The Environmental Challenge Facing Gold Mining', *Journal of Mines, Metals and Fuels*, Vol. 47(12), pp. 352-357.

Mittenthal, R.A. (1999), *Effective Philanthropy: the Importance of Focus for Foundations, a Smaller Field of Vision Can Mean Greater Effectiveness*, The Conservation Company. Briefing paper, URL: http://www.consco.com (accessed on 19/04/01).

MJ (1997), 'Diamonds', *Metals & Minerals Annual Review 1999, Supplement to Mining Journal*, pp. 27-30.

MJ (1998), *Diamond supplement to Mining Journal*, Vol. 331(8505).

MJ (1999), 'Diamonds: De Beers' Perspective', *Precious Metals and Minerals: Annual Review, Supplement to Mining Journal*, pp. 31-33.

MJ (2001a), 'Gold in 2000', *Mining Journal*, Vol. 336(8634), 18 May 2001, pp. 373-376.

MJ (2001b), *Mining Journal*, London, Vol. 336(8618), 26 January 2001, p. 77.

MJ (2002), 'Sustainable Development and the Minerals Industry', *Mining Journal*, Vol. 338(8672), 15 February 2002, pp. 121-123.

Moneva, J. M. and Llena, F. (2000), 'Environmental Disclosure in the Annual Reports of Large Companies in Spain', *The European Accounting Review*, Vol. 9(1), pp. 7-29.

Morris, S. D. (1996), 'Angola Diamond Trade, Environment and Conflict', *TED Case Studies*, Vol. 5(2), URL: http://www.american.edu/TED/class/all.htm (accessed on 29/05/02).

Mostakhov S.E., Nekrasov I.A., Dmitrieva Z.M., Kalmykova A.I. (1980), *Yakutskaya ASSR (Yakut ASSR)* [Yakut Autonomous Soviet Socialist Republic], Knizhnoje izdatelstvo, Yakutsk.

Motyashov, A.P. and Rastorguyev, G.N. (eds) (1991), *Zolotaya Yakutia* [Golden Yakutia], Moscow.

Mullen, J. (1997), 'Performance-based Corporate Philanthropy: How 'Giving Smart' Can Further Corporate Goals', *Public Relations Quarterly*, Vol. 42(2), pp. 42-49.

Murray, S. (1997), 'Gold', *Metals and Minerals Annual Review, Supplement to Mining Journal*.

Murray, S. (1998a), 'Gold', *Metals and Minerals Annual Review, Supplement to Mining Journal*.

Murray, S. (1998b), 'Gold', *Precious Metals and Minerals: Annual Review Supplement, London*, Vol. 330(8481), pp. 2-9.

Murray, S. (1999), 'Gold', *Metals & Minerals Annual Review, Supplement to Mining Journal*.

Namdeb (1998), *Annual Review*, Namdeb Diamond Corporation (Pty) Limited.

Namdeb (1999), *Annual Review*, Namdeb Diamond Corporation (Pty) Limited.

Namdeb (2000), *Annual Review*, Namdeb Diamond Corporation (Pty) Limited.

Namdeb [online] URL: www.namdeb.com (accessed 01/05/01).

Nasledie [online], URL: http://nasledie.ru/ (accessed on 4/05/02).

Natarov, O. (2000), 'Zolotodubycha v Rossii: Tekuschie Problemy i Blizhaishie Perspektivy' [Gold Mining in Russia: Current Problems and Perspectives for Near Future], *Rynok Zennykh Bumag* [Journal of Stock Market], No. 3, pp. 12-20.

Neu, P., Warsame, H. and Pendwell, K. (1998), 'Managing Public Impressions: Environmental Disclosures in Annual Reports', *Accounting, Organizations and Society*, 23(3), pp. 265-282.

Newcombe, K. (1999), 'Building a New Natural Resources Management Strategy', *Environment Matters*, No. 5, Annual Review, June 1998-June 1999, pp. 50-53, URL: http://www.worldbank.org/ (accessed on 12/06/02).

Newmont (1999), *Newmont Annual Report*, Newmont, Denver.

Newmont (2000), *Annual Report*, Newmont, Denver.

Newmont Gold [online] URL: http://www.newmont.com (accessed 08/04/02).

Niskala, M. and Pretes, M. (1995), 'Environmental Reporting in Finland: a Note on the Use of Annual Reports', *Accounting, Organization and Society*, Vol. 20(6), pp. 457-466.

Niskanen, J. and Nieminen, T. (2001), 'The Objectivity of Corporate Environmental Reporting: a Study of Finnish Listed Firms' Environmental Disclosures', *Business Strategy and the Environment*, Vol. 10(1), pp. 29-37.

Normandy Mining Limited [online] URL: http://www.normandy.com.au (accessed 01/05/01).

Normandy Mining Limited (1998a), *Annual Shareholder Report*, Normandy Mining Limited, Adelaide.

Normandy Mining Limited (1998b), *Environment Report: Summary*, Normandy Mining Limited, Adelaide.

Normandy Mining Limited (1999a), *Annual Shareholder Report*, Normandy Mining Limited, Adelaide.

Normandy Mining Limited (1999b), *Environment Report: Australasia*, Normandy Mining Limited, Adelaide.

Normandy Mining Limited (2000a) *Annual Shareholder Report*, Normandy Mining Limited, Adelaide.

Normandy Mining Limited (2000b), *Safety & Health Environment Community Report*, Normandy Mining Limited, Adelaide.

Norton, M. (ed.) (1981), *Raising Money from Trusts*, Directory of Social Change, London.

NPO Yakutalmaz [Yakutalmaz Corporation] (1992), *Okhrana Prirody 1992* [Nature Protection 1992], Unpublished Report, Mirny.

Obshetvenny Registr Certificatsii System Ecologicheskogo Menedgmena [Public Register for Certification of Environmental Management Systems] [online], URL: http://www.14000.ur/register (accessed on 5/07/02).

OECD (Organisation for Economic Co-operation and Development) (2000), *Survey of Existing Activities for the Promotion of Environmental Management in Enterprises in the NIS, CCNM/ENV/EAP(2000)92*, Organisation for Economic Co-operation and Development, URL: http://www.oecd.org/ (accessed on 21/05/01).

Otto, J.M. (1997), 'A National Mineral Policy as a Regulatory Tool', *Resources Policy*, Vol. 23(1-2), pp. 1-7.

Padmore, G. (1992), 'Government Participation in Mining Projects', *Natural Resources Forum*, Vol. 16(2), pp. 132-140.

Parsons, C.J. and Kraus, P.D. (eds) (1969), *Practical Gem Knowledge for the Amateur*, Lapidary Journal Inc., San Diego, California.

Pastizzi-Ferencic, D. (1992), 'Natural Resources and Environmentally Sound Sustainable Development', *Natural Resources Forum*, Vol. 16(1), pp. 3-10.

Patrizi, P. and McMullen, B. (1998), *Evaluation in Foundations: the Unrealized Potential*, Prepared for W.K. Kellogg Foundation Evaluation Unit, URL: http://www.wkkf.org/ (accessed on 24/01/05).

Patten, D.M. (1991), 'Exposure, Legitimacy, and Social Disclosure', *Journal of Accounting and Public Policy*, No. 10, pp. 297-308.

Pavlov, N.P. and Afanasieva, V.M. (1997), *Tragediya i Bol' Sedogo Vilyuya* [Tragedy and Pain of Grey Vilyuy], SAPI-Torg-Kniga, Yakutsk.

Peattie, K. (1992), *Green Marketing*, The M& Handbook Series, Pitman Publishing, London.

Pershing, J. (1999), 'Fossil Fuel Implications of Climate Change Mitigation Responses', *IEA Climate Change Studies, 25 October-5 November 1999, Bonn*, URL: http://www.iea.org/ (accessed on 24/04/01).

Peterson, D. J. (1995), 'Russia's Environmental and Natural Resources in Light of Economic Regionalisation', *Post-Soviet Geography*, Vol. 36(5), pp. 291-309.

Petrova, P.G. and Volozhin, L.I. (1996), *Ekologiya Cheloveka v Usloviyakh Severa: Yakutia, Respublika Sakha* [Human Ecology in Conditions of the North: Yakutia, the Republic of Sakha], Tipografiya Voenno-inzhinernoi Akademii.

Picton, J. (1996), 'Africa's Position', *Diamond Supplement to Mining Journal*, Vol. 328(8403), 22 November 1996, pp. 14-19.

Picton, J. (1997), 'De Beers' Place in the Sun', *Diamond Supplement to Mining Journal*, Vol. 329(8452), 24 October 1997, pp. 5-6.

Pine, A.H., Thayer, G.E. and Coghlan, A., Innovation network (1998), *Program Evaluation Practice in the Non-profit Sector*, URL: http://www.innonet.org/ (accessed on 20/04/01).

Placer Dome (1998), *Annual Report*, Placer Dome, Canada.

Placer Dome (1999a), *Annual Report*, Placer Dome, Canada.

Placer Dome (1999b), *Sustainability Report*, Placer Dome Inc, Vancouver.

Placer Dome (ud.), *Sustainability policy*, Placer Dome Inc, Vancouver.

Placer Dome [online] URL: http://placerdome.com (accessed 01/05/01).

Placer Dome Asia Pacific Limited (2000), *Sustainability Report for Calendar Year 2000.*

Placer Dome Group (1998), *Sustainability Report*, Placer Dome Inc, Vancouver.

Placer Dome Inc (2000), *Business Report and Financial Results*, Placer Dome Inc, Canada.

Plantz, M.C., Greenway, M.T., and Hendricks, M. (1997), *Outcome Measurement: Showing Results in the Nonprofit Sector, New Directions for Evaluation: Using Performance Measurement to Improve Public and Nonprofit Programs*, URL: http://national.unitedway.org/outcomes/ndpaper.htm (accessed on 10/05/01).

Poiseev, I.I. (1999), *Ustoichivoe Razvitie Severa* [Sustainable Development of the North], Nauka, Novosibirsk.

Pokshishevsky, V. (1974), *Geography of the Soviet Union*, Progress Publishers, Moscow.

Porfiriev, B. (1997), 'Environmental Policy in Russia: Economic, Legal, and Organisational Issues', *Environmental Management*, Vol. 21(2), pp. 147-157.

Porter, M.E. and Kramer, M.R. (1999), 'Philanthropy's New Agenda: Creating Value', *Harvard Business Review*, Vol. 77(5), pp. 121-131.

Poulin, R. and Sinding, K. (1993), 'Mining Economics and the Environment', *Natural Resources Forum*, Vol. 17(2), pp. 157-163.

Pravitelstvo RS(Y) [Government of the Republic of Sakha (Yakutia)] (1995), *Gosudarstvennyy Doklad o Sostoyanii Okruzhayuschei Sredy Respubliki Sakha (Yakutia) v 1994 Godu* [State Report on the Environmental Situation in the Republic of Sakha in 1994], Pravitelstvo Respubliki Sakha (Yakutia), Yakutsk.

Pravitelstvo RS(Y) [Government of the Republic of Sakha (Yakutia)] (1996), *Gosudarstvennyy Doklad o Sostoyanii Okruzhayuschei Sredy Respubliki Sakha (Yakutia) v 1995 Godu* [State Report on the Environmental Situation in the Republic of Sakha in 1995], Pravitelstvo Respubliki Sakha (Yakutia), Yakutsk.

Pravitelstvo RS(Y) [Government of the Republic of Sakha (Yakutia)] (1997), *Gosudarstvennyy Doklad o Sostoyanii Okruzhayuschei Sredy Respubliki Sakha (Yakutia) v 1996 Godu* [State Report on the Environmental Situation in the Republic of Sakha in 1996], Pravitelstvo Respubliki Sakha (Yakutia), Yakutsk.

Pravitelstvo RS(Y) [Government of the Republic of Sakha (Yakutia)] (1998), *Gosudarstvennyy Doklad o Sostoyanii Okruzhayuschei Sredy Respubliki Sakha (Yakutia) v 1997 Godu* [State Report on the Environmental Situation in the Republic of Sakha in 1997], Pravitelstvo Respubliki Sakha (Yakutia), Yakutsk.

Pravitelstvo RS(Y) [Government of the Republic of Sakha (Yakutia)] (1999), *Gosudarstvennyy Doklad o Sostoyanii Okruzhayuschei Sredy Respubliki Sakha (Yakutia) v 1998 Godu* [State Report on the Environmental Situation in the Republic of Sakha in 1998], Pravitelstvo Respubliki Sakha (Yakutia), Yakutsk.

Pravitelstvo RS(Y) [Government of the Republic of Sakha (Yakutia)] (2001), *Gosudarstvennyy Doklad o Sostoyanii Okruzhayuschei Sredy Respubliki Sakha (Yakutia) v 1999 Godu* [State Report on the Environmental Situation in the Republic of Sakha in 1999], Pravitelstvo Respubliki Sakha (Yakutia), Yakutsk.

Pressend, M., Mthethwa, D., Hlabane, M. and Mara, I. (2001), 'Breaking Communication Barriers between Communities Affected by Mining and the Mining Industry in South Africa', Witbank Office: Group of Environmental Monitoring (GEM) and Environmental Monitoring Group (EMG), URL: http://www.mineralresourcesforum.org/docs/pdfs/Communication.pdf (accessed on 24/01/05).

PricewaterhouseCoppers and MMSD (2001), *Mining and Minerals Sustainability Survey 2001*, PricewaterhouseCoppers, URL: http://www.iied.org/ (accessed on 18/07/02).

Prokopiev, V. (1998), 'Almaznaya Voina kak Prolog Bitvy za Mineral'no-syr'evye Resursy Yakutii' [Diamond War as a Prologue of the Battle for Mineral Resources of Yakutia], *Yakutia*, No.20 (n28872), 20 May 1998, URL: http://www.yakutia.ru/ (accessed on 29/05/98).

Protasov, V. F. and Molchanov, A.V. (1995), *Ecology, Health and Management of Nature in Russia*, Finance and Statistics, Moscow.

Protokol No. 58 (ud.) *Protokol No. 58: O Komplekse Pervoocherednykh Mer po Ozdorovleniyu Ekologicheskoi Obstanovki v Basseine Reki Vilyuy* [Protocol 58: About Complex of Primary Measures for Improvement of Environmental Situation on Basin of the River Vilyuy], Unpublished Document, Yakutsk.

Pryde, P.R. (1976), 'Protection and Restoration of Surface-mined Land in the Soviet Union', in F. Singleton (ed.), *Environmental Misuse in the Soviet Union*, Praeger Publishers, New York, pp. 60-71.

Pryde, P.R. (1994), 'Observations on the Mapping of Critical Environmental Zones in the Former Soviet Union', *Post-Soviet Geography*, Vol. 35(1), pp. 38-49.

Puddephatt, R.J. (1978), *The Chemistry of Gold*, Elsevier Scientific Publishing Company, Amsterdam.

Rae, M. (1998), 'Environmental Reporting: Ore or Overburden?' *Groundwork*, Vol. 1(3), pp. 18-19.

Rae, M. and Rouse, A. (2001), *Mining Certification Evaluation Project: Independent Certification of Environmental and Social Performance in the Mining Sector*, A WWF-Australia Discussion Paper, January, World Wide Fund for Nature, Melbourne.

Ramani, R.V. (1995), 'Mining Disasters Caused and Controlled by Mankind: the Case for Coal Mining and Other Minerals. Part 1: Causes of Mining disasters', *Natural Resources Forum*, Vol. 19(3), pp. 233-242.

Ranganathan, J. (1999), 'Signs of Sustainability: Measuring Corporate Environmental and Social Performance', in M. Bennett and P. James (eds), *Sustainable Measures: Evaluation and Reporting of Environmental and Social Performance*, Greenlead Publishing, Sheffield, pp. 475-495.

Rasmuson, E. (1993), 'A Founder's Reflection on the Early Days of the Alaska Permanent Fund Corporation, Presented to the Alaska Permanent Fund Corporation Board of Directors, September 28, 1993' in APFC (Alaska Permanent Fund Corporation) (1997)

The Early History of the Alaska Permanent Fund: Perspectives on the Origin of Alaska's Oil Savings Account, Trustees Papers, 5, pp. 1-9.

Regelbrugge, L. (2001), 'Corporate Foundation Boards and Evaluation', *Foundation News and Commentary*, Vol. 42(1), URL: http://www.foundationnews.org/ (accessed on 6/03/01).

Rex Diamond Mining Corporation (2000), *Annual Report*, Rex Diamond Mining Corporation.

Rex Mining (1998), *Rex Diamond Mining Corporation Annual Report.*

Rex Mining (1999), *Rex Diamond Mining Corporation Annual Report.*

Rex Mining [online] URL: http://www.rexmining.be (accessed 4/04/01).

RIA Novosti (1999), 'ALROSA-De Beers: Novye Gorizonty' [ALROSA-De Beers: New Horizons], *Yakutia*, No. 29212, URL: http://www.yakutia.ru/ (accessed on 5/10/99).

Rikhardson, P.M. (1999), 'Statutory Environmental Reporting in Denmark: Status and Challenges', in M. Bennett and P. James (eds), *Sustainable Measures: Evaluation and Reporting of Environmental and Social Performance*, Greenlead Publishing, Sheffield, pp. 344-353.

Rio Tinto (1998a), *Annual Report and Accounts*, St. Ives plc, England.

Rio Tinto (1998b), *Social and Environment Report*, Westerham Press on Evolve Era, England.

Rio Tinto (1999a), *Annual Report and Accounts*, St. Ives plc, England.

Rio Tinto (1999b), *Social and Environment Report*, St. Ives plc, England.

Rio Tinto (2000a), *Annual Report and Financial Statements*, St. Ives plc, England.

Rio Tinto (2000b), *Social and Environment Review*, Beacon Press, UK.

Rio Tinto [online] URL: http://riotinto.com (accessed 01/05/01).

Roberts, C.B. (1991), 'Environmental Disclosures: a Note on Reporting Practices in Mainland Europe', *Accounting, Auditing & Accountability Journal*, Vol. 4(3), pp. 62-71.

Rombouts, L. (1999), 'Diamonds', *Precious Metals and Minerals: Annual Review, Supplement to Mining Journal*, pp. 27-30.

Rombouts, L. (2000), 'The Diamond-mining Scene', *Mining Journal*, Vol. 334(8587), 16 June 2000, pp. 464-466.

Rondinelli, D. and Berry, M.A. (2000), 'Environmental Citizenship in Multinational Corporations: Social Responsibility and Sustainable Development', *European Management Journal*, Vol. 18(1), pp. 70-84.

Rondinelli, D.A. and Vastag, G. (1998), 'Private Investment and Environmental Protection: ALCOA-Köfém's Strategy in Hungary', *European Management Journal*, Vol. 16(4), pp. 422-430.

Rossi, P.H., Freeman, H.E. and Wright, S.R. (1979), *Evaluation: a Systematic Approach*, Sage Publications, London.

Rubchenko, M. and Shokhina, E. (2001), 'Chastny Shurf' [Private Bore Pit], *Expert*, No. 5, 5 February 2001, pp. 26-29.

Sagers, M. (1997), 'Regional Trends in Russian Gold Production', *Post-Soviet Geography and Economics*, Vol. 38(6), pp. 315-356.

Sánchez, L.E. (1998), 'Industry Response to the Challenge of Sustainability: the Case of the Canadian Nonferrous Mining Sector', *Environmental Management*, Vol. 22(4), pp. 521-531.

Santema, S. and Van De Rijt, J. (2001), 'Strategy Disclosure in Dutch Annual Reports', *European Management Journal*, Vol. 19(1), pp. 101-108.

Santos-Reyes, J. and Beard, A.N. (2002), 'Assessing Safety Management Systems', *Journal of Loss Prevention in the Process Industries*, Vol. 15(2), pp. 77-95.

SAPI (1995), *Ustav Fonda SAPI (Sakhaalmazprovinvest)* [Charter of the SAPI Foundation], Unpublished Document, Yakutsk.

SAPI (1996) *Polozhenie o Mirninskom Ulusnom Filiale Fonda SAPI (Sakhaalmazprovinvest)* [Regulation on the Mirny District Division of the SAPI Foundation], Unpublished Document, Yakutsk.

SAPI (1998a), *Regional'naya Zelevaya Programma 'Deti Severa: Obrazovanie, Kultura, Sport na 1999-2001 gg'* [Regional Target Programme 'Children of the North: Education, Culture, Sport for 1999-2001'], Unpublished Document, Yakutsk.

SAPI (1998b), *Programma Povysheniya Effektivnosti Almazogranil'nogo Proizvodstva na 1999-2001 gg* [Programme for Improvement of the Diamond-cutting Effectiveness for 1999-2001], Unpublished Document, Yakutsk.

SAPI (1998c), *Programma Vnedreniya Spezial'nogo Mekhnizma Finansirovaniya Sel'skogo Khozyaistva, Ukrepleniya Vnutrennego Rynka Sekslokhozyaistvennoi Produkzii v 8 Ulusakh Almaznoi Provinzii RS(Y) na 1999-2001 Gody* [Programme for Introduction of Special Mechanism for Financing Agriculture, Strengthening Internal Market of Agricultural Products in Eight Districts of the Diamond province for 1999-2001], Unpublished Document, Yakutsk.

SAPI (1998d), *Regional'naya Zelevaya Programma 'Ozdorovlenie Ekologicheskoi Situazii v Almaznoi Provinzii v 1999-2001 gg'* [Regional Target Programme 'Improvement of the Environmental Situation in the Diamond Province for 1999-2001'], Unpublished Document, Yakutsk.

SAPI (1998e), *Poyasnitel'naya Zapiska k Budzhetu Fonda SAPI na 1998 God* [Explanation Note for the Budget of the SAPI Foundation in 1998], Unpiblished Document, Yakutsk.

SAPI (1998f), *Otchet po Vypolneniyu Programmy Fonda SAPI za 1993-1997 gg* [Report on Implementation of Programme of the SAPI Foundation for 1993-1997], Unpublished Report, Yakutsk.

SAPI (1998g), *Programma 'Ekologia' v 1998 Godu* [Programme 'Ecology' in 1998], Unpublished Document, Yakutsk.

SAPI (1999a), *Programma Investizii v Neftyanuyu i Gazovuyu Promyshlennost' Respubliki Sakha (Yakutia)* [Programme for Investment in the Oil and Gas Industry of the Republic of Sakha (Yakutia)], Unpublished Document, Yakutsk.

SAPI (1999b), *Svodny Budzhet Fonda SAPI na 1999 God* [Summary Budget of the SAPI Foundation for 1999], Unpublished Document, Yakutsk.

SAPI (2000a), *Utverzhdenie Zaklucheniya Popechitel'skogo Soveta Fonda SAPI po Itogam Finansovo-khozyaistvennoi Deyatel'nosti Fonda SAPI za 1999 God* [Ratification of Conlcusions of the Supervisory Board of the SAPI Foundation on Results of Financial and Economic Operations of the SAPI Foundation in 1999], Unpublished Document, Yakutsk.

SAPI (2000b), *Otchet Finansirovaniya Programm Budzheta SAPI za 2000 God za Period s Yanvarya po Mai* [Report on Financing Programmes of the SAPI Foundation in 2000 from January to May], Unpublished Report, Yakutsk.

SAPI (ud. a), *SAPI*, Kuduk, Yakutsk.

SAPI (ud. b), *5 let Fondu SAPI* [Five Years of the SAPI Foundation], SAPI, Yakutsk.

Sarkar, S. (1999), *Eco-socialism or Eco-capitalism?* Zed-Books Ltd, London.

Savvinov, D. (1999), 'Almazy i Ecologia' [Diamonds and the Environment], *Yakutia*, No. 31, 23 February.

Savvinov, D. (ed.) (1993), *Ekologiya r. Vilyuy* [Ecology of the River Vilyuy], RAN Institut Prikladnoi Ekologii Severa [Russian Academy of Sciences, Institute of Applied Ecology of the North], Yakutsk.

Savvinov, D.D. (1994), 'Ustoichivost' Ekosystem Severa pri Technogennom Vozdeistvii na Nikh' [Sustainability of Ecosystems in the North in Case of Industrial Impact], in -, *Tolerantnost: Materialy Regional'noi Nauchno-prakticheskoi Konferenzii 'Tolerantnost Kak Sposob Vyzhyvaniya Narodov v Usloviyakh Severa', Yakutsk 28-29 Aprelya 1994*

[Tolerance: Materials of the Regional Scientific and Practical Conference 'Tolerance as the Way of Survival of People in the Conditions of the North', Yakutsk 28-29 April 1994], YNZ RAN SO [Yakutsk Scientific Centre, Russian Academy of Sciencesm Siberian Branch, Yakutsk, pp. 55-62.

Savvinov, D.D. (2000), *Zagryaznenie Prirodnoi Sredy pri Promyshlennom Osvoenii (na Primere Yakutii)* [Environmental Pollution of the Industrial development: a Case of Yakutia], *Nauka i Obrazobanie* [Journal of Science and Education], No. 1, Yakutsk.

Savvinov, D.D., Teptirgyanov, M.M., Marshintsev, V.K. (1992), *Ekologiya Basseina r. Vilyuy: Promyshlennoe Zagryaznenie* [Ecology of the Basin of the River Vilyuy: Industrial Contamination], YNZ RAN SO [Yakutsk Scientific Centre, Russian Academy of Sciences, Siberian Branch], Yakutsk.

Schroeder, G. (1998), 'Dimensions of Russia's Industrial Transformation, 1992-1998: an Overview', *Post-Soviet Geography and Economics*, Vol. 39(5), pp. 243-270.

Schwartz, R. (2000), 'The Politics of Evaluating Government Third Sector Collaborations', *4th European Society Conference, Lausanne, 12-14 October 2000*, URL: http://www.europeanevaluation.org/pdf/1-3_schwartz.pdf (accessed on 24/05/01).

Scott, P. (2000), 'Reporting in the Mining sector', *Mining Environmental Management*, Vol. 8(2), pp. 10-12.

Sengupta, M. (1993), *Environmental Impacts of Mining: Monitoring, Restoration and Control*, Lewis Publishers Inc, London.

Seymoar, N., Roca, C. and Lavergne, R. (1998), 'Lima Workshop on Mining and Sustainable Development in Americs', *Report of Proceedings of Lima Workshop in Mining and Sustainable Development in Americas, 27-29 June 1999, Lima, Peru*, URL: http://www.iisd.ca/pdf/miningrep.pdf (accessed on 5/10/99).

Shearer, T. (2002), 'Ethics and Accountability: From the For-self to For-the-others', Accounting, Organizations and Society, Vol. 27 (6), pp. 541-573.

Shipkov, R.Ju. (1999), *Ekologicheskaya Obstanovka Almaznoi Provinzii: Problemy i Puti Resheniya: Doklad Zamestitelya Predsedatelya Praviletstva RS(Y) na Parlamentskikh Slushaniyakh 26 Marta 1999* [Environmental Situation in the Diamond Province: Problems and Solutions: Report of the Deputy Prime Minister of the Government of RS(Y) at the Parliament Hearings on 26th March 1999], Unpublished Report, Yakutsk.

Shrivastava, P. (1995), 'Industrial/environmental Crises and Corporate Social Responsibility', *The Journal of Socio-Economics*, Vol. 24(1), pp. 211-227.

Sinding, K. (1999), 'Environmental Impact Assessment and Management in the Mining Industry', *Natural Resources Forum*, Vol. 23(1), pp. 57-63.

Singleton, F. (1976), *Environmental Misuse in the Soviet Union*, Praeger Publishers, New York.

Singleton, F. (1987), *Environmental Problems in Soviet Union and Eastern Europe*, Lynne Reinner Publishers, London.

Sipkins, S. (1995), 'Sierra Leone Mining, Exports and Pollution', *TED Case Studies*, Vol. 4(1), URL: http://www.american.edu/TED/class/all.htm (accessed on 29/05/02).

Sobolev, N.V. (1981), 'Siberian Diamonds', *World Mining*, No. 34, pp. 55-57.

Soldatenko, M.N. (2001), 'Razrabotka i Vnedrenie Systemy Ekologicheskogo Menedjmenta na ZAO 'Sibkabel' [Development and Introduction of Environmental Management System in the 'Sibkabel Company], in -, *Ekologichesky Menedjement i Obshestvenny Dialog: Perspektivy v Rossii, Materialy Mezhdunarodnoi Konferenzii, Moskva, 15-16 Noyabrya 2001* [Environmental Management and Public Dialogue: Perspectives for Russia, Proceedings of International conference, Moscow, 15-16 November 2001], GEOPLAN-Ecoline, Moscow, URL: http://www.ecoline.ru/ (accessed on 02/07/02).

Solomov, N. (1998), 'Dostoyanie Respubliki: My v Otvete za Nego' [Heritage of the Republic: We Are Responsible for It], *Yakutia*, No. 94 (28876), 27 May 1998, URL: http://www.yakutia.ru/ (accessed on 31/01/00).

SouthernEra Resources Limited (1999), *Annual Report*, Stratagem Marketing & Design.

SouthernEra [online] URL: http://www.southernera.com (accessed 4/04/01).

SouthernEra (2000), *Annual Report*, Walter J. Mashko & Co. Inc, Toronto.

SouthernEra Resouces Limited (1998), *Annual Report*, SouthernEra, Toronto.

Stagliano, A. and Walden, W. (1998), 'Assessing Quality of Environmental Disclosure Themes', *2nd Asian-Pacific Interdisciplinary Research on Accounting, 4-6 August 1998, Osaka, Japan*, URL: http://www3.bus.osaka-cu.ac.jp/apira98/archives/pdfs/75.pdf (accessed on 9/05/02).

Stevenson, M.G. (1996), 'Indigenous Knowledge in Environmental Assessment', *Arctic*, Vol. 40(3), pp. 278-291.

Stigson, B. (1999), 'Sustainable Development for Industry and Society, *Building Research & Information*, Vol. 27(6), pp. 424-430.

Strongman, J. (1998), 'Mining and the Community from Enclave to Sustainable Development', *Asia/Pacific Mining and the Community Conference, July 26-30 1998, Madang, PNG*, URL: http://www.worldbank.org/ (accessed on 15/07/02).

Tanin, S. (1999), 'Trubka Mira' [Mir Pipe], *Nezavisimaya Gazeta*, 16 November 1999, No. 17(42), URL: http://www.ng.ru/(accessed on 10/01/00).

Tanner, A. (1999), *Russian Crisis has Helped Environment*, URL: http://www.planetark.org/ (accessed on 20/01/99).

Tarras-Wahlberg, H. and O'Toole, M.J. (2000), 'Namibian Marine Diamond Mining', *Mining Environmental Management*, Vol. 8(4), pp. 12-14.

Taylor, B., Hutchinson, C., Pollack, S. and Tapper, R. (1994), *The Environmental Management Handbook*, Pitman Publishing, London.

Teslenko, N. (1997), *Organizazia Torgovli Dragozennymi Kamnyami* [Organisation for Trade of Precious Stones], Infra-M, Moscow.

Thayer, C. E. and Fine, A.H. (2001), 'Evaluation and Outcome Measurement in the Non-profit Sector: Stakeholder Participation', *Evaluation and Program Planning*, Vol. 24(1), pp. 103-108.

Thoresen, J. (1999), 'Environmental Performance Evaluation – a Tool for Industrial Improvement', Journal of Cleaner Production, Vol. 7 (5), pp. 365-370.

Tichotsky, J. (2000), *Russia's Diamond Colony: the Republic of Sakha*, Harwood Academic Publishers, Malaysia.

Tilling, M. (2001), *Corporate Social Reporting: a Cook's Tour*, School of Commerce Research Paper Series, No. 9.

Tilt, C.A. (1994), 'The Influence of External Pressure Groups on Corporate Social Disclosure: Some Empirical Evidence', *Accounting, Auditing & Accountability Journal*, Vol. 7(4), pp. 47-72.

Tilt, C.A. (2001), 'The Content and Disclosure of Australian Corporate Environmental Policies', *Accounting, Auditing & Accountability Journal*, Vol. 14(2), pp. 190-212.

Tilt, C.A. and Symes, C.F. (1999), 'Environmental Disclosure by Australian Mining Companies: Environmental Conscience or Commercial Reality?' *Accounting Forum*, Vol. 23(2), pp. 137-154.

Toryannikov, B.N. (1999), 'Zoloto: Vse Zoloto Mira, Zoloto Rossii' [Gold: World Gold, Russian Gold], *Kreditreforma*, January-March, pp. 18-22, URL: http://www.creditreform.spb.ru/globalworld.htm (accessed on 27/10/99).

Tysyachnyuk, M. (1998), 'Ustoichivoe Razvitie v Ramkakh Pravitel'stvennykh Programm Rossii' [Sustainable Development within Governmental Programmes in Russia],

EcoNews, Vol. 4(138), 31 October 1998, URL: http://www.friends-partners.org/oldfriends/valery/econews/1998/eco138.html (accessed on 10/02/00).

UNCED (United Nations Conference on Environment and Development) (1992) *Agenda 21: Programme of Action for Sustainable Development. Rio Declaration on Environment and Development. Statement of Forest Principles; the Final Text of Agreements Negotiated by Governments at the United Nations Conference on Environment and Development (UNCED), 3-14 June 1992, Rio de Janeiro, Brazil*, New York: United Nations Department of Public Information.

UNEP (1999), *UNEP's Activities in Mining and the Environment*, United Nations Environmental Programme, URL: http://natural-resources.org/ (accessed on 17/04/00).

US EPA (1994), *Technical Resource Document, Extraction and Beneficiation of Ores and Minerals, Vol. 2, Gold*, United States Environmental Protection Agency, URL: http://www.epa.gov/ (accessed on 5/04/01).

Varnakov, A.N. (2001), 'Resultaty Proekta 'Razvitie Obschestvennogo Dialoga i Sotrudnichestva v Oblasti Ekologicheskogo Menedjmenta v Promyshlennom Sektore RF' v Ryazanskoi Oblasti' [Results of the Project 'Development of Public Dialogue and Co-operation in the area of Environmental Management in the Industrial Sector of the Russian Federation], in -, *Ekologichesky Menedjement i Obshestvenny Dialog: Perspektivy v Rossii, Materialy Mezhdunarodnoi Konferenzii, Moskva, 15-16 Noyabrya 2001* [Environmental Management and Public Dialogue: Perspectives for Russia, Proceedings of International Conference, Moscow, 15-16 November 2001], GEOPLAN-Ecoline, Moscow, URL: http://www.ecoline.ru/ (accessed on 2/07/02).

Vassie, L., Tomas, J.M. and Oliver, A. (2000), 'Health and Safety Management in UK and Spanich SMEs: a Comparative Study', *Journal of Safety Research*, Vol. 31(1), pp. 35-43.

Veber, A. (1999), 'Poyavit'sya li u Rossii Strategiya Ustoichivogo Razvitiya' [Will Russia Have a Strategy for Sustainable Development], *Informazionno-analitichesky Zhurnal 'Obozrevatel'* [Information and Analytical Journal 'Viewer'], No. 10, URL: http://www.incom-svyaz.com/~dn/observer/N10_99/10_07.HTM (accessed on 10/10/00).

Vecherin, P.P. (1997), *Ot Tresta do Kompanii, ot Palatok do Gorodov: Khronologiya 'Yakutalmaza' 1957-1992 gg* [From Trust to Company, from Tents to Cities: Chronology of 'Yakutalmaz' 1957-1992], Mirninsky Rabochy, Mirny.

Vinichenko, V.N., Guseva, T.V. and Daiman, S.Ju. (2001), 'Opyt Razvitiya Sistem Ekologicheskogo Menedzhmenta v Rossii: Resultaty Mezhdunarodnykh Kommercheskykh Projektov' [Experience of Developing Environmental Management Systems in Russia: Results of International Non-profit Projects], in -, *Ekologichesky Menedzhement i Obschestvenny Dialog: Perspektivy v Rossii, Materialy Mezhdunarodnoi Konferenzii, Moskva, 15-16 Noyabrya 2001* [Environmental Management and Public Dialogue: Perspectives for Russia, Proceedings of International Conference, Moscow, 15-16 November 2001], GEOPLAN-Ecoline, Moscow, URL: http://www.ecoline.ru/ (accessed on 2/07/02).

Volostnov, D.V. and Bareisha, V.M. (1999), *Opyt Tomskoi Oblasti: Resultaty Prinyatiya Zakona ob Ekologicheskom Audite, Doklad na Seminare v Tomske, 19-21 Avgusta 1999* [Experience of Tomsk Oblast: Results of Adoption of the Law about Environmental Audit, Report at the Seminar in Tomsk, 19-21 August 1999], URL: http://www.ecoline.ru/ (accessed on 5/07/02).

Volotovsky, K.A. and Kuznetsova, L.V. (1993), 'Nekotorye Problemy Ratsional'nogo Prirodopol'zovaniya v Oblasti Prozhivaniya Aborigenov Jugo-Vostoka Yakutii [Some Problems of Rational Natural Resource Use in the Region of Livelihood of Indigenous People in South-east of Yakutia], in D.D. Savvinov (ed.), *Ekologicheskie Tradizii*

Aborigenov Severa [Ecological Traditions of Indigenous People of the North], Institut Prikladnoi Ekologii Severa [Institute of Applied Ecology of the North], Yaktusk, p. 45.

Vukina, T., Beghin. J.C., Solakoglu, E.G. (1999), 'Transition to Markets and the Environment: Effects of the Change in the Composition of Manufacturing Output', *Environment and Development Economics*, Vol. 5(4), pp. 582-598.

Walden, W.D. and Schwartz, B.N. (1997), 'Environmental Disclosures and Public Policy Pressure', *Journal of Accounting and Public Policy*, Vol. 16(2), pp. 125-154.

Walker, J. and Howard, S. (2002), *Finding the Way Forward: How Could Voluntary Action Move Mining towards Sustainable Development?* International Institute for Environment and Development and WBCSD, London, URL: http://www.iied.org (accessed 10/02/05).

Walton, J. and Alabaster, T. (1996), 'Institutional Environmental Reporting Using the World Wide Web', in J.P. Ulhøi and H. Madsen (eds), *Industry and the Environment: Practical Applications of Environmental Management Approaches in Business*, The Aarhus School of Business, pp. 29-39.

Warhurst, A. (1998), 'Corporate Social Responsibility & the Mining Industry', *MERN Research Bulletin & Newsletter*, No. 13-14, pp. 81-97.

Warhurst, A. (ed.) (1999), *Mining and the Environment: Case Studies from the Americas*, International Development Research Centre, Ottawa.

Warhurst, A. (2001), 'Corporate Citizenship and Corporate Social Investment: Drivers of Tri-sector Partnerships', *Journal of Corporate Citizenship*, No.1, pp. 57-73.

Warhurst, A. and Bridge, G. (1997), 'Economic Liberalisation, Innovation and Technology Transfer: Opportunities for Cleaner Production in the Mineral industry', *Natural Resources Forum*, Vol. 21(1), pp. 1-12.

Warhurst, A. and Insor, R. (1996), 'Environmental Issues for Developing Countries Arising from Liberalized Trade in the Mining industry', *Natural Resources Forum*, Vol. 20(1), pp. 27-35.

Warhurst, A. and Mitchell, P. (2000), 'Corporate Social Responsibility and the Case of Summitville Mine', *Resources Policy*, Vol. 26(2), pp. 91-102.

Warhurst, A. and Noronha, L. (2000), 'Corporate Strategy and Viable Future Land Use: Planning for Closure from the Outset of Mining', *Natural Resources Forum*, Vol. 24(2), pp. 153-164.

Warner, M. (2000a), *Guidance Note for 'Getting Started' Tri-sector Partnerships for Managing Social Issues in the Extractive Industries*, Business Partners for Development Natural Resources Cluster, Working Paper No. 6, URL: http://bpd-naturalresources.org (accessed on 25/10/01).

Warner, M. (2000b), *Tri-sector Partnerships for Social Investment within the Oil, Gas and Mining Sectors: an Analytical Framework*, Business Partners for Development Natural Resources Cluster, Working Paper No. 2, URL: http://bpd-naturalresources.org (accessed on 25/10/01).

Watkins, M. (1998), 'Maximising Environmental Management Capacity through Stakeholder Participation', *MERN Research Bulletin & Newsletter*, No. 13/14, pp. 98-101.

Watts, P. and Holme, L., WBCSD (1998), *Corporate Social Responsibility: Meeting Changing Expectations*, World Business Council for Sustainable Development, Switzerland, URL: http://www.wbcsd.org (accessed on 8/10/00).

WCED (World Commission on Environment and Development) (1987), *Our Common Future*, Oxford University Press.

Welford, R. (ed.) (1998), *Corporate Environmental Management 1: Systems and Strategies*, 2nd ed., Earthscan Publications Ltd, London.

Wells, D. (2001), 'Voluntary Initiatives for the Mineral Sector: the Australian Experience', *MMSD Workshop, Santa Fe, USA, 18 July 2001*, URL: http://mineralresourcesforum.unep.ch/docs/pdfs/Wells_Santa_Fe_Speech.pdf (accessed on 20/05/02).

Wheeler, D. and Elkington, J. (2001), 'The End of the Corporate Environmental Report? Or the Advent of Cybernetic Sustainability Reporting and Communication', *Business Strategy and the Environment*, Vol. 10(1), pp. 1-14.

Wheeler, D. and Sillanpää, M. (1997), *The Stakeholder Corporation: a Blueprint for Maximising Stakeholder Value*, Pitman Publishing, London.

White, A.L. and Zinkl, D. (1999), 'Standardisation: the Next Chapter in Corporate Environmental Performance Evaluation and Reporting', in M. Bennett and P. James (eds), *Sustainable Measures: Evaluation and Reporting of Environmental and Social Performance*, Greenleaf Publishing, Sheffield, pp. 117-131.

Williams, S.M. and Ho Wern Pei, C.A. (1999), 'Corporate Social Disclosure by Listed Companies on Their Web Sites: an International Comparison', *The International Journal of Accounting*, Vol. 34(3), pp. 389-419.

Wilmshurst, T.D. and Frost, G.F. (2000), 'Corporate Environmental Reporting: a Test of Legitimacy Theory', *Accounting, Auditing & Accountability Journal*, Vol. 13 (1), pp. 10-26.

Wood, D. (1991), 'Corporate Social Performance Revisited', *Academy of Management Review*, Vol 16 (4), pp. 758-769.

Wood, H. (1998), 'A New Face for Environmental Action: Community Action and Involvement', *International Minerals and Mining*, Vol. 1(10), pp. 276-280.

Woodward, D., Edwards, P. and Birkin, P. (2001), 'Some Evidence on Executive's View of Corporate Social Responsibility', *British Accounting Review*, Vol. 33, pp. 357-397.

World Bank (1992), *World Development Report 1992: Development and the Environment*, Oxford University Press, New York.

World Bank Group (2003), Russian Federation at a Glance, URL: http://www.worldbank.org (accessed on 29/04/04).

World Gold Analyst (2000), *World Gold Analyst, June 2000, Quarter Review*, September, URL: http://www.mining-journal.com/ (accessed on 2/03/01).

World Gold Council (2001), *World Official Gold Holdings, March 2001*, URL: http://www.gold.org (accessed on 10/05/02).

WWF Australia (2000), *Ore or Overburden II? WWF's 2nd Annual Scorecard on Mining Company Environmental Reports*, World Wild Fund for Nature-Australia, URL: http://www.wwf.org.au (accessed on 10/04/01).

YaGAP (Yakutskoe Gosudarstvennoe Aerogeodezicheskoe Predpriyatie) (1998), *Respublika Sakha(Yakutia)* [Republic of Sakha (Yakutia], Roskartographia, Yakutsk.

Yakovleva, N.P. and Alabaster, T. (2001), 'Ecological Modernisation: Critical Analysis of the SAPI Foundation in the Republic of Sakha (Yakutia)', in I. Massa and V. Tynkkynen (eds), *The Struggle for Russian Environmental Policy*, Kikimora Publications, Helsinki, pp. 107-122.

Yakovleva, N.P., Alabaster, T. and Petrova, P.G. (1999), 'Environmental and Health Impacts of Diamond Mining in the Republic of Sakha (Yakutia)', *World mining Environment Congress, National Mining Research Centre-Skochinsky Institute of Mining, 7-11 September 1999.*

Yakovleva, N.P., Alabaster, T. and Petrova, P.G. (2000), 'Natural Resource Use in the Russian North: a Case Study of Diamond Mining in the Republic of Sakha', *Environmental Management and Health*, Vol. 11(4), pp. 318-336.

Yakutia (2002), 'Vokrug ALROSA i ee Presidenta' [Around ALROSA and its President], *Yakutia*, 5 January 2002, URL: http://www.gazetayakutia.ru (accessed 5/01/02).

Young, J.E. (2000), *Gold: at What Price?* Project Underground, URL: http://www.moles.org/ProjectUnderground/reports/goldprice.html (accessed on 12/06/02).

Yuen, C.P. and Yip, D. (2002), 'Corporate Environmental Reporting - the CLP Power Experience', *Corporate Environmental Strategy*, Vol. 9(1), pp. 95-100.

Zakharova, T. (1996), 'Pravovye Osnovy Okhrany Okruzhayuschei Sredy v Respublike Sakha' [Legal Basis for Environmental Protection in the Republic of Sakha], in -, *Ekologicheskie Problemy Yakutii* [Ecological Problems of Yakutia], Yakutsk, pp. 166-170.

Zank, N.S. (1995), 'Privatizing the Mineral Sector', *Natural Resources Forum*, Vol. 19(3), pp. 215-221.

Ziegler, C.E. (1987), *Environmental Policy in the USSR*, Frances Pinter Publishers, London.

Zoloto: Stranizy Zolotopromyshlennika [Gold: Pages of Gold Miners] [online], URL: http://www.aurum79.ru (accessed on 12/06/02).

Legal Documents of the Russian Federation in Order of their Legal Status

Constitution of the Russian Federation12 December 1992.

Law of RSFSR 'Protection of Atmospheric Air', 14 June 1982.

Law of RSFSR 'Sanitary and Epidemiological Well-Being of Population', 19 April 1991.

Law of RSFSR 'Land Code', 25 April 1991.

Law of RSFSR 'Protection of Natural Environment', 19 December 1991.

Law of the Russian Federation 'Mineral Resources', 21 February 1992.

Law of the Russian Federation 'Responsibilities for Breach of the Submission Order of State Statistical Reports', 13 May 1992, No. 2761-1.

Law of the Russian Federation 'Specially Protected Areas', 14 March 1995.

Law of the Russian Federation 'Animals', 24 April 1995.

Law of the Russian Federation 'Warer Code', 16 November 1995.

Law of the Russian Federation 'Ecological Expertise', 23 November 1995.

Law of the Russian Federation 'Specially Protected Natural Areas', 15 February 1995, No. 33-ФЗ.

Law of the Russian Federation 'Forest Code', 29 January 1997.

Decree of the Central Committee of the Communist Party of the Soviet Union and the Council of Ministers of the USSR 'Operations for Enforcement of Protection of Seas, Rivers and Other Watercourses of The Arctic Basin From Contamination', 15 January 1981, No. 65.

Decree of the President of the Russian Federation 'Basics of State Strategy of the Russian Federation on Environmental Protection and Provision of Sustainable Development', 4 February 1994.

Decree of the President of the Russian Federation 'Conception for Transition to Sustainable Development', 1 April 1996.

Decision of the Government of the Russian Federation 'Establishment of the United State System of Environmental Monitoring', 23 November 1993, No. 1229.

Decree of the Government of the Russian Federation 'Plans for Actions of the Government of the Russian Federation in the Field of Environmental Protection for 1994-1995', 18 May 1994.

Decree of the Government of the Russian Federation 'Plans for Actions of the Government of the Russian Federation in the Field of Environmental Protection for 1996-1997', 19 February 1996.

Decision of the Government of the Russian Federation 'Conception of Federal Target Program "Supplying Population of Russia with Drinking Water" and Implementation of Actions of Primary Importance for the Improvement of Water Supply for Population', 6 March 1998, No. 292.

Decision of the State Committee for Statistics of the Russian Federation 'Instruction for Completion of Statistical Report about Formation and Disposal of Toxic Waste form N2 -ТП - Toxic Waste', 14 September 1993, No. 180.

Decree of the State Committee for Statistics of the Russian Federation 'Adoption of Forms for Federal State Statistical Monitoring over Traumatism and Instructions for Their Completion', 26 July 1995, No. 105.

Decree of the State Committee for Statstics of the Russian Federation 'Instruction for Completion of the Form of Federal State Statistical Monitoring for Injuries at Workplaces, Occupational Diseases and Related Material Costs (Form No. 7 – Injuries)', 25 February 1997, No. 13.

Decree of the State Committee for Statistics of the Russian Federation 'Form No. 2-тп (Toxic Waste): Information on Creation, Delivery, Use and Allocation of Toxic Waste of Production and Consumption', 11 September 1998, No. 95.

Decree of the State Committee for Statistics of the Russian Federation 'Instruction for Completion of the Form of Federal State Statistical Monitoring No. 2-тп (Toxic Waste): Information on Creation, Delivery, Use and Allocation of Toxic Waste of Production and Consumption', 21 October 1998, No. 101, with revision on 26.11.99 No. 106.

Decree of the State Committee for Statistics of the Russian Federation 'Form No. 4-ос: Information on Current Expenses for Environmental Protection, Ecological and Natural Resource Payments', 24 July 2000, No. 68.

Decree of the State Committee for Statistics of the Russian Federation 'Form No. 2-тп (Air): Information on Protection of Atmospheric Air', 24 July 2000, No. 68.

Decree of the State Committee for Statistics of the Russian Federation, 'Instruction for Completion of the Form of Federal State Statistical Monitoring No. 4-ос: Information on Current Expenses for Environmental Protection, Ecological and Natural Resource Payments', 29 September 2000, No. 90.

Decree of the State Committee for Statistics of the Russian Federation 'Instruction for Completion of Form of Federal State Statistical Monitoring No. 2-тп (Air): Information on Protection of Atmospheric Air', 29 September 2000, No. 90.

Decree of the State Committee for Statistics of the Russian Federation 'Form No. 2-тп (Water Use): Information on Water Use', 13 November 2000, No. 110.

Legal Documents of the Republic of Sakha (Yakutia) in Order of their Legal Status

Constitution of the Republic of Sakha, 27 April 1992.

Law of the Republic of Sakha (Yakutia) 'Environmental Protection in Yakut-Sakha Soviet Socialistic Republic', 17 October 1991, No. 648-XII.

Law of the Republic of Sakha (Yakutia) 'Administrative Responsibility for Environmental Offences', 27 March 1993, No. 1426-XII.

Law of the Republic of Sakha (Yakutia) 'Specially Protected Natural Areas of the Republic of Sakha (Yakutia)', 6 May 1996, 3 No. 113-I.

Law of the Republic of Sakha (Yakutia) 'Hunting and Hunting Economy', 14 June 1998, 3 No. 34-II.

Law of the Republic of Sakha (Yakutia) 'Fishing, Fishing Economy and Protection of Water Biological Resources', 17 June 1999, 3 No. 104-II.

Law of the Republic of Sakha (Yakutia) 'State Budget of the Republic of Sakha (Yakutia) in 2000', 24 December 1999, 3 No. 140-II (in force).

Law of the Republic of Sakha (Yakutia) State Budget of the Republic of Sakha (Yakutia) for 2001, 22 December 2000, 3 No. 239-II.

Decree of Supreme Soviet of the Republic of Sakha (Yakutia) 'Set of Primary Measures for Improvement of Environmental Situation in the River Vilyuy Basin', 20 March 1992.

Resolution of the Chamber of Republic of the Legislative Assembly (Il Tumen) of the Republic of Sakha (Yakutia) 'Certain Measure for the Enforcement of State Support for Oil and Gas Industry in the Republic of Sakha', 24 December 1998, ПП No. 84-II.

Decree of the President of the Republic of Sakha (Yakutia) 'Creation of "Sakhaalmazprovinvest" Financial Corporation', 30 June 1993, No. 481 (article 4 cancelled).

Decree of the President of the Republic of Sakha (Yakutia) 'Introduction of Special Regime for Exploitation and Protection of Unique Lakes in the Republic of Sakha (Yakutia)', 16 August 1994, No. 836.

Decree of the President of the Republic of Sakha (Yakutia) 'Actions for the Establishment of Specially Protected Natural Areas', 16 August 1994, No. 837.

Decree of the President of the Republic of Sakha (Yakutia) 'Establishment of the Department for Biological Resources under the Ministry for Nature Protection of the Republic of Sakha (Yakutia)', 29 October 1995, No. 118.

Decree of the President of the Republic of Sakha (Yakutia) 'Program for Implementation of State Environmental Policy of the Republic of Sakha (Yakutia) in the Field of Enrichment of Biodiversity of Flora and Fauna', 7 May 1996, No. 1376.

Decree of the President of the Republic of Sakha (Yakutia) 'Actions for the Completion of the Establishment of the System of Specially Protected Areas in the Republic of Sakha (Yakutia)', 31 December 1998, No. 654.

Decree of the President of the Republic of Sakha (Yakutia) 'Registration and Usage of the Part of Lease Payment from Joint-stock Company "ALROSA" in the Amount of 2 % from its Main Production Sales', 15 March 1999, No. 711 (in force).

Decree of the President of the Republic of Sakha (Yakutia) 'Order for Delivery, Registration and Usage of 2% Assignments, a Part of Lease Payments from ALROSA Company', 8 July 1999, No. 804 (cancelled).

Decree of the President of the Republic of Sakha (Yakutia) 'Inclusion of Kobyai District in Environmental and Socio-economic Programs in the Diamond Province', 4 November 1999, No. 902.

Decree of the President of the Republic of Sakha (Yakutia) 'Additional Measures for Completion of the Establishment of the Republican System of Specially Protected Natural Areas – Ytyk Kere Sirder', 28 December 1999, No. 932.

Decree of the President of the Republic of Sakha (Yakutia) 'Order for Delivery, Registration and Usage of 2% Assignments from Joint-stock Company "ALROSA"', 17 January 2000, No. 948 (in force).

Order of the President of the Republic of Sakha (Yakutia) 'Additional Measure for Regulation of Obligatory Payments from Almazy Rossii-Sakha Company to the Special Fund', 11 September 1996, No. 491-РП (cancelled).

Order of the President of the Republic of Sakha (Yakutia) 'Creation of Regional Center for Socio-economic Research in Western Yakutia', No. 176-RP, 18 July 1998.

Decree of the Government of the Republic of Sakha (Yakutia) 'Emergency Socio-economic and Environmental Protection Measures for Rehabilitation of Environmental Situation in the River Vilyuy Basin', No. 201, 27 April 1992.

Decision of the Government of the Republic of Sakha (Yakutia) 'Adoption of the United State System of Environmental Monitoring in the Republic of Sakha (Yakutia)', 4 March 1996, No. 86.

Decision of the Government of the Republic of Sakha (Yakutia) 'Adoption of Regulations for National Nature Parks (Aan Aiylgy), Resource Reserves (Erkeeyi Sirder), Natural Monuments (Aiylga Menelere)', 27 November 1997, No. 515.

Decision of the Government of the Republic of Sakha (Yakutia) 'Development of Republican Target Program "Supplying Population of Yakutia with Drinking Water"', 12 April 1999, No. 191.

Decision of the Government of the Republic of Sakha (Yakutia) 'Implementation of the Decree of the President of the Republic of Sakha (Yakutia) of 17 January 2000 No. 948 'Order for Delivery, Registration and Usage of 2% Assignments from ALROSA', 22 March 2000, No. 176 (in force).

Decision of the Government of the Republic of Sakha (Yakutia) 'Adoption of Republican Target Program "Supplying Population of Yakutia with Drinking Water"', 12 February 2000, No. 70.

Decision of the Government of the Republic of Sakha (Yakutia) 'Budget of the Target Budget Fund for Implementation of Socio-economic and Environmental Programs in Districts of the Diamond Province of the Republic of Sakha (Yakutia) in 2000', 15 December 2000, No. 647.

Decision of the Government of the Republic of Sakha (Yakutia) 'Rules for Hunting in the Republic of Sakha (Yakutia)', 25 December 2000, No. 660.

Decision of the Government of the Republic of Sakha (Yakutia) 'Budget of the Target Budget Fund for Implementation of Socio-economic and Environmental Programs in Districts of the Diamond Province of the Republic of Sakha (Yakutia) in 2001', 16 April 2001, No. 216.

Index